Süßwasserfauna von Mitteleuropa

Begründet von A. Brauer

Herausgegeben von

J. Schwoerbel (†) und P. Zwick

Stellvertretender Herausgeber für Band 7

R. Gerecke

Band 7/2-2

Antonio Di Sabatino
Reinhard Gerecke
Terence Gledhill
Harry Smit

Chelicerata: Acari II

Süßwasserfauna von Mitteleuropa

7/2-2

Spektrum
AKADEMISCHER VERLAG

Wichtiger Hinweis für den Benutzer
Der Verlag, der Herausgeber und die Autoren haben alle Sorgfalt walten lassen, um vollständige und akkurate Informationen in diesem Buch zu publizieren. Der Verlag übernimmt weder Garantie noch die juristische Verantwortung oder irgendeine Haftung für die Nutzung dieser Informationen, für deren Wirtschaftlichkeit oder fehlerfreie Funktion für einen bestimmten Zweck. Der Verlag übernimmt keine Gewähr dafür, dass die beschriebenen Verfahren, Programme usw. frei von Schutzrechten Dritter sind. Die Wiedergabe von Gebrauchsnamen, Handelsnamen, Warenbezeichnungen usw. in diesem Buch berechtigt auch ohne besondere Kennzeichnung nicht zu der Annahme, dass solche Namen im Sinne der Warenzeichen- und Markenschutz-Gesetzgebung als frei zu betrachten wären und daher von jedermann benutzt werden dürften. Der Verlag hat sich bemüht, sämtliche Rechteinhaber von Abbildungen zu ermitteln. Sollte dem Verlag gegenüber dennoch der Nachweis der Rechtsinhaberschaft geführt werden, wird das branchenübliche Honorar gezahlt.

Bibliografische Information der Deutschen Nationalbibliothek
Die Deutsche Nationalbibliothek verzeichnet diese Publikation in der Deutschen Nationalbibliografie; detaillierte bibliografische Daten sind im Internet über http://dnb.d-nb.de abrufbar.

Springer ist ein Unternehmen von Springer Science+Business Media
springer.de

Spektrum Akademischer Verlag ist ein Imprint von Springer

10 11 12 13 14 5 4 3 2 1

Planung und Lektorat: Dr. Ulrich G. Moltmann, Bettina Saglio
Satz: Mitterweger & Partner, Plankstadt
Umschlaggestaltung: SpieszDesign, Neu-Ulm

ISBN 978-3-8274-1894-4

8. Acari: Hydrachnidia II

By Antonio Di Sabatino, Reinhard Gerecke, Terence Gledhill and Harry Smit

List of authors

Dr. Antonio Di Sabatino
Dipartimento di Scienze Ambientali
University of L'Aquila
Via Vetoio 20
I-67100 Coppito, Italy
Email: antonio.disabatino@univaq.it

Dr. Reinhard Gerecke
Biesingerstraße 11
D-72070 Tübingen, Germany
Email: reinhard.gerecke@uni-tuebingen.de

Terence Gledhill CBiol, MSB, FLS
Freshwater Biological Association
Windermere Laboratory
The Ferry Landing, Far Sawrey
Ambleside Cumbria, LA 22 OLP, United Kingdom
Email: tgledhill@fba.org.uk

Drs. Harry Smit
Zoological Museum, University of Amsterdam
Plantage Middenlaan 64
NL-1018 DH Amsterdam, The Netherlands
Email: smit.h@wolmail.nl

Contents

Superfamily Hydryphantoidea PIERSIG, 1896 1
Key to families, subfamilies and genera of the European fauna 1

Family Hydrodromidae K. VIETS, 1936 8
Genus *Hydrodroma* KOCH, 1837 9

Family Hydryphantidae PIERSIG, 1896 12
Subfamily Diplodontinae K. VIETS, 1936 13
Genus *Diplodontus* DUGÈS, 1834 13
Subfamily Euthyadinae K. VIETS, 1931 15
Genus *Euthyas* PIERSIG, 1898 15
Genus *Panisellus* K. VIETS, 1925 17
Genus *Panisopsis* K. VIETS, 1926 18
Genus *Panisus* KOENIKE, 1896 21
Genus *Parathyas* LUNDBLAD, 1926 24
Genus *Parathyasella* K. VIETS, 1949 32
Genus *Thyasella* K. VIETS, 1926 33
Genus *Thyasides* LUNDBLAD, 1926 34
Genus *Thyopsis* PIERSIG, 1899 36
Genus *Trichothyas* K. VIETS, 1926 37
Subgenus *Lundbladia* K. VIETS, 1929 37
Genus *Vietsia* LUNDBLAD, 1926 39
Genus *Zschokkea* KOENIKE, 1892 40
Subfamily Hydryphantinae PIERSIG, 1896 42
Genus *Georgella* KOENIKE, 1907 42
Genus *Hydryphantes* KOCH, 1841 44
Subgenus *Hydryphantes* KOCH, 1841 48
Subgenus *Polyhydryphantes* K. VIETS, 1926 54
Subfamily Protziinae KOENIKE, 1909 56
Genus *Partnunia* PIERSIG, 1896 56
Genus *Protzia* PIERSIG, 1896 59
Subfamily Pseudohydryphantinae K. VIETS, 1926 69
Genus *Pseudohydryphantes* K. VIETS, 1907 69
Subfamily Tartarothyadinae K. VIETS, 1934 70
Genus *Tartarothyas* K. VIETS, 1934 70
Subfamily Wandesiinae SCHWOERBEL, 1961 72
Genus *Wandesia* SCHECHTEL, 1912 72
Subgenus *Pseudowandesia* HABEEB, 1958 73
Subgenus *Wandesia* SCHECHTEL, 1912 74

Superfamily Lebertioidea THOR, 1900 76
Key to families 76

Family Anisitsiellidae KOENIKE, 1910 79
Genus *Bandakia* THOR, 1913 81
Genus *Nilotonia* THOR, 1905 83
Subgenus *Dartia* SOAR, 1917 85
Subgenus *Dartiella* K. VIETS, 1929 87
Genus *Sigthoria* KOENIKE, 1907 89

Family Lebertiidae THOR, 1900 90
Genus *Lebertia* NEUMAN, 1880 91
Subgenus *Brentalebertia* GERECKE, 2008 93
Subgenus *Eolebertia* GERECKE, 2009 96
Subgenus *Lebertia s. str.* NEUMAN, 1880 97
Subgenus *Mixolebertia* THOR, 1906 117
Subgenus *Pilolebertia* THOR, 1900 140

Family Oxidae K. Viets, 1926 145
Genus *Oxus* Kramer, 1877 145
Subgenus *Oxus* Kramer, 1877 148
Subgenus *Gnaphiscus* Koenike, 1898 151
Family Rutripalpidae Sokolow, 1934 152
Genus *Rutripalpus* Sokolow, 1934 152
Family Sperchontidae Thor, 1900 154
Subfamily Sperchontinae Thor, 1900 154
Genus *Sperchon* Kramer, 1877 154
Subgenus *Hispidosperchon* Thor, 1901 162
Subgenus *Sperchon* Kramer, 1877 167
Genus *Sperchonopsis* Piersig, 1896 171
Family Teutoniidae Koenike, 1910 173
Genus *Limnolegeria* Motaş, 1928 173
Genus *Teutonia* Koenike, 1889 175
Family Torrenticolidae Piersig, 1902 177
Subfamily Torrenticolinae Piersig, 1902 177
Genus *Monatractides* K. Viets, 1926 179
Genus *Pseudotorrenticola* Walter, 1906 184
Genus *Torrenticola* Piersig, 1896 185
Subgenus *Megapalpis* Halbert, 1944 185
Subgenus *Torrenticola* Piersig, 1896 187
Hydrachnidia, general morphology and terminology 203
Hydrachnidia, key to superfamilies 207

Acknowledgements 212
Glossary 213
References 216

Introductory remarks

This volume deals with European species of two out of the eight superfamilies of true fresh water mites generally accepted world-wide, namely Hydryphantoidea and Lebertioidea. The superfamilies Stygothrombioidea, Hydrovolzioidea, Eylaoidea and Hydrachnoidea, were treated in the first Chelicerata volume of this series (Davids et al. 2007), the remaining superfamilies, Hygrobatoidea and Arrenuroidea will be subject of the third volume. Unfortunately, the plan to produce a fourth volume dealing with larvae of all the superfamilies (as announced in Vol. I) has had to be abandoned.

An explanation of morphological terminology and abbreviations (page 203–205, Figs 8-75-77), an illustrated key to superfamilies (page 206–211, Figs 8-78-81) and a glossary (page 213) are found at the end of this volume.

At genus level, all keys refer to Europe as a whole. At species level, the area treated includes Central and Northern Europe extending in the West to the Southwestern margin of the Pyrenees, in the South to the Southern margin of the Alps and in the East to the Eastern state borders of Austria, Slovakia, Poland and the Baltic countries.

Superfamily Hydryphantoidea PIERSIG, 1896

1896c Hydryphantinae PIERSIG, Sitzungsberichte naturf. Ges. Leipzig 22: 48.

Diagnosis: Colour red, rarely brown, orange or yellow. Membranous parts of integument papillate. Lateral eyes located at anterolateral edge of idiosoma, not on a common eye plate. Coxae typically in four groups, on each side, Cx-I+II and Cx-III+IV fused to form two coxal plates, these plates separated. Palp five-segmented, often chelate, never uncate. Chelicera two-segmented. Larvae as far as known, except for one subfamily, (Wandesiinae) aerial.

Discussion: Without any doubt, this is a para- or polyphyletic clade, including taxa branched off before the evolutionary level of aquatic larvae was reached in the stem line of water mites. In the design of the genital field and mouth parts, many genera show a high degree of similarity to terrestrial parasitengone mites, but there are also evolutionary lines that have developed outstanding new standards and distinct morphological adaptations. Some larvae in the genus *Wandesia* display an aquatic life style, but their legs are hardly adapted for swimming. It would not be surprising to find further, and better, adaptations to larval aquatic life within this clade, achieved in convergence to the "higher" water mites (superfamilies Lebertioidea + Hygrobatoidea + Arrenuroidea). Future systematic research will most probably result in the splitting of Hydryphantoidea into several separate clades and the transfer of some of its elements to other superfamilies is also possible.

Of the seven families included by COOK (1974, 1992), only Hydryphantidae and Hydrodromidae, both characterized by a chelate palp, are represented in the European fauna. Other families, all with non-chelate palps, are Teratothyadidae (Africa, Sumatra, two genera), Rhynchohydracaridae (North and South America, three subfamilies), Ctenothyadidae (E Asia, New Zealand), Thermacaridae (Asia, North and South America, only genus *Thermacarus*) and Zelandothyadidae (New Zealand, Tasmania).

TUZOVSKIJ (2007) proposed to re-establish as a family Thyadidae K. VIETS, 1926l, a taxon ranked as a subfamily of Hydryphantidae here following COOK's proposal (1974). This taxonomic change, however, does not provide any clearer insight into phylogenetic relationships. However, following ÖZDIKMEN (2006), the correct name of former Thyadidae is Euthyadidae K. VIETS, 1931c, and this is followed here but at subfamily level, i. e. Euthyadinae. The only character state in common to adults of all genera, is the probably plesiomorphic absence of swimming setae. All other potential synapomorphies listed in TUZOVSKIJ's revision regard the larval stage only – a stage unfortunately still unknown in 50 % of the genera attributed to this clade. In view of such an unsatisfactory state, we retain here the traditional system. After further study of preadult stages, some euthyadine genera may be shown to be related to representatives of other hydryphantid subfamilies.

Key to families, subfamilies and genera of the European fauna

Some statements in this key are only valid for taxa represented in the European fauna. Excluded as *genus incertum* (DI SABATINO et al. 2009): *Pseudothyas* THOR, 1899c.

1 Pointed dorsodistal projection of P-4 reaching the tip level of P-5 (*8-1 a*; *8-6 b*), often slightly bent dorsally; numerous Ac on genital plates (*8-1 c*); numerous densely-arranged long setae near margins of coxae (*8-6 a*); lateral eye lenses not enclosed in capsules (*8-1 b*); dorsal and posteroventral idiosoma completely smooth; legs II-IV with swimming setae (*8-7 a-d, f*). . . . . Family Hydrodromidae (only genus: *Hydrodroma*, page 8)

\- Pointed dorsodistal projection of P-4 not reaching tip of P-5, not bent dorsally (*8-1 d*; *8-4f*); morphology of lateral eyes, Ac, setation of coxae and legs various (Family Hydryphantidae). 2

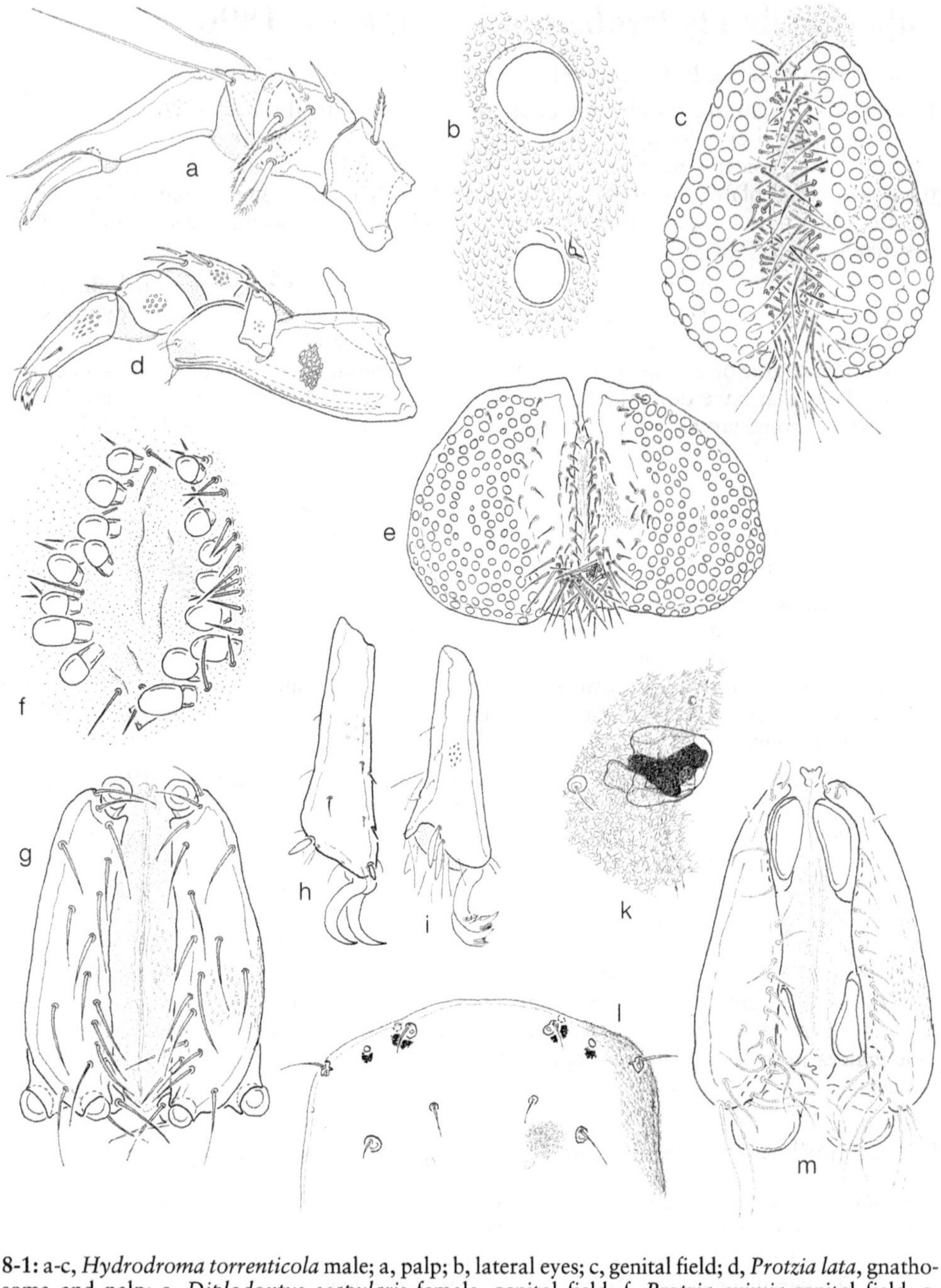

8-1: a-c, *Hydrodroma torrenticola* male; a, palp; b, lateral eyes; c, genital field; d, *Protzia lata*, gnathosoma and palp; e, *Diplodontus scapularis* female, genital field; f, *Protzia eximia* genital field; g, *Zschokkea oblonga* female, genital field; h, *Partnunia steinmanni* male, leg claws; i, *Protzia invalvaris* male, leg claws; k, *Pseudohydryphantes parvulus* female, integument structure and lateral eye capsule; l-m, *Tartarothyas romanica* male, l, frontal dorsum; m, genital field.

2 (1) Postgenital idiosoma vermiform (*8-34 a*); lateral eyes often little developed; genital plates reduced to irregular strips or absent (*8-34 b-d*). . . . . Subfamily Wandesiinae (only genus in the study area: *Wandesia*, page 72)

\- Postgenital idiosoma not conspicuously elongated; lateral eyes and genital plates various. 3

3 (2) Dorsal idiosoma completely soft (*8-22 a*, in some cases small sclerotized spots indicate position of muscle insertions); median eye, if present, lying free in the integument, not on a sclerotized plate. 4

\- Dorsal idiosoma with a frontal shield (e. g., *8-2 a, 8-3 c-d*) and/or regularly-arranged plates (e. g., *8-2 f, 8-4 b*), varying in size from small, roundish platelets (in diameter similar to glandularia – *8-5 b*) to extended shields (*8-4 d*); median eye, if present, often surrounded by a sclerotized plate. (In juveniles with little sclerotization, platelets may be faint, observe in tangential view at higher magnification in case of doubt). 9

4 (3) Genital field with four to more than 40 pairs of Ac (*8-1 e-f*). 5

\- Genital field with three pairs of Ac (*8-1 g, m*). 7

5 (4) Acetabula arranged on a pair of genital plates, not stalked; genital flaps well-developed, half-moon-shaped (*8-1 e, 8-8 a*); legs with swimming setae (*8-8 d*). . . . . . . Subfamily Diplodontinae (only genus: *Diplodontus*, page 13)

\- Acetabula in the soft integument surrounding gonopore, in most species stalked; genital flaps strap-shaped or reduced (*8-1 f, 8-28 b, 8-29 b*); legs without swimming setae (subfamily Protziinae). 6

6 (5) Legs with simple claws (*8-1 h*); genital flaps strongly sclerotized, their medial margins with numerous regularly-arranged setae (*8-28 b*). . . . . . *Partnunia* (page 56)

\- Legs with pectinate claws bearing fine lateral clawlets (*8-1 i*); genital flaps more or less reduced, sometimes irregularly-arranged setae the only remnants of flaps (*8-1 f, 8-29 b*). . . . . . *Protzia* (page 59)

7 (4) Legs with swimming setae; integument of the only species known from the area iridescent due to stellate papillae ending in fine tips (*8-1 k*); Ac oval in shape, in area between gonopore and genital flaps (*8-32 b*); lateral eyes in capsules (*8-1 k*). . . . . Subfamily Pseudohydryphantinae (only genus: *Pseudohydryphantes*, page 69)

\- Legs without swimming setae; integument with simple papillae; Ac and lateral eyes various. 8

8 (7) Lateral eyes reduced in size, not encapsulated (*8-1 l*); Ac oval, in the membranous area between gonopore and genital flaps (*8-1 m*); suture line separating Cx-I/II nearly parallel to longitudinal idiosomal axis; chelicerae with very strong claws (*8-33 c*, basal segment/claw ratio < 1.8). . . . . . Subfamily Tartarothyadinae (only genus: *Tartarothyas*, page 70)

\- Lateral eyes in capsules (*8-22 a*); Ac round; Ac-1 and -3 embedded in genital flaps, Ac-2 surrounded by posterior genital flap margin and appearing embedded in the flap (*8-1 g*); cheliceral claws less strong (basal segment/claw ratio > 1.8). . . . . . . Subfamily Euthyadinae partim, genus *Zschokkea* (page 40)

9 (3) Dorsum with a characteristic frontal shield which includes a pair of anterolateral preocularia, and with concave posterior margin between two posterolateral projections (*8-2 a*); other dorsalia unsclerotized or (rarely) developed as small, roundish spots or absent; legs generally with swimming setae (subfamily Hydryphantinae). 10

\- If an isolated frontal shield present (e. g., *8-3 c-d*), it neither includes the preocularia anterolaterally, nor forms a pair of projections posterolaterally; often with large, symmetrically-arranged dorsal and ventral plates; legs without swimming setae (subfamily Euthyadinae partim). 1

10 Genital field lateral margin straight in anterior, rounded in posterior part (*8-2 b*); if more than three pairs of Ac present, restricted to posterior part. . . . . . *Hydryphantes* (page 48)

\- Genital field lateral margin equally rounded or with a central constriction; numerous Ac both in the anterior and posterior part (*8-2 c*). . . . . . *Georgella* (page 42)

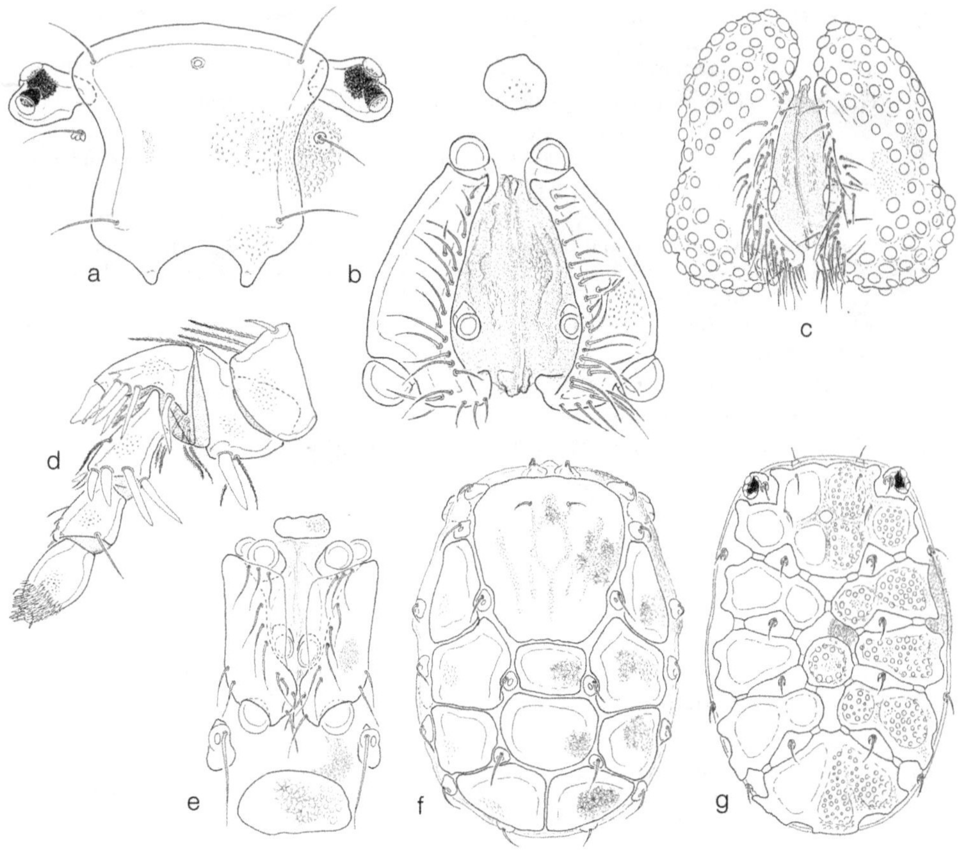

8-2: a-b, *Hydryphantes ruber* male; a, frontal shield; b, genital field; c, *Georgella helvetica* male genital field; d-f, *Dacothyas savulescui*; d, male I-L-3; e, female genital field; f, female dorsal idiosoma; g, *Tadjikothyas fibulata, male* dorsum.

11 (9) Dorsum with three large medial plates (anterior plate: frontal shield including frontal plates and Dc-1; central plate: medially-fused Dc-2; posterior plate: medially-fused Dc-3), these plates surrounded by isolated platelets and glandularia (*8-2 f*). 12

\- Arrangement of dorsal plates various, not with three medial plates in series (e. g., *8-4 b, d, 8-5 a-b*). 13

12 (11) Three pairs of Ac; males (pregenital sclerite longer than wide or reduced): Cx-III medial margins not close, with one pair of slender genital flaps not divided into anterior and posterior platelets (*8-20 a*); I-L-3 without a lateral extension (*8-20 d*); females (pregenital sclerite wider than long, crescent-shaped): genital flaps narrow, maximum width similar to diameter of posterior Ac (*8-20 b*). . . . . *Trichothyas* (page 37)

\- Four pairs of Ac (*8-2 e*); males: Cx-III close, with medial margins nearly in touch, dividing genital flaps into anterior and posterior pairs of platelets; I-L-3 medially unsclerotized, with a strongly developed lateral extension bearing an enlarged seta (*8-2 d*); females: Genital flaps large, maximum width about twice diameter of posterior Ac (*8-2 e*).
. . . . *Dacothyas* Motaş, 1959
(Not recorded from the study area. Two species known from Romania and Turkey – Özkan & Bader 1988)

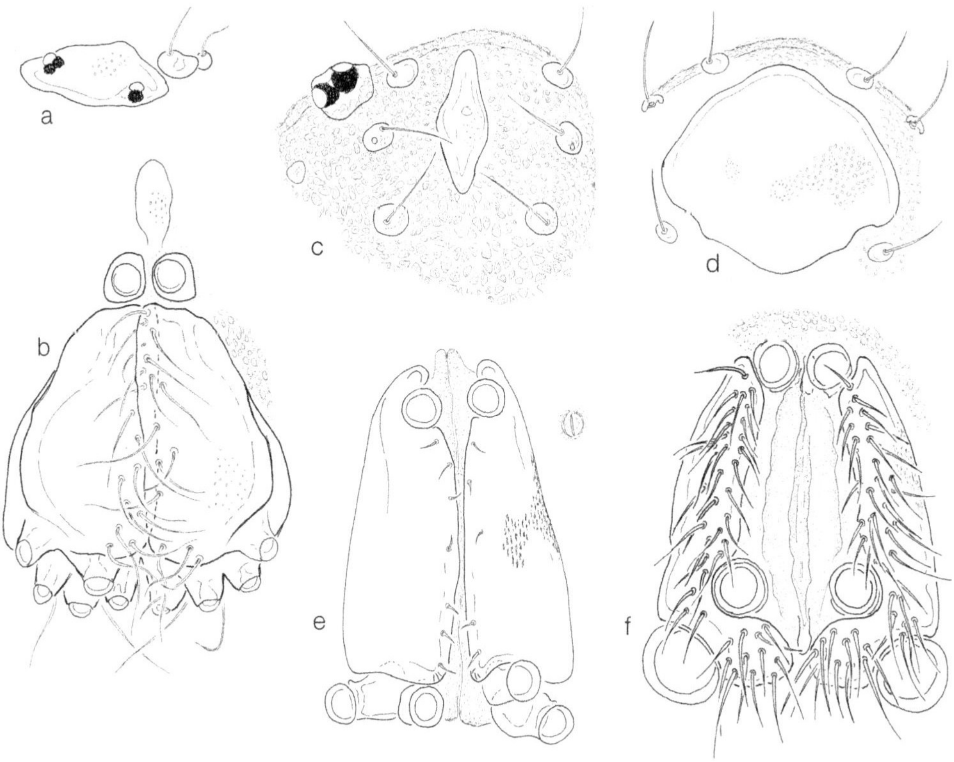

8-3: a-b, *Panisellus thienemanni* male; a, lateral eye platelets; b, genital field; c, *Euthyas truncata* female, frontal shield and integument; d, *Vietsia scutata* male, frontal shield; e, *Thyasella mandibularis* female, genital field (anterior setae lacking, but insertion points visible); f, *Panisus michaeli* male, genital field.

13 (11) More than three pairs of Ac (*8-3 b*); lateral eyes with small lenses, lying in the frontal area on a pair of platelets (*8-3 a, 8-10 a*). . . . . *Panisellus* (page 17)

\- Three pairs of Ac (*8-3 e-f*); lateral eyes with larger lenses, lying more laterally in roundish capsules (*8-2 g, 8-3 c*). 14

14 (13) Dorsal plates linked by subcuticular strips (*8-2 g*) or forming a complete latticed shield (*8-17 a, d*). 15

\- Dorsal plates, if present, separate or fused in various ways, but not linked by subcuticular strips – if a large dorsal shield is present, its surface with normal porosity, not latticed (*8-4 d*). 16

15 (14) Dorsum covered by a latticed shield with 'annual ring'-like marginal growth strips, only lateral eyes surrounded by small membranous areas (*8-17 a, d*). . . . . *Thyopsis* (page 36)

\- Dorsal plates linked by subcuticular strips, but separated by membranous areas; frontal shield including Dl-1 and Dc-1; Dc-2 and Dc-4 on each side fused with neighbouring Dl-2 and -4; Dc-3 fused medially forming a round central platelet, posterior shield including Dc-5 (*8-2 g*). . . . . *Tadjikothyas* Sokolow, 1948

(Not recorded from the study area. Two species known from Tadzikistan, Romania and Turkey – Özkan 1988a).

16 (14) Papillae of integument triangular or quadrangular, arranged in rosettes on a lattice-like structure (*8-3 c*); a small rhomboid frontal plate bearing pigmented median eye, all other muscle insertion sclerites small and knob-shaped (*8-9 a*). . . . . *Euthyas* (page 15)

\- Papillae of integument normal, roundish, not arranged in particular patterns; if a frontal shield present, not small and rhomboid (e. g. *8-3 d*); dorsal plates various. 17

17 (16) A large frontal shield present, all other muscle insertions unsclerotized (*8-3 d, 8-21 d*); genital flaps with a dense cover of fine, similar setae; Ac small (*8-21 b*). . . . . *Vietsia* (page 39)

\- If a large frontal shield present, idiosoma bearing further symmetrically-arranged dorsal and ventral plates (e. g., *8-4 b, 8-5 a*); genital flap setation various, often with a row of stronger setae at medial margin; Ac may be large. 18

18 (17) Anterior genital flap margin with an extension embracing Ac-1; at least one genital flap seta inserting anterior to this Ac (*8-3 e, 8-4 a*). 19

\- Anterior genital flap margin without an extension embracing Ac-1; no setae located anterior to this Ac (*8-3 f, 8-5 c-d*). 23

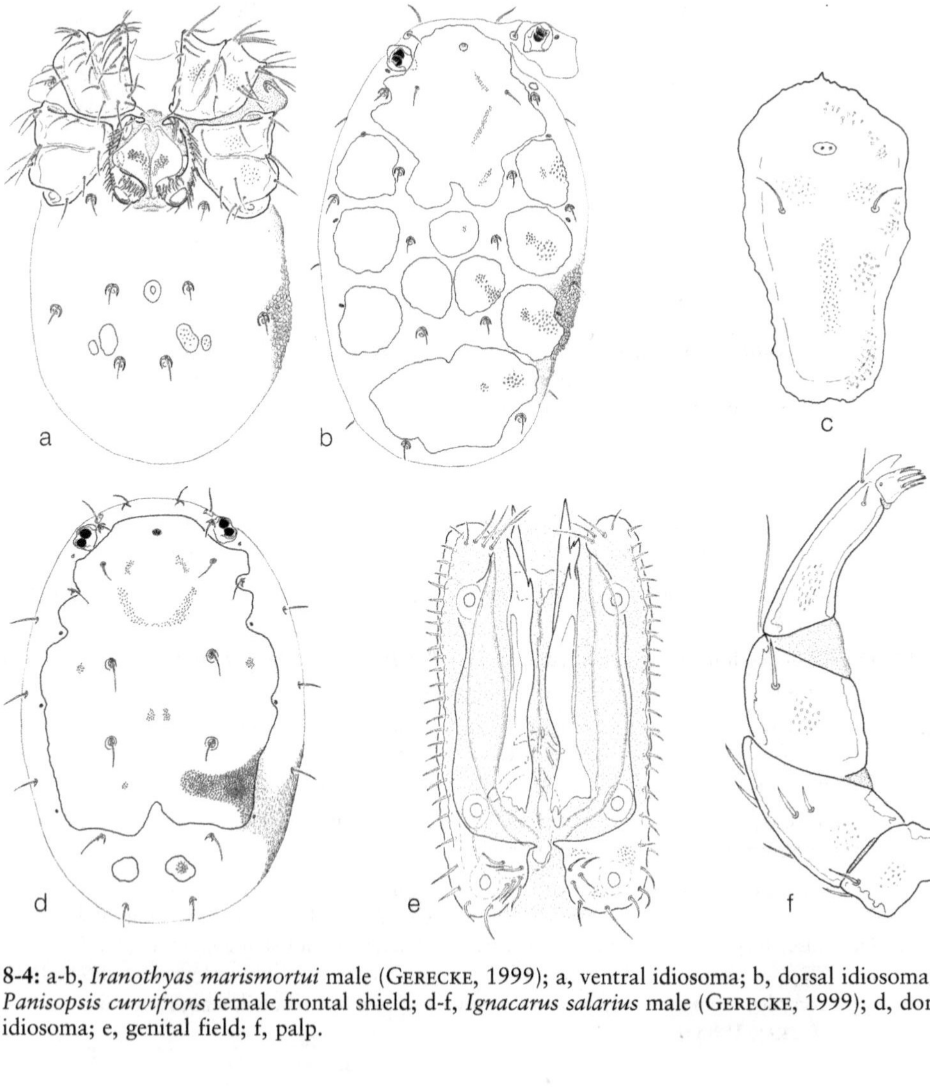

8-4: a-b, *Iranothyas marismortui* male (GERECKE, 1999); a, ventral idiosoma; b, dorsal idiosoma; c, *Panisopsis curvifrons* female frontal shield; d-f, *Ignacarus salarius* male (GERECKE, 1999); d, dorsal idiosoma; e, genital field; f, palp.

19 (18) P-4 relatively short, P-5 relatively long (*8-15 b*); frontal shield roundish, not including postocularia (*8-15 a*). 20

\- P-4 relatively long, P-5 relatively short (*8-4 f*); frontal shield various. 21

20 (19) Body sclerites weakly sclerotized, Dc-1 separate from frontal shield (*8-15 a*); frontal eye not visible; genital flaps with less than 20 pairs of setae, single seta anterior to Ac-1 nearly invisible. . . . . *Thyasella* (page 33)

\- Body sclerites strongly sclerotized, Dc-1 fused with frontal shield (*8-15 c*); frontal eye well visible; genital flaps with more than 20 pairs of setae, single seta(e) (n = 1-3) anterior to Ac-1 clearly visible. . . . . *Parathyasella* (page 32)

21 (19) Frontal shield slender, laterally not reaching level of eye capsules (*8-4 c*); posterior Ac lying free in postgenital membranous integument. . . . . *Panisopsis* (page 18)

\- Frontal shield large, laterally reaching, or extending beyond, level of eye capsules (*8-4 b, d*), occasionally making part of a shield covering the whole dorsum; posterior Ac embedded into posterior margin of genital flaps (*8-4 a, e*). 22

22 (21) Posterior dorsum with paired plates Dc-5; P-4 relatively slender (*8-4 f*); Ac small (diameter Ac-3 < 20 µm, *8-4 e*); males: dorsum nearly completely covered by large shield (*8-4 d*); genital field equiped with strong, pointed, blade-like structures (*8-4 e*).
. . . . *Ignacarus* GERECKE, 1999

(Not recorded from the study area. Monotypic genus known only from a hypersaline spring in Southern Spain – GERECKE 1999)

\- Posterior dorsum with an un paired plate formed by medially fused Dc-5; P-4 stouter; Ac large (diameter Ac-3 > 30 µm, *8-4 a*); males: frontal shield including dorsocentralia 1, often also dorsocentralia 2, further dorsocentralia and –lateralia lying separate (*8-4 b*); genital field with a pair of tongue-shaped lips densely denticulate on inner surfaces (*8-4 a*).
. . . . *Iranothyas* BADER, 1984

(Not recorded from the study area. In the West Palaearctic only one species found in a spring in Israel – GERECKE 1999, PEŠIĆ et al. 2009)

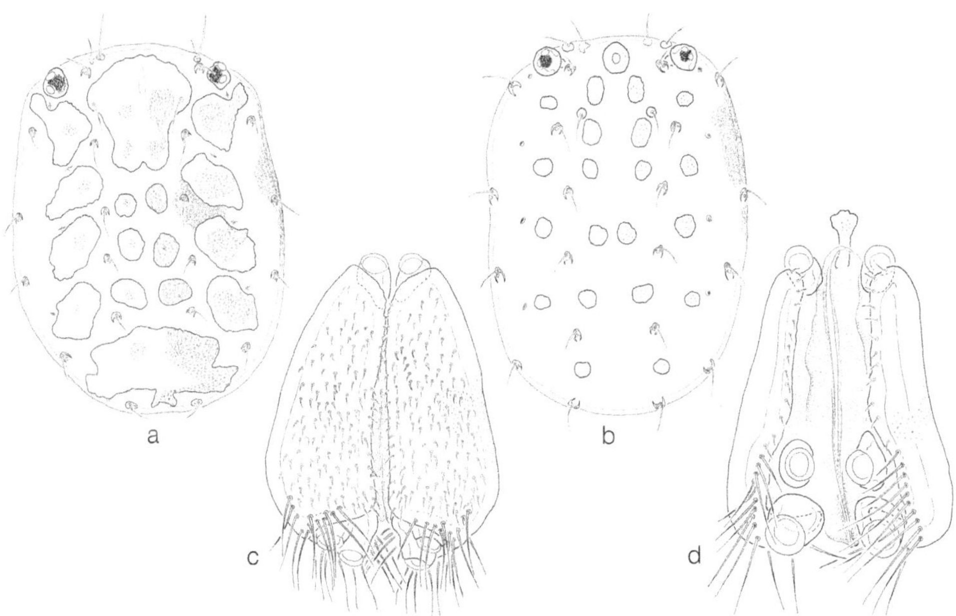

8-5: a, *Panisus michaeli* male, dorsal idiosoma; b, *Parathyas palustris* male, dorsal idiosoma; c, *Thyasides dentatus* female, genital field; d, *Parathyas thoracata*, genital field.

23 (18) Dorsal plates large, frontal shield formed by fusion of pre- and postfrontalia, postocularia and Dc-1; Dc-5 medially fused to a large posterodorsal plate (*8-5 a*). . . . . *Panisus* (page 21)

\- Dorsal plates smaller, frontal shield including only pre- and postfrontalia or completely lacking; Dc-5 not fused medially (*8-5 b*). 24

24 (23) Frontal shield typically formed by pre- and postfrontalia; median eye pigment forming two spots (*8-16 a*); genital flap setae densely dispersed all over the surface (*8-5 c*). . . . . *Thyasides* (page 34)

\- Frontal area various, often frontal eye not fused with surrounding plates; median eye pigment, if present, forming one spot; genital flap setae restricted to the medial margin (*8-5 d* - in species from outside the area treated, with setae located on flap surface, frontal eye separate from frontalia). . . . . *Parathyas* (page 24)

Family Hydrodromidae K. VIETS, 1936

1936d Hydrodromidae K. VIETS, Die Tierwelt Deutschlands und angrenzender Meeresteile 31/32: 140.

Diagnosis: Colour red, orange or yellow. Idiosoma soft, without sclerotized dorsalia or ventralia. No frontal eye. Lateral eyes not encapsulated, on each side with two lenses separate from each other, associated with the integument. Cx-I+II medially in touch, only separated by a fine membranous line. Rows of characteristic long fine setae at medial margin of Cx-I, and posterior margins of Cx-II, -III and -IV. Genital flaps with more or less equally rounded lateral and slightly concave medial margins, bearing fine setae and numerous small Ac arranged in several rows along medial edge. In immature specimens, Ac appearing as lying on irregularly-shaped sclerotized rings fused to each other in various degrees, locally remaining separated from each other by irregular slits. Often a slight sexual dimorphism in shape and setation (males with more slender flaps, setae more numerous and longer). Legs bearing swimming setae. Leg claws without claw blade, with a dorsal clawlet.

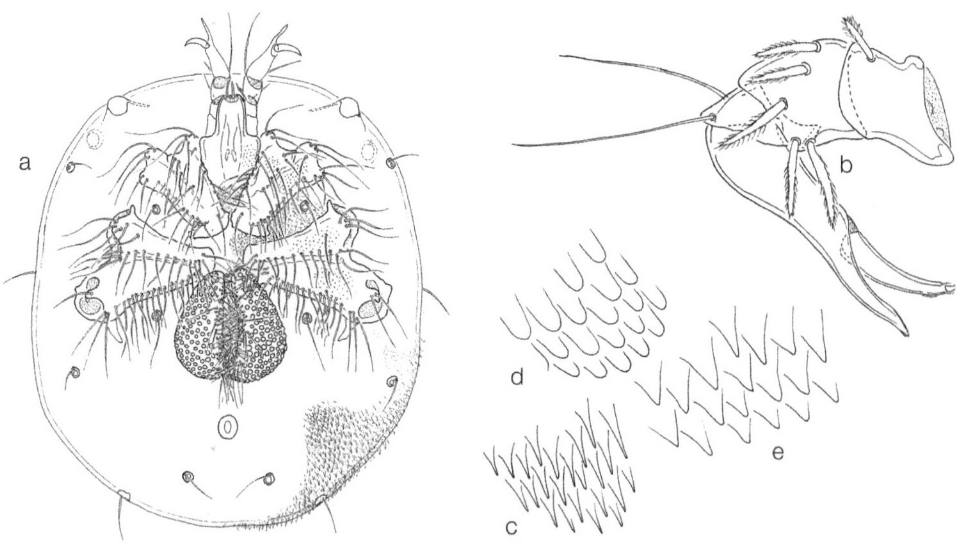

8-6: a-b, *Hydrodroma pilosa*; a, male ventral idiosoma; b, palp; c-e, *Hydrodroma* integument structure; c, *H. pilosa*; d, *H. despiciens*; e, *H. torrenticola*.

Gnathosoma rostrum short. Palp five-segmented; P-2 with pectinate mediodistal setae; P-4 with a long, pointed dorsodistal extension reaching tip of the slender and elongated P-5.

Remarks: The family is highly distinct and probably monotypic. Apart from the genus *Hydrodroma*, represented nearly worldwide by numerous species, only one other monotypic genus, *Oxopsis* NORDENSKIÖLD, 1905, has been recorded from Sudan. COOK (1974) suggested that the characters of the single specimen on which this taxon is based (lateral compression of idiosoma, presence of a median dorsal furrow) resulted from distortion during preservation.

Genus *Hydrodroma* KOCH, 1837

1837 *Hydrodroma* KOCH, Deutschlands Crustaceen, Myriapoden und Arachniden 14: 7.

Type species: *Hydrodroma despiciens* (MÜLLER, 1776) (sub nom. *H. umbrata* Koch, 1837), designated by KOCH (1842).

Diagnosis: As given for the family. Idiosoma not laterally compressed, wider than high and in dorsal view often subquadratic.

Preadult stages: WILES (1985) and the bibliography cited there. Deutonymphs: similar to adults, but with only two Ac in the provisional genital field.

Biology: Life cycle known for one central European species only (see below). Much wider variety of adaptations to be expected in tropical faunas.

Habitat: All European species in standing waters or pool areas of streams, with well-developed swimming setae, species-specific differences in setae numbers. In tropical species, various degrees of swimming setae reduction, up to a total loss in species living in stream riffles and cascades.

Distribution: All continents except Antarctica.

Key to species

(In normal position, anterior leg segment surface directed ventrally, posterior surface directed dorsally)

1 II-L-5 with more than four swimming setae, IV-L-5 posteriorly with 10 or more swimming setae; integument papillae acutely pointed (*8-6 c*); genital flaps with 6-8 Ac lying along an imaginary transverse line crossing at maximum width. . . *Hydrodroma pilosa* (page 11)

- II-L-5 with one or without swimming setae, IV-L-5 posteriorly with less than 10 swimming setae; integument papillae not acutely pointed (*8-6 d-e*); genital flaps with 4-6 Ac lying along an imaginary transverse line crossing at maximum width. 2

2 (1) III-L-4 with more than eight swimming setae, IV-L-5 anteriorly without swimming setae; integument papillae rounded (*8-6 d*). . . . . *Hydrodroma despiciens* (page 9)

- III-L-4 at most with eight swimming setae, IV-L-5 anteriorly with 2-5 swimming setae; integument papillae bluntly pointed (*8-6 e*). . . . . . *Hydrodroma torrenticola* (page 12)

1. *Hydrodroma despiciens* (MÜLLER, 1776) (*8-6 d, 8-7 c-d*)

1776 *Hydrachna despiciens* MÜLLER, Zoologiae Danicae prodromus, Havniae (Hallager): 190.

Description: Colour red. Integument papillae long, distally rounded. Idiosoma length 700-2000 µm. Coxal setae numbers Cx-I, 20-24; Cx-II, 20-26; Cx-III, 22-25; Cx-IV, 19-22. Leg setae numbers: II-L-5 posterior 1; III-L-4 posterior 9-13, III-L-5 posterior 6-10; IV-L 4 anterior 9-14, posterior 8-15, IV-L-5 posterior 6-9. Genital plate length 225-305 µm, with at most 6 longitudinal rows of Ac. Males: Genital plate more slender

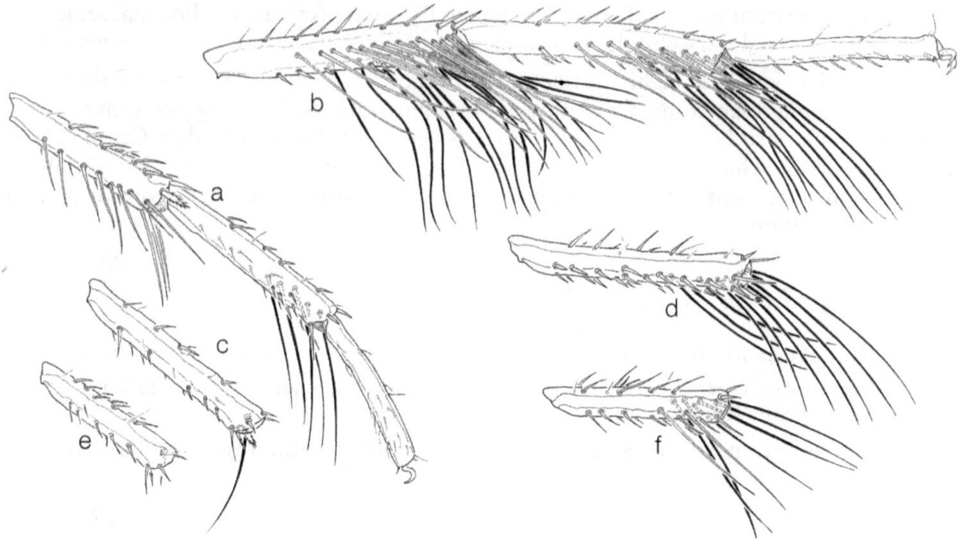

8-7: *Hydrodroma* legs in anterior view; a-b, *H. pilosa*; a, II-L-4-6; b, IV-L-4-6; c-d, *H. despiciens*; c, II-L-5; d, IV-L-5; e-f, *H. torrenticola*; e, II-L-5; f, IV-L-5.

setae more numerous than in females, all hollow. Females: Genital plate larger; Ac number 38-75, with 20-44 setae, which in the anterior part are solid and flat, in the posterior part hollow; Ac number 48-90.

Similar species: *Hydrodroma reinhardi* Pešić, 2002e, recorded from streams in Montenegro, Central Italy, Sardinia and Corsica, is similar both in the presence of only one swimming seta on II-L-5, and in the absence of swimming setae from anterior IV-L-5. It differs from *H. despiciens* in reduced swimming setae numbers on III/IV-L (less than 10 on both posterior III-L-4 and anterior IV-L-4). For differences to *H. pilosa*, see below.

Preadult stages: Larva: Wiles (1985). Deutonymph: Numerous descriptions under this name uncertain (the detailed description published under this name by Tuzovskij (1990) due to integument structure is clearly referable to *H. pilosa*). Distinguished from *H. pilosa* in shape of integument papillae and complete absence of setae from II-L-5.

Biology: Unclear. Due to a long history of taxonomic confusion, resulting in late detection of two locally coexisting taxa (see *H. pilosa*), interpretation of most published data impossible. In all cases of possible taxonomic attribution, life cycle studies refer to the sister taxon *H. pilosa*. This is true also for all references to *H. despiciens* made in the chapter introducing Hydrachnidia in volume 1 of this book. Adults feed on chironomid egg masses (Wiles 1982).

Habitat: Standing waters of various dimensions, with a preference for lower alkalinity (Wiles 1985, Gerecke 1991a).

Distribution: One of the few water mites considered previously cosmopolitan. However, since *H. pilosa* has been distinguished from *H. despiciens*, evidence increases that the latter is much less widely distributed than supposed in former times. Under Mediterranean and semi-arid climate in Southern Italy, *H. despiciens* is rare and mostly restricted to mountain ponds at higher elevations (Gerecke 1991a). Pešić & Smit (2007a, b) demonstrated that Australian populations formerly attributed to *H. despiciens* represent several clearly distinct autochthonous species. Therefore, new taxonomic evidence is needed concerning all other extra-European records.

2. *Hydrodroma pilosa* BESSELING, 1940 (*8-6 a-c, 8-7 a-b*)

1940 *Hydrodroma despiciens pilosa* BESSELING, Ent. Ber. 10: 246.

1991a *Hydrodroma pilosa* GERECKE, Lauterbornia 7: 66.

Description: Colour red. Integument papillae sharply pointed. Idiosoma length 700-2000 µm. Coxal setae numbers Cx-I, 17-22; Cx-II, 19-25; Cx-III, 21-25; Cx-IV, 24-27. Leg setae numbers: II-L-5 posterior 8-10; III-L-4 posterior 15-19, III-L-5 posterior 10-16; IV-L-4 anterior 14-18, posterior 15-20, IV-L-5 anterior 5-9, posterior 11-16. Genital plate length 265-350 µm. Males: Genital plate more slender, setae more numerous than in females, all hollow; Ac number 55-90. Females: Genital plate a little larger, with 33-42 setae, which are in the anterior part solid and flat, in the posterior part hollow; Ac number 60-100.

Similar species: Obviously a sister taxon of *H. despiciens*. In larvae, WILES (1985) found larger general dimensions as the only differences in comparison with *H. despiciens*. The coexistence of populations of both taxa in several sites induced GERECKE (1991a) to attribute species rank to what formerly was considered two subspecies or "forms" of uncertain taxonomic significance. Since then, the differences mentioned were verified with surprising stability in several geographical areas (KOWALIK 2002, PEŠIĆ 2002a). Also a study on chromosomal numbers, published under the name *H. despiciens* by ONRAT et al. (2006), refers to *H. pilosa*.

Preadult stages: Larva: WILES (1985). Deutonymph: WILES (1985); under the name *H. despiciens*: TUZOVSKIJ (1990).

Biology: Under temperate conditions, adults (possibly also nymphal stages) hibernating in the bottom sediment of lakes and appearing first in March/April in deeper layers (MEYER & SCHWOERBEL 1981). From May to November, adults and deutonymphs in the riparian macrophyte belt, where oviposition takes place on the underside of *Nymphaea alba* or *Nuphar lutea* leaves. Fertilization by indirect spermatophore transfer. Ovigerous females from end of May till late November, but main oviposition period from May to July only. A single female may produce numerous egg clutches with a total number of up to 800 eggs. Larval emergence from June to September, with a peak in late June. Larvae hatch preferably at dusk, climb to the water surface and attach to thorax, rarely abdomen, of emerging insects. Parasitism observed on at least 40 different species of the Dipteran families Chaoboridae, Chironomidae and Tipulidae, and the Trichopteran families Leptoceridae and Limnephilidae (WILES 1987, SMUKALLA & MEYER 1988). Host preference varying geographically: populations in a S German lake with *Chaoborus flavicans* (MEIGEN, 1830) as the preferred host (MEYER 1985), but neglecting the chironomid *Cricotopus sylvestris* (FABRICIUS, 1794), a species heavily parasitized in the British Isles (WILES 1987). Duration of parasitic phase and postparasitic protonymphal stage only a few days each, but with an increase in idiosoma size by 100 % during parasitism. At summer temperatures, deutonymphal stage normally taking only about four weeks, tritonymphal resting stage only a few days (MEYER 1985, WILES 1987 – see there for many further details on the life cycle).

Habitat: Lakes, ditches, ponds and pools of streams.

Distribution: Western Palaearctic. Widely distributed in temperate and warm climates, with a tolerance against eutrophication and elevated electrolyte contents. Possibly by far more than 50 % of the records under the name of *H. despiciens* in reality refer to *H. pilosa*.

3. *Hydrodroma torrenticola* (WALTER, 1908) (*8-1 a-c, 8-6 e, 8-7 e, f*)

1908b *Diplodontus torrenticolus* WALTER, Arch. Hydrobiol. 4: 1.

1956a *Hydrodroma danuviensis* SCHWOERBEL, Arch. Hydrobiol., Suppl. 24 (1): 69.

1964a *Hydrodroma torrenticola danuviensis* SCHWOERBEL, Arch. Hydrobiol., Suppl. 27: 393.

Description: Colour orange to brown. Integument papillae bluntly pointed. Idiosoma length 600-1500 µm. Coxal setae numbers Cx-I, 23-33; Cx-II, 26-36; Cx-III, 21-25; Cx-IV, 20-23. Medial margin of Cx-II and Cx-III shorter, posterior margin of Cx-IV less concave than in the preceding species. Leg setae numbers: II-L-5 posterior 1; III-L-4 posterior 5-8, III-L-5 posterior 4-8; IV-L 4 anterior 7-10, posterior 7-9, IV-L-5 anterior 2-4, posterior 4-7. Genital plate length 225-275 µm, sexual dimorphism as in *H. despiciens*. Males: Genital plate more slender, setae more numerous than in females, all hollow; Ac number 39-58. Females: Genital plate a little larger, with 25-45 setae, which are in the anterior part are solid and flat, in the posterior part, hollow; Ac number 40-70.

Similar species: *Hydrodroma torrenticola* differs from the two other Central and Western European species not only in colour and distinctly lower swimming setae numbers, but also in generally smaller dimensions. *Hydrodroma reinhardi*, a species similar in having reduced general swimming setation (e. g., III-L-4 with less than five swimming setae), differs in the absence of swimming setae from anterior IV-L-5.

Preadult stages: Larva and deutonymph: WILES (1985).

Habitat and Biology: Pools of low order streams, slowly flowing riparian reaches of higher order streams. Deutonymphs occasionally in the hyporheic. Life cycle unknown.

Distribution: Central, Western and Southern Europe.

Family Hydryphantidae PIERSIG, 1896

1896c Hydryphantinae PIERSIG, Sitzungsberichte naturf. Ges. Leipzig 22: 48.

Diagnosis: Membranous integument papillate, dorsal and ventral sclerites various, from extended and fusing to large shields to poorly developed or completely absent. Frontal eye absent or present. Lateral eyes often in capsules, but in some hyporheophilic taxa not encapsulated or reduced. Genital field plesiotypically with movable flaps and three pairs of Ac flanking the gonopore, but development to polyacetabulism and shifting of Ac on the flaps in several clades. Legs with or without swimming setae. Palp chelate, with a pointed dorsodistal extension on P-4 flanking the proximal half of P-5 (occasionally, this extension formed as a strong seta).

Remarks: This is obviously a para- or polyphyletic family, with a high number of genera which do not fit the diagnosis of one of the remaining six, probably monophyletic hydryphantoid families (see also above, discussion of Hydryphantoidea). In addition to the seven subfamilies represented in Europe and treated here, COOK (1974), K. O. VIETS (1987), MITCHELL (2003) and GOLDSCHMIDT & GERECKE (2003) include Ankelothyadinae (Chile), Chimerathyadinae (N America, only genus *Chimerathyas*), Cowichaniinae (N America, only genus *Cowichania*), Cyclothyadinae (E Asia), Mamersinae (Africa, S and E Asia, Australia) and Eupatrellinae (E Africa, S America). In view of the phylogenetically heterogeneous character of this family, no general statement on biology and distribution is made.

Subfamily Diplodontinae K. VIETS, 1936

1936d Diplodontinae K. VIETS, Die Tierwelt Deutschlands und angrenzender Meeresteile 31/32: 140.

Diagnosis: Lateral eyes in capsules and these with well developed apodemes, median eye free in integument; dorsalia and ventralia small or reduced; if swimming setae present, these arranged in longitudinal lines on posterior surface of II-L-5, III-L-4/5 and IV-L-4/5, one to three additional swimming setae distally on IV-L-3; genital plates bearing numerous Ac (on each side from 15 to more than 300 – an additional pair between the genital plates and anterior part of gonopore in one African species); numerous setae on genital plates, mostly lined up along their medial margins, some of them scattered between Ac, a characteristic group of more densely-arranged setae at posteromedial genital plate margin; cheliceral claw straight; P-2 mediodistally with one to six strong, sometimes curved setae.

Remarks: A well-defined monotypic taxon separated from all other hydryphantids by large morphological gaps. Diplodontines are most similar to species of the monotypic Eupatrellinae (GERECKE 2004a, GOLDSCHMIDT & GERECKE 2003) in: (1) multiplied, laterally arranged secondary Ac (in *Eupatrella* on stalks in the soft integument); (2) numerous setae flanking the gonopore in two groups, with shorter anterior, and longer posterior setae (in *Eupatrella* distributed on two pairs of platelets); (3) simple leg claws, and (4) presence of small, but distinct muscle insertion sclerites in some probably early derivative species. The first two characters could indicate a common origin, the latter two character states are plesiomorphic.

Distribution: Palaearctic, Ethiopian, Oriental, Australian.

Genus *Diplodontus* DUGÈS, 1834

1834b *Diplodontus* DUGÈS, Ann. Sci. Natur. Paris (s. 2) Zool. 1: 148.

Type species: *Diplodontus scapularis* DUGÈS, 1834, designated by LUCAS (1844).

Diplodontus scapularis DUGÈS, 1834 (*8-1 e, 8-8 a-d*)

1834b *Diplodontus scapularis* DUGÈS, Ann. Sci. Natur. Paris (s. 2) Zool. 1: 148.

Description: Colour black in the anterior, red in the posterior part, a coloration unique in water mites. Dorsal and posteroventral idiosoma completely membranous in juveniles, with small sclerotized muscle insertion spots in older specimens. Integument papillae rounded or slightly pointed, bent posteriorly, in vertical view giving a squamose aspect to skin surface. Cx-I+II forming posteromedial extensions embracing the gnathosomal bay, but medially separated by a membranous strip. P-2 with one or two long and strong medial setae. Chelicera large and strong, basal segment about twice as long as claw (basal segment/claw ratio 1.7-2.1). Males: Idiosoma length/width 1000/700. Genital flaps relatively small, posteromedially extending into a distinct projection; genital setae fine and short, widely scattered along the edge flanking gonopore (in this area less than 10 setae), more densely arranged on posteromedial projections. Ejaculatory complex with *brachia distalia* longer than *brachia proximalia*. Legs, in particular III/IV-L, robust and with strong, shortened claws (e. g. length/height ratio IV-L-5 < 4.0). Females: Idiosoma length/width 1000-2500/900-1800. Genital field with larger flaps, posteromedially with an area of thickened sclerotization, but without projections; genital setae in anterior and central part of medial edge in shape as in males, but more numerous, in the posterior strongly

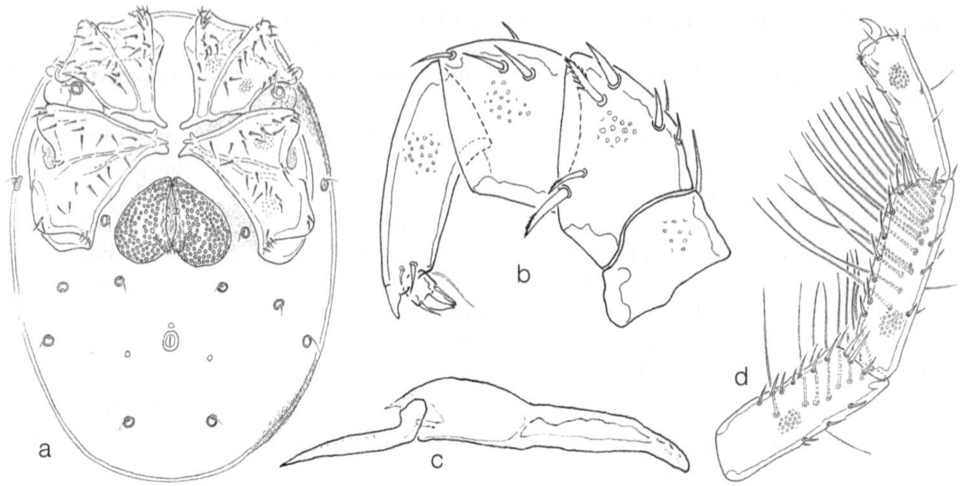

8-8: *Diplodontus scapularis* male; a, ventral idiosoma; b, palp; c, chelicera; d, IV-L-4-6.

sclerotized area thickened and longer. Legs more slender (e. g., length/height ratio IV-L-5 > 4.0), leg claws of III-/IV-L less robust than in males.

Similar species: A sister species is *D. semiperforatus* WALTER, 1925a, recorded from Western Mediterranean saline inland waters (one doubtful record from the S French coast, see GERECKE 1991a). Common features not found in other species of the genus are the stout shape of the palp, and a marked sexual dimorphism in genital field and IV-L (GERECKE 2004). *Diplodontus semiperforatus* differs in shape of chelicerae (claw relatively longer, length ratio basal segment/claw < 1.7) and male ejaculatory complex (smaller in size, with less developed *brachia distalia*, GERECKE 1991a). Due to its preference for saline and hypersaline waters, and a larval host preference for halobiont beetles, it is restricted to semi-arid and arid areas of the SW Palaearctic (MORENO et al. 2004).

Preadult stages: Larva: MORENO et al. (2004). Deutonymph: With about 30-60 pairs of Ac, on both sides arranged in a smaller medial group and a larger lateral group, separated by a setae-bearing sclerotized bow (GERECKE 1991a, 2004a).

Biology: In Morocco, larvae were found in July attached to intersegmental membranes of the subelytral dorsal abdomen of hydrophilid beetles (*Laccobius moraguesi* RÉGIMBART, 1898; *Hemisphaera guignoti* SCHÄFER, 1975; *Hydrochus grandicollis* KIESENWETTER, 1870). They develop on the host to the protonymphal stage and leave at the deutonymphal moult (MORENO et al. 2004). The widely scattered distribution of the species suggests that a considerable dispersal capacity is maintained at least in some parasitized beetle host specimens.

Habitat: Remnant pools of summer-dry streams, small standing water bodies such as farmland drinking troughs, dune ponds and ditches; often at elevated electrolyte concentrations (conductivity 500-2.800 µS/cm: GERECKE 1991a).

Distribution: SW Palaearctic, only a few, mostly old, records from coastal reaches of the area treated (France, Var, Pas-de-Calais, Hérault: DUGÈS 1834b, MOTAŞ 1928f, 1928g; BARROIS 1889b; The Netherlands: DAVIDS 1979; England (GLEDHILL, personal records); Germany: KOENIKE 1896b).

Subfamily Euthyadinae K. VIETS, 1931

1926l Thyasinae K. VIETS, Zool. Anz. 69: 190.

1931c Euthyasinae K. VIETS, Zool. Anz. 93: 210.

2007 Thyadidae TUZOVSKIJ, Acarina 15 (2): 181.

Diagnosis: Lateral eyes on each side in capsules or on a common eye plate. Median eye generally present, but often not pigmented. Dorsal and posteroventral muscle insertions typically sclerotized, often developing characteristic patterns of large plates, but in some taxa sclerotization reduced to a frontal plate; a sclerotized ring around the frontal eye, or completely absent. Genital field plesiotypically with large flaps and three pairs of Ac, in several taxa flaps reduced to small sclerotized strips and / or polyacetabulism occurring. Legs without swimming setae, leg claws simple.

Remarks: Originally, the subfamily Euthyadinae was introduced for those genera bearing a pigmented frontal eye, supposing a sister relation to Thyadinae including all genera lacking pigment in the frontal eye. As this system was evidently highly artificial – frequently, specimens with and without frontal eye pigmentation were found within a single population – COOK (1974) proposed the synonymization of the two subfamilies. Since ÖZDIKMEN (2006) pointed out that *Thyas* and Thyadinae, were preoccupied names in Lepidoptera, Euthyadinae became the valid subfamily name for all genera formerly included in Thyadinae + Euthyadinae.

The subfamily is surely a polyphyletic reservoir including several phylogenetically distant clades with uncertain relationships among each other and to members of other subfamilies. This taxon is held together by plesiomorphies such as the presence of a frontal eye, the absence of swimming setae, and a strong development of idiosoma sclerotized plates. Better taxonomic solutions are required on a large taxonomic scale: an appropriate phylogenetic assessment of the euthyadine genera will probably result in changes at the family, possibly also at the superfamily level. Concerning TUZOVSKIJ's (2007) proposal to re-establish this clade at family rank, see discussion of Hydryphantoidea. In view of taxonomic heterogeneity, it is senseless including topics such as habitat preference, distribution or life cycles at subfamily level.

Genus *Euthyas* PIERSIG, 1898

1875 *Bradybates* NEUMAN (nec TSCHUDI, 1838), Öfvers. Vetensk.-Akad. Förhandl., Stockholm 2: 104.

1898e *Euthyas* PIERSIG, Zool. Anz. 21: 575.

Type species: *Bradybates truncatus* NEUMAN, 1875, by original designation.

Diagnosis: Integument papillae on a subepidermal lattice with subtriangular openings, which are over large parts of the dorsal and posteroventral idiosoma arranged in a rosette pattern. Lateral eyes in capsules. Median eye with two integument spots, on a rhombic frontal shield made up of the frontal eye sclerite and frontalia. Dc-1+2 of each side fused to a rod-shaped plate, all other dorsalia small and rounded. Pre- and postocularia on large bases. Glandularia sclerites rounded, with a small gland pore. Dgl-3 shifted laterally. Coxae and legs with dense, uniform setation. Posteromedial margin of Cx-I+II not projecting. Genital field nearly circular, genital flaps densely covered along medial margins and over the surface by numerous setae rather uniform in shape and size. Three or four pairs of Ac on spindle-shaped basal sclerites. Gnathosoma slender, rostrum long, palp with dense, rather uniform setation, P-4 bearing numerous dorsal setae.

Remarks: The plesiomorphic, dense uniform setation of appendages and genital flaps suggests that this genus represents a clade which branched off early from the stem line of Hydrachnidia. The integument structure is an autapomorphy strongly distinguishing this line from all other members of Hydryphantidae. In the study area only one species.

Euthyas truncata (NEUMAN, 1874) (*8-3 c, 8-9 a-c*)

1875 *Bradybates truncatus* NEUMAN, Öfvers. Vetensk.-Akad. Förhandl., Stockholm 2: 104.

Description: Frontal shield rhombic, variable in shape from stout to slender. Genital field without pregenitale, but postgenitale present. Genital flaps densely covered with uniform, fine setae. Three pairs of small Ac located respectively at the anterior edge, in the centre, and at the posterior edge of gonopore. Excretory pore sclerite pear-shaped. Gnathosoma rostrum highly variable in shape (more or less curved, diameter of mouth opening 30-50). Chelicera length/height 560/96. Palp rather stout, with one dorsal seta on P-1, and dense setation on segments 2-4. Males: Idiosoma length up to 1700; gnathosoma length/width 550-590/170, length/height P-1, 75/98; P-2, 171/96; P-3, 129/104; P-4, 243/61; P-5, 57/29 µm. Females: Idiosoma length up to 2500 µm; genital flap setation sparser than in males; gnathosoma length/width 440/150, length/height P-1, 64/96; P-2, 169/93; P-3, 118/96; P-4, 203/43; P-5, 57/25 µm.

Similar species: Two other species known from the Palaearctic, *E. elephantula* GERECKE, 1996b (Sardinia) and *E. tazzecensis* GERECKE, 1999 (Morocco), differ in the presence of four pairs of Ac which are distinctly larger in size.

Preadult stages: Larva: Many descriptions – e. g., PIERSIG (1892a), LUNDBLAD (1927c), BIESIADKA & CICHOCKA (1990), WAINSTEIN (1980). Deutonymph: With two pairs of Ac; inconfusable due to the unique integument structure (WALTER 1907b, TUZOVSKIJ 1990).

Biology: Larvae parasitic on Culicidae (*Ochlerotatus*: MULLEN 1974), as well as various genera of Tipulidae and Mycetophilidae (SMITH & OLIVER 1986) – all data from N America. In The Netherlands found from the beginning of April until mid-May. The species reappears in late summer and autumn (SMIT & VAN DER HAMMEN 2000).

Habitat: Small, often temporary and semi-aquatic water bodies such as coastal dune waters, quagfens, marshes and terrestrializing waters. Occasionally in lowland streams.

Distribution: Holarctic. Widespread in Europe.

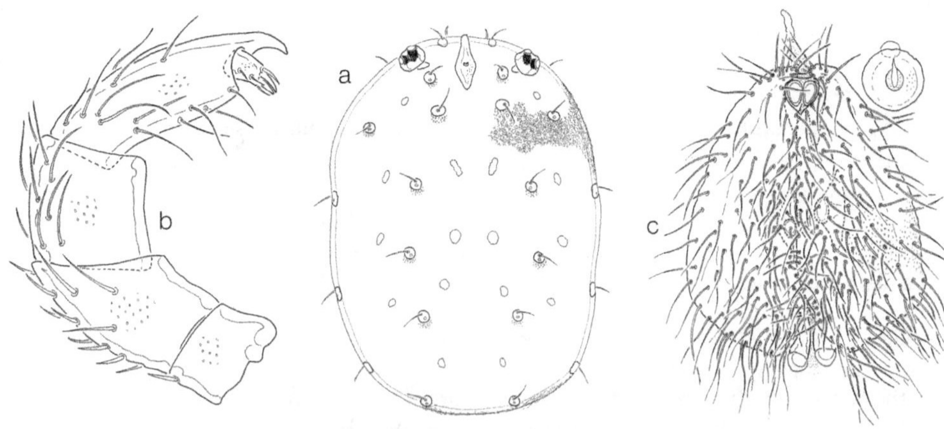

8-9: *Euthyas truncata*; a, dorsal idiosoma; b, palp; c, genital field; upside right: excretory pore.

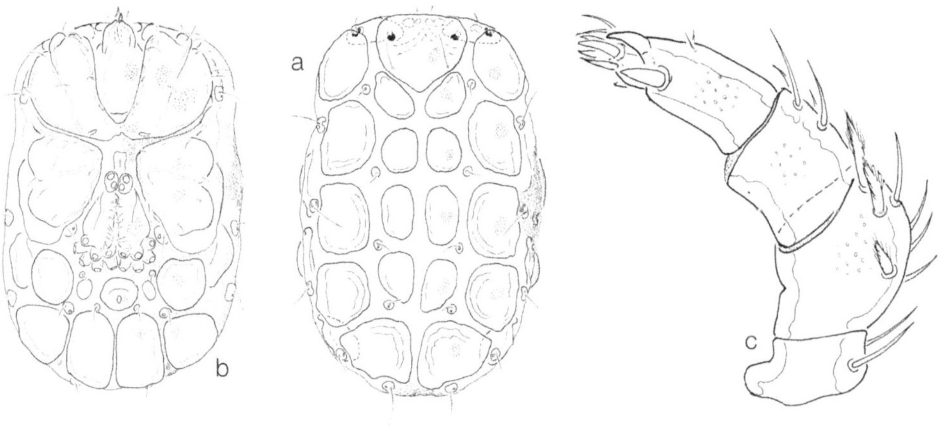

8-10: *Panisellus thienemanni* female; a, dorsal view; b, ventral view; c, palp.

Genus *Panisellus* K. VIETS, 1925

1925g *Panisellus* K. VIETS, Zool. Jbch. Syst. 50: 479.

Type species: *Panisellus thienemanni* (K. VIETS, 1920b), by original designation.

Diagnosis: Lateral eye lenses separate from each other on transverse-oval flattened platelets directed anteriorly (in dorsal view extensively covered by frontal sclerites, usually two small pigment dots visible). Frontale + frontalia fused to a subtriangular frontal shield, median eye not developed. All other dorsal and posteroventral sclerites not fused but expanded and occupying most of the idiosoma surface, membranous interspaces papillate. Coxae with very weak setation, posteromedial margins of Cx-I+II fused to a small bridge posterior to gnathosomal bay. Genital field with one or two pairs of Ac anterior to genital flaps and a further three to six pairs located at posterior edge of genital flaps. Chelicera with very strong claw (ratio basal segment/claw about 2.0). Palp short and stout, P-4 distally bearing two strong peg-like setae, one located at the lateral, the other at the medial segment edge.

Remarks: A highly distinct monotypic genus.

Panisellus thienemanni (K. VIETS, 1920) (*8-3 a-b, 8-10 a-c*)

1920b *Panisus thienemanni* K. VIETS, Arch. Hydrobiol. 12: 806.

1925g *Panisellus thienemanni* K. VIETS, Zool. Jbch. Syst. 50: 479.

Description: Colour brownish-yellow. All dorsal sclerites with a slight central depression surrounded by a particularly thickened border. Genital field on each side with a group of 4-6 posterior Ac on volcano-like extensions of a common basal sclerite that is fused to the posterior genital flap edge. Setae restricted to medial border of genital flaps in both sexes – the statement of K. VIETS (1936d) concerning a sexual dimorphism with setae extending to flap surface in males not correct. Gnathosoma robust, in lateral view with equally S-shaped ventral margin and short rostrum. Palp setation: P-1, 2, dorsally; P-2, 8-10, laterally and dorsally; P-3, 4 dorsally; P-4 one fine hair-like seta dorsally, two very strong, peg-like setae, one mediodistally, the other laterodistally. Male idiosoma length/width 830-850/520-615 µm. Genital field with one pair of Ac at anterior edge. Female idiosoma length/width 1000/630 µm. Genital field with two pairs of Ac at anterior edge.

Preadult stages: Larva: BOEHLE (1996). Deutonymph: (K. VIETS 1936d).

Biology: Larvae emerging from April till June, parasitic on Collembola (BOEHLE 1996).

Habitat: Helocrenes.

Distribution: British Isles, Central and South-eastern Europe including the Alps.

Genus *Panisopsis* K. VIETS, 1926

1926f *Panisopsis* K. VIETS, Zool. Anz. 66: 147.

1926b *Panisoides* LUNDBLAD, Ent. Tidskr. 47: 208.

1984b *Walterothyas* BADER, Entomol. Basil. 9: 7 – GERECKE (1996b).

1992 *Sindacoides* BADER, Entomol. Basil. 15: 5 – DI SABATINO et al. (2009).

Type species: *Panisopsis vigilans* (PIERSIG, 1896a), by monotypy.

Diagnosis: Median eye well developed, pigmentation well or poorly developed, or completely absent, in Western Palaearctic species surrounded by a frontal plate including frontalia and dorsocentralia 1 and postocularia. In some Nearctic and Oriental species, it is surrounded by membranous integument bearing variously fused sclerites. Coxae with long fine setae lined up along transverse margins and suture lines of Cx-I+II, less numerous and scattered on Cx-III+IV. Legs very stout and densely covered with strong setae, III-/IV-L with very strong, slightly elongate claws. Genital field with three Ac on each side lying more or less in a line. Genital flaps anteriorly embracing Ac-1 with an extension bearing 2-6 setae, posterior part, on the level of Ac-2, greatly narrowed, but not interrupted and embracing Ac-3 with a seta-bearing extension. Genital setae distinctly stronger in females than in males. Excretory pore with crescent-shaped anterior and posterior sclerites.

Remarks: Notwithstanding the considerable variation of the frontal area, *Panisopsis* appears to be a well-defined monophyletic clade. This is also confirmed from the analysis of larval characters (MARTIN 2003).

Biology: Larvae parasitic on Diptera (Culicidae ? BESSELING 1964; Chironomidae: MARTIN et al. 2010).

Preadult stages: Larvae treated with species. Deutonymphs with two pairs of Ac.

Key to species

1 Frontal shield with a large central opening surrounding the frontal eye, and a bifurcated posterior extension – *8-11 b*); remaining dorsal sclerites very small, similar in size to glandularia. . . . . *Panisopsis vigilans* (page 21)

- Frontal shield with a small central pore including the frontal eye, posterior margin without a bifurcated extension – *8-4 c, 8-11 a*); remaining dorsal sclerites distinctly larger than glandularia. 2

2 (1) Integument papillae longer than broad, in lateral view forming spine like tips directed posteriorly and extending over the following papilla; leg setation extremely developed – e. g., in III-L-4/5 covering the whole distal segment half (*8-11 c*); frontal shield (*8-11 e*) large (length/width 0.9-1.2); palp (*8-12 b*) more slender (length/height P-4 > 3.8). . . . . *Panisopsis setipes* (page 20)

- Integument papillae broader than long, in lateral view slightly bent posteriorly but not extending over the following papilla; leg setation less strongly developed – e. g., in III-L-4/5 restricted to the distal third (*8-11 d*); frontal shield (*8-4 c*) more slender (length/width > 1.2); palp (*8-12 c-d*) stouter (length/height P-4 < 3.8). 3

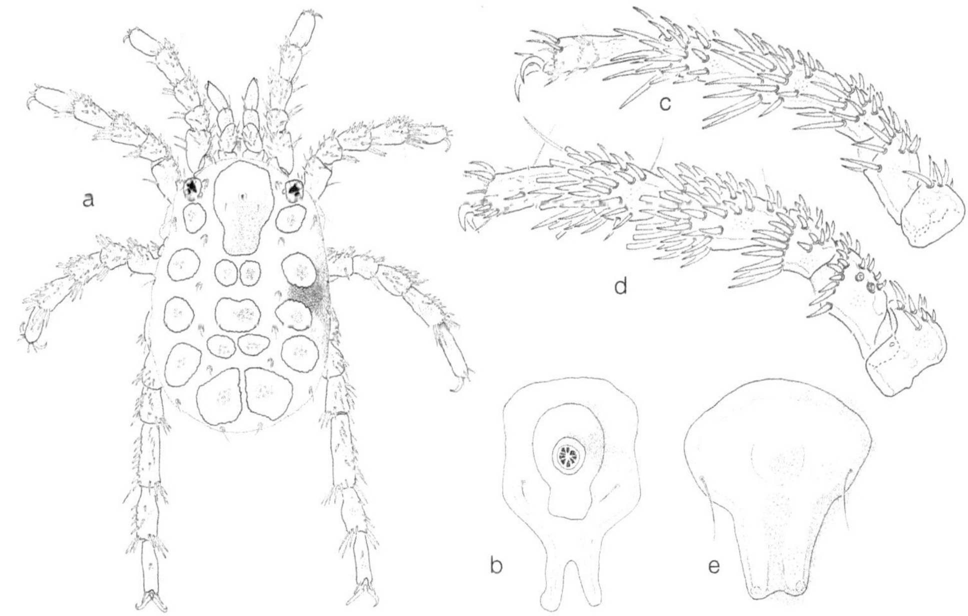

8-11: *Panisopsis*; a, *P. thori* male, dorsal view (from DI SABATINO et al. 2000); b, *P. vigilans* frontal shield; c, *P. setipes* III-L; d, *P. curvifrons* III-L; e, *P. setipes* frontal shield.

3 (2) Dc-5 medially fused (*8-12 c*); frontal shield (*8-4 c*) minor in size (length/width < 300/220 µm); P-2 (*8-12 a*) more slender (length/height 1.6-1.9). . . . . *Panisopsis curvifrons* (page 19)

\- Dc-5 medially separate (*8-11 a*); frontal shield major in size (length/width > 300/220 µm); P-2 (*8-12 d*) stouter (length/height 1.3-1.6). . . . . *Panisopsis thori* (page 21)

1. *Panisopsis curvifrons* (WALTER, 1907) (*8-4 c, 8-11 d, 8-12 a, c*)

1907a *Thyas curvifrons* WALTER, Zool. Anz. 31: 299.

1984b *Walterothyas curvifrons* BADER, Entomol. Basil. 9: 7 – GERECKE (1996b).

Description: Colour deep red. Integument with flat papillae directed slightly posteriorly. Frontal shield slender, maximum width near anterior margin, posteriorly equally narrowed. Dorsocentralia 2-4 separate or medially fused. Dc-5 forming a large, inversely trapezoidal plate. Cx-I medial margin anteriorly straight, posteriorly equally convex (not embracing gnathosomal base). Palp robust (P-4 length/height 3.4-3.7). Males: Idiosoma length/width 870-1050/600-700; frontal shield length/width 260/135; genital flap length 190 µm, with 22 marginal setae, pregenitale small, wider than long; gnathosoma length 200 µm. Females: Idiosoma length/width 1000-1150/650-820; frontal shield length/width 200-275/135-180; genital flap length 200-240 µm, with 20-30 setae; pregenitale longer than wide.

Preadult stages: Larva: WALTER (1915a, 1922a), MARTIN (2003). Deutonymph: undescribed.

Biology: Larva parasitic on Chironomidae (GERECKE & MARTIN 2006).

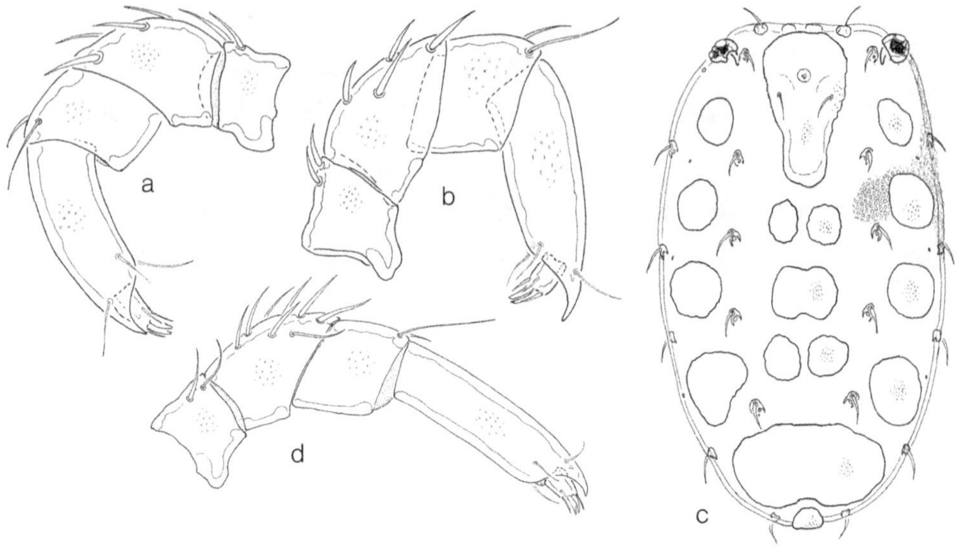

8-12: *Panisopsis*; a, *P. curvifrons*, palp; b, *P. setipes*, palp; c, *P. curvifrons* dorsal idiosoma; d, *P. thori* palp.

Habitat: Springs on mineral bedrock, preferably in strong-flowing rheocrenes between decaying wood and mosses or on hygropetric surfaces. Often above the tree line.

Distribution: Western Palaearctic: Mediterranean, Alps.

2. *Panisopsis setipes* (K. Viets, 1911) (*8-11 c, e, 8-12 b*)

1911f *Thyas setipes* K. Viets, Zool. Anz. 38:332.

1926b *Panisoides setipes* Lundblad, Ent. Tidskr. 47: 208.

1992 *Sindacoides ticinensis* Bader, Entomol. Basil. 15: 5 – Di Sabatino et al. (2009).

Description: Colour red. Integument with strongly pointed papillae, strongly bent posteriorly. Frontal shield large, anteriorly convex, posteriorly strongly narrowed. Frontal eye unpigmented. Dorsolateralia large, extending to flanks. All Dc not fused medially, Dc-2-4 small, circular, Dc-5 large, triangular. Cx-I medial margin anteriorly slightly concave, posteriorly embracing gnathosomal base with a slight convex projection. Palp relatively slender (P-4 length/height 3.9-4.9). Males: Idiosoma length/width 900-1300/550-1000; frontal shield length/width 290-360/270-330; genital flap length 190-220 µm, with 22-33 marginal setae, pregenitale small, wider than long; gnathosoma length 250-300 µm. Females: Idiosoma length/width 850-1370/400-1000; frontal shield length/width 320-350/370; genital flap length 210-250 µm, with 20-30 setae; pregenitale longer than wide.

Preadult stages: Larva: Biesiadka & Cichocka (1990). Deutonymph undescribed.

Habitat and Biology: Shaded forest springs, in the Mediterranean area restricted to higher elevations (*Fagus* belt). Life cycle unknown.

Distribution: Western Palaearctic: Northern, Central and Southern Europe.

3. *Panisopsis thori* (WALTER, 1907) (*8-12 d*)

1907a *Thyas thori* WALTER, Zool. Anz. 31: 299.

Description: Colour red. Integument with flattened papillae directed slightly posteriorly. Frontal shield slender, pear-shaped, maximum width posterior to frontal eye, posteriorly abruptly narrowed. Frontal eye pigmented. Dorsalia roundish to transverse-oval. All dorsocentralia not fused medially, Dc-2-4 smaller than lateralia, Dc-5 large, triangular. Cx-I medial margin anteriorly concave, posteriorly embracing gnathosomal base with a convex projection. Genitalia poorly developed, mostly invisible in both sexes. Palp robust (P-4 length/height 3.4-3.5). Males: Idiosoma length/width 1000/750; frontal shield length/width 270-330/170-245; genital flap length 200 µm, with 24 marginal setae; gnathosoma length 200 µm. Females: Idiosoma length/width 1100/800 µm; in dimensions of frontal shield and genital flaps similar to male.

Preadult stages: Larva and deutonymph: HUBAULT (1927).

Habitat and Biology: Rhithrobiont. Preferably in cascades and turbulent sectors of low order streams, also in hygropetric rheocrenes. In the High Atlas up to 2300 m asl. Life cycle unknown.

Distribution: Mediterranean, Alps, Central Europe.

4. *Panisopsis vigilans* (PIERSIG, 1896) (*8-11 b*)

1896a *Thyas vigilans* PIERSIG, Zool. Anz. 19: 441.

Description: Colour red. Integument with flattened papillae directed slightly posteriorly. Anterior part of frontal shield subquadratic or pentagonal, posterior part with two long divergent extensions. Frontal eye strongly pigmented, in a narrow ring surrounded by a large membranous area. Dorsalia knob-shaped and small, never fused with neighbouring sclerites. Cx-I medial margin anteriorly slightly concave, posteriorly embracing gnathosomal base with slightly convex medial projections. Genital flaps with 8-10 setae each on the anterior and posterior extension. Palp robust (P-4 length/height 3.5-3.7). Males: Idiosoma length/width 1150/950; genital flap length 235 µm, with about 40 marginal setae; pregenitale small, roundish. gnathosoma length 250 µm. Females: Idiosoma length/width 1800/1000; frontal shield length/width 360/250 µm; pregenital of sclerite long, anteriorly enlarged; genital flap length 275 µm, with about 30 marginal setae.

Preadult stages: Larva: PIERSIG (1896-1899), K. VIETS (1928a). Deutonymph: K. VIETS (1928a).

Habitat and Biology: Temporary standing waters, peat bogs. Life cycle unknown, larvae possibly parasitic on Culicidae (BESSELING 1964, uncertain association).

Distribution: Northern and Central Europe – a record from Turkey (ÖZKAN & BOYACI (1990) refers to *P. thori* (DI SABATINO et al. 2009).

Genus *Panisus* KOENIKE, 1896

1896b *Panisus* KOENIKE, Zool. Anz. 19: 356.

Type species: *Panisus cataphractus* (KOENIKE, 1895), designated by WOLCOTT (1905b).

Diagnosis: Frontal eye unpigmented, roundish or slit-shaped, sometimes hardly visible, in the anterior part of a plate made up of frontalia, Dc-1 and postocularia. Lateral eyes encapsulated. Dc-3 may fuse medially, more rarely also Dc-2 and Dc-4, in old specimens fusion of further neighbouring plates may occur. Dc-5 fused to a characteristic large posterodorsal plate. All glandularia free. Genital field with three pairs of Ac lying more or less in a longitudinal line. In Palaearctic species genital flaps with anterior margin obliquely truncated, their tips extending over Ac-1 only in exceptional cases, thus in general no setae lying anterior to Ac-1, posterior part not narrowed on level of Ac-2, embracing Ac-3 with a setae-bearing extension. Genital setae in males finer and in several rows, in females thicker and in one single row. Pregenital sclerite well developed, often elongated, in females, roundish or absent in males. Palp stout, with few setae, P-4 with medioventral hair-like seta away from distal margin, at about 2/3 of total segment length.

Remarks: The genital field of the Nearctic *P. cataphractus* Koenike, 1896b bears an anterior extension of genital flaps completely embracing Ac-1, with two setae located anterior to Ac-1 (Cook 1974). The significance of this character for a global definition of the genus (that obviously grades into *Panisopsis*) needs further investigation.

Preadult stages: Larva: Lundblad (1927c), Tuzovskij (2005: *Panisus major* Tuzovskij, 1994)

Key to species

1 Frontal shield wider than long, rarely as long as wide; dorsolateralia with strongly developed lateral projections extending to the flanks; pores of sclerites fine in the centre, gradually enlarged peripherally (*8-13 d*). *Panisus michaeli* (page 22)

\- Frontal shield as long as wide or longer than wide; dorsolateralia transversely oval or with weakly developed lateral projections; pores of sclerites fine in the centre, very large in a circular zone surrounding the centre, again rather fine near the edge (*8-13 c*). . . . . *Panisus torrenticolus* (page 23)

1. *Panisus michaeli* Koenike, 1896 (*8-3 f, 8-5 a, 8-13 d*)

1896b *Panisus michaeli* Koenike, Zool. Anz. 19: 356.

1981a *Panisus sarasini* Bader, Entomol. Basil. 6: 56 – Gerecke (1996b).

Description: Pores of idiosoma plates uniform, very fine in the centre, gradually increasing in size towards the margin. Occasionally a circular area may have pores larger in size than both medially and laterally, but this area never having the aspect of a spoke-wheel. Extensions of Dc-l and Dc-5 often coming in contact or fusing with those of neighbouring

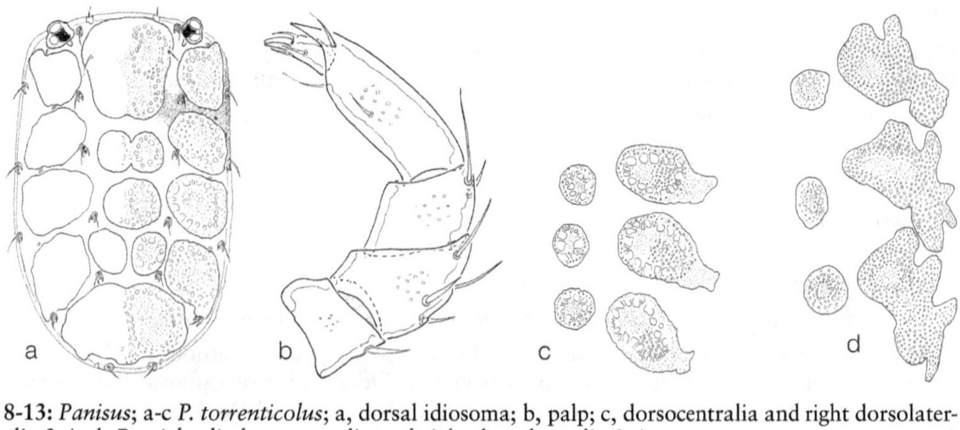

8-13: *Panisus*; a-c *P. torrenticolus*; a, dorsal idiosoma; b, palp; c, dorsocentralia and right dorsolateralia 2-4; d, *P. michaeli*, dorsocentralia and right dorsolateralia 2-4.

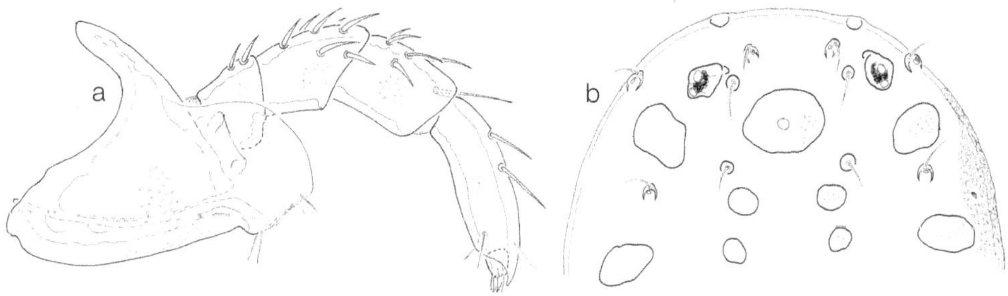

8-14: *Parathyas thoracata*; a, gnathosoma and right palp; b, frontal dorsum.

plates. Dc-3 rarely medially fused. Anterior margin of genital flaps normally ending laterally to Ac-1, rarely extending to anterior margin of, but not embracing Ac-1 (occasionally here one seta, but normally setation ending posterior to level of Ac-1). Palp stout. Males: Idiosoma length/width 780-1400/470-1000; frontal shield length/width 230-290/340-360; genital flap length 160-190; P-4 length 120 µm. Females: Idiosoma length/width 780-1800/460-1400; frontal shield length/width 230-390/230-520; genital flap length 220-290; P-4 length 160-180 µm.

Preadult stages: Larva: LUNDBLAD (1927c), MARTIN (2003). Deutonymph: K. VIETS (1936d), TUZOVSKIJ 1990), characterized by the sclerite porosity as described for adults.

Similar species: Following the definition given here, *P. michaeli* is highly variable. Occasionally, specimens with character combinations intermediate between this and the following species may be found. They could be attributable to the little known *P. lundbladi* BADER, 1981a, recorded only from Spain and not sufficiently defined. Also, populations from southern Italy recorded by GERECKE (1996b) as *P. michaeli*, in dimensions, sclerite porosity and shape of frontal shield intermediate between this species and *P. torrenticolus*, may eventually represent *P. lundbladi* or another undescribed species. Consequently, measurement data given by GERECKE (1996b) are not considered here.

Biology: Larvae according to LUNDBLAD (1927c) parasitic on Hymenoptera (Braconidae) – a host association doubted by GERECKE & MARTIN (2006).

Habitat: Helocrenes, preferably in montane areas, also in sites with moderate organic pollution.

Distribution: Western Palaearctic.

2. *Panisus torrenticolus* PIERSIG, 1898 (*8-13 a-c*)

1898b *Panisus torrenticolus* PIERSIG, Zool. Anz. 91: 451.

1909 *Thyas clypeolata* MAGLIO, Atti Soc. ital. Sci. natur., Pavia 48: 253 – DI SABATINO et al. (2009).

Description: Pores of idiosoma plates various in shape, very fine in the centre, very large in a circular area surrounding the centre and abruptly again finer near the margin. The pores in the circular area often subtriangular, giving the plate the aspect of a spoke-wheel. Margins of Dc-l and Dc-5 equally convex or with slightly-developed lateral extensions. Dc-3 often medially fused. Anterior margin of genital flaps ending laterally to Ac-1. Palp a little stouter than in *P. michaeli*. Males: Idiosoma length/width 620-850/450-650; frontal shield length/width 180-200/180-200; length genital flap 140-150; P-4, 95-110 µm. Females: Idiosoma length/width 780-1300/620-780; frontal shield length/width 200-270/190-270; length genital flap 150-180; P-4, 110-120 µm.

Preadult stages: Larva: MOTAŞ (1928f: "*P. clypeolatus*"). Deutonymph: not described, characterized by a sclerite porosity as described for adults.

Similar species: See *P. michaeli*.

Habitat and Biology: Low- and middle-order streams, occasionally in rheocrenes. Larvae parasitic on chironomids (MOTAŞ 1928f).

Distribution: Western Palaearctic, not recorded from Fennoscandia.

Genus *Parathyas* LUNDBLAD, 1926

1836 *Thyas* KOCH, Deutschlands Crustaceen, Myriapoden und Arachniden 5: 18 *nec* HÜBNER, 1824.

1926b *Parathyas* LUNDBLAD, Ent. Tidskr. 47: 208.

1974 *Todothyas* COOK, Mem. Amer. Ent. Inst. 21: 74 - DI SABATINO et al. (2009, 2010).

2006 *Acerbitas* ÖZDIKMEN, Munis Entomology & Zoology 1: 51 - DI SABATINO et al. (2010)

2007 *Thyas* KOCH, 1836 – TUZOVSKIJ, Acarina 15: 182.

Type species: *Parathyas thoracata* (PIERSIG, 1896a), by original designation.

Diagnosis: Lateral eyes encapsulated. Integument papillate. Median eye (lacking pigment or forming one diffuse mark) free in the integument or fused with the neighbouring frontalia to form a rounded or subtriangular frontal plate. Muscle attachment sclerites small or medium-sized. Except for Dc-1+2 (occasionally fused, but medially remaining separate) no fusion of neighbouring dorsalia. Legs relatively slender, distal margins of III-/IV-L-4/5 with characteristic strong setae. Inner margin of genital flap with fine setae anteriorly, stronger and longer setae posteriorly, in general without setae on surface and laterally. Genital field with 3 pairs of Ac; Ac-1 lying anterior to anterior edge, Ac-2 and -3 near posterior edge of flaps, their basal sclerites to various degree fused to each other and to flap margin, their locations forming angular points of a more or less elevated trapezoid shape. Excretory pore surrounded by a sclerotized ring. Sexual dimorphism weak: pregenital sclerite generally away from genital field in males, but located at anterior gonopore edge in females. Male genital flaps, especially anterior to Ac-2, relatively shorter and larger than in females.

Remarks: As both in preadult and adult stages, no significant differences can be detected between the type species of *Parathyas* and the character states diagnostic for *Thyas* KOCH, 1863, TUZOVSKIJ (2007) proposed to synonymize *Parathyas* Lundblad, 1926 with this genus. Previously, ÖZDIKMEN (2006), reporting on the homonymy of *Thyas* KOCH, 1836 with *Thyas* HÜBNER, 1924 (Lepidoptera), incorrectly applied the rules of the International Code of Zoological Nomenclature, introducing *Acerbitas nom. nov.* for the water mite genus *Thyas*. As stated by DI SABATINO et al. (2009), the first available name for *Thyas* KOCH, 1836 was *Todothyas* COOK, 1974. However, TUZOVSKIJ (2007) was obviously correct in the synonymization of *Parathyas* with *Thyas sensu* KOCH (DI SABATINO et al. 2010). Thus, as *Parathyas* is the oldest available name for the genus, and DI SABATINO et al. (2010) refused the former subdivision of the genus into subgenera (based on fusion, or not, of Ac-2+3 to the medioposterior genital flap margins), all species formerly placed in *Thyas* KOCH *should now be transferred to Parathyas.*

As all common character states of *Parathyas* are possibly plesiomorphies, it is uncertain if the species included have common phylogenetic roots. Some characters previously used for species discrimination are hardly suitable: integument papillae are very dense, appearing hexagonal in juveniles, in older specimens are far apart and round; palps of one and the same specimen may show strong differences (P-1 ventrally bulging or not, P-2 ventral margin straight or with a kink, P-3 maximum height in the centre or near the distal end). A thorough morphological analysis, in most cases including descriptions of preadult stages, is given by TUZOVSKIJ (2007) for all species recorded from Russia. The frontal eye of all species recorded from the area treated is unpigmented. As in some other hydryphantid genera, in addition to the pregenital sclerite a so-called anterior sclerite is located below the integument surface at the anterior end of the gonopore. If it is fused to, or away from, the pregenital sclerite, it may be of importance for recognition of species.

Biology: Deutonymphs and adults of many species feeding on chironomid larvae and microcrustaceans (TUZOVSKIJ 2007).

Preadult stages: Larva: Numerous descriptions – e.g., PRASAD & COOK (1972), TUZOVSKIJ (1980, 2007). Deutonymph: With two pairs of Ac in the provisional genital field (TUZOVSKIJ 1990). RACK (1976) reports parasitism by undetermined *Parathyas* species on Limoniidae (*Elliptera*, *Gonomyia*, *Erioptera* and *Molophilus*), with the abdomen and the preabdominal intersegmental membrane as the preferred attachment site. Other host taxa, with varying attachment sites, are reported mostly from outside the treated area (Diptera: Chironomidae, Culicidae, Ptychopteridae, Tipulidae; Trichoptera: SMITH & OLIVER 1976, MÜNCHBERG 1935b, 1936b).

Key to species

Excluded as *species incertae* (DI SABATINO et al. 2009): *Parathyas extendens* (GEORGE, 1901d); *P. valvata* (THOR, 1899a).

1 Frontal eye on a round frontal shield originating from fusion of frontalia and frontal eye sclerite (*8-14 b*); palp segment P-4 with several strong dorsal setae (*8-14 a*) . . . . . *Parathyas thoracata* (page 32)

- Frontal eye surrounded by a small sclerotized ring, frontalia visible as two pairs of sclerotized dots, occasionally on each side fused to a small longish platelet (*8-5 b*); palp segment P-4 generally without dorsal setae (*8-15 a-c*). 2

2 (1) Basal sclerites of Ac-2/3 fused to each other and generally also to posterior margin of genital flaps (*8-16 b-c*). Ac-3 may be located far lateral to Ac-2 (in this case, imaginary lines crossing centres of Ac-2/3 would meet in an obtuse angle in the posterior part of genital field (note that this character is visible only when flaps are closed - *8 16 f*). 3

- Basal sclerites of Ac-2/3 separate from each other, not fused to posterior margin of genital flaps (*8-15 d*). Ac-3 lying posterior to Ac-2 (imaginary lines crossing the centres of Ac-2/3 would meet in an acute angle in the anterior part of genital field). 5

3 (2) Mouth parts slender (*8-16 a*): gnathosoma with long, pointed rostrum; chelicera length/height > 5.0; P-3 length/height 1.4, P-4 length/height > 4.5; in males, numerous setae on genital flap surface. . . . . . *Parathyas bruzelii* (page 27)

- Mouth parts more robust (*8-16 e*): gnathosomal rostrum shorter, chelicera length/height < 5.0, P-3 length/height 1.1-1.3, P-4 length/height < 4.5; both sexes without setae on genital flap surface. 4

4 (3) Acetabula small (*8-16 b* - diameter Ac-3 < 25 µm). Anterior edge of genital flaps narrow, often oblique. Excretory pore with separate anterior and posterior sclerites or with a laterally narrowed ring. Dorsocentralia and dorsolateralia in diameter like glandularia or slightly larger. . . . . . *Parathyas dirempta* (page 29)

- Acetabula larger (*8-16 c* - diameter Ac-3 > 30 µm). Anterior edge of genital flaps larger, truncated. Excretory pore with circular sclerotization. Dorsocentralia and dorsolateralia larger than diameter of glandularia. . . . . . *Parathyas barbigera* (page 26)

5 (2) Gnathosoma with very short rostrum (in lateral view, a right-angled kink immediately behind the mouth opening – *8-15 a*). Diameter of frontal eye platelet 50 % or less the length of a lateral eye capsule. 6

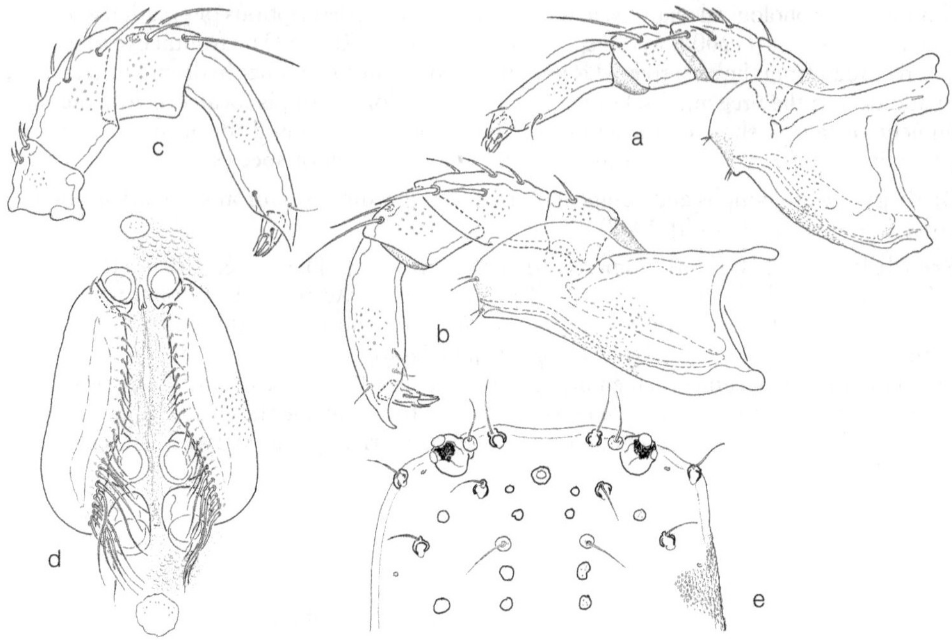

8-15: *Parathyas*; a, *P. pachystoma*, gnathosoma and right palp; b, *P. palustris*, gnathosoma and palp; c, *P. colligera*, palp; d-e, *P. inepta* male; d, genital field; e, frontal idiosoma.

- Gnathosoma with longer rostrum (in lateral view, ventral margin smoothly curved, without a kink – *8-15 b*). Diameter of frontal eye platelet approximately equal to length of lateral eye capsule. 7

6 (5) Frontalia generally fused (as in *8-5 b* – if separate, placed close to each other). Female pregenital sclerite separate from anterior genital sclerite (similar to the situation given in *8-16 c* for a male of *P. barbigera*, but closer). Setae of Dgl-1 stronger and shorter. . . . . *Parathyas pachystoma* (page 30)

- Frontalia separate, interspace wider than diameter of one frontale (*8-15 e*). Female pregenital sclerite fused with anterior genital sclerite (as in *8-16 b* for *P. dirempta*). Setae of Dgl-1 more slender and longer. . . . . *Parathyas inepta* (page 30)

7 (5) P-1 with one seta, P-3 with two to three setae nearly identical in length (*8-15 b*). Dorsal plates larger, in particular pre- and postfrontale, occasionally fused to the frontal eye ring. . . . . *Parathyas palustris* (page 31)

- P-1 with three to four setae, P-3 with two long and two or three short setae (*8-15 c*). Dorsal plates smaller, frontalia never fused to frontal eye ring. . . *Parathyas colligera* (page 28)

1. *Parathyas barbigera* (K. Viets, 1908) (*8-16 c-e*)

1908e *Thyas barbigera* K. Viets, Zool. Anz. 33: 670.

Description: Integument papillae flat. Median eye platelet small (diameter 40-70 µm, about 50 % length of eye capsule), occasionally with an asymmetrical border of secondary sclerotization. Frontalia similar in dimensions to median eye, occasionally pre- and postfrontale of one or both side(s) fused to a longish platelet. Only four pairs of dorsocentral plates due to fusion of Dc-1+2 (Lundblad 1927c). Legs slender, with characteristic three-pointed robust setae at distal segment margins (principal setal tip flanked by a pair of lateral tips). Genital flaps with broad transverse anterior and strongly oblique posteromedial margins, completely embracing, and locally fused to, basal sclerites Ac-2+3. Ac-3 located posterolaterally to Ac-2. Gnathosoma with short rostrum, in dorsal view length/width 1.5. Palp setae: P-1, 1-3; P-2, 9-16, 4-7 of them on medial surface, strong and pennate; P-3, 4-10 - 2-3 of them located distally, long and fine, P-4, 4 hair-like setae - 2 dorsodistally, 2 ventromedially (distal one stronger). P-4 rather slender, with slight sexual dimorphism (length/height ratio males 3.6-4.0, females 3.4-3.7). Excretory pore surrounded by a sclerotized ring with a knob-shaped anterior extension. Males: Frontalia generally fused; idiosoma length/width 1000-1600/750-1200; genital flap length 190-240 µm, with 24-27 setae; gnathosoma length 210-240 µm. Females: Frontalia generally separate; idiosoma length/width 1200-1700/900-1300; genital flap length 195-270 µm; pregenital plate large, fused with pregenital sclerite; gnathosoma length 230-300 µm.

Similar species: Mouthparts rather variable in shape, tending to grade into the situation described for *P. bruzelii*. The latter differing in P-4 being narrowed near segment base, distally slightly inflated, and presence of setae on male genital flaps.

Preadult stages: Larva: Numerous descriptions, e.g., LUNDBLAD (1927c), MÜNCHBERG (1956), BIESIADKA & CICHOCKA (1990), MITCHELL (2003), TUZOVSKIJ (2007). Deutonymph: LUNDBLAD (1927c), TUZOVSKIJ (2007).

Biology: Eggs with a particular envelope, which to a certain degree, protects against evaporation (MULLEN 1977). Newly hatched larvae move to the water surface and attach to membranous body parts of egg-laying dipterans. Preferred hosts Culicidae (*Ochlerotatus*: IMAMURA 1951b, 1954b, MULLEN 1974, 1977, NIELSEN 1998), occasionally Tipulidae (MULLEN 1977) and Limoniidae (*Dicranomyia, Erioptera, Limonia, Pilaria, Molophilus*: MÜNCHBERG *1956*, RACK *1976, 1977). Duration of parasitic phase 5-8 days, protonymphal stage one-two weeks. Deutonymph and adults can survive situations of drought hidden in humid crevices of sediment for more than 12 months (aestivation, hibernation). Adults feed on culicid eggs (*MULLEN *1977). In The Netherlands found in springtime, rarely collected after mid-May (*SMIT *&* VAN DER HAMMEN *2000).*

Habitat: Temporary vernal pools (e.g. coastal dune waters, carr, temporary ponds). Occasionally in other waters such as lowland streams.

Distribution: Holarctic. Widespread in Europe.

2. *Parathyas bruzelii* (LUNDBLAD, 1926) (*8-16 a*)

1926b *Thyas bruzelii* LUNDBLAD, Ent. Tidskr. 47: 205.

Description: Median eye small (diameter 30-50 µm, about 1/3 length of lateral eye capsule), Dc-1+2 fused; Dc and Dl in size similar to, or larger than, lateral eye capsules. Legs slender, with characteristic three-pointed robust setae at distal segment margins (the principal tip flanked by a pair of lateral tips). Genital flaps with transverse anterior and strongly oblique posteromedial margins. Genital flap setae small at medial margins, very long and strong at posterior margins. Excretory pore with anterior and posterior sclerotization, not forming a continuous ring. Gnathosoma in lateral and dorsal view more slender than in all other species of the genus (length/width ratio 2.7). Palp setae: P-1, 1-3; P-2, 7-10; P-3, 4-7, 2-3 of these located distally, long and fine, P-4, 3-4. P-4 proximally narrowed, distally slightly thickened. Males: Frontalia fused; idiosoma length 900-1500; genital flap length/width 250-290/90-100 µm, bearing numerous setae scattered over the surface. Females: Frontalia separate; idiosoma length 1100-1700; genital flap length/width 270-285/120-130; gnathosoma length 310-385 µm.

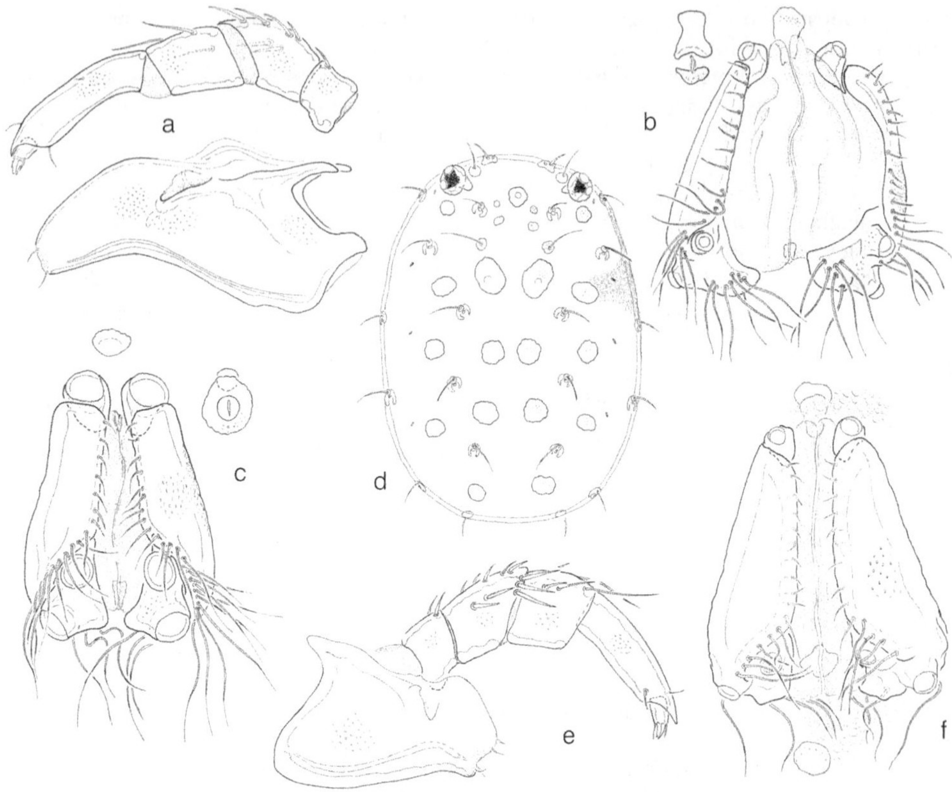

8-16: *Parathyas*; a, *P. bruzelii* gnathosoma and palp; b, *P. dirempta*, female genital field with opened flaps – at top left: excretory pore; c-e, *P. barbigera* male; c, genital field – at top right: excretory pore; d, dorsum; e, gnathosoma and palp; f, *P. dirempta*, female genital field with closed flaps.

Preadult stages: Larva: WAINSTEIN (1980), BIESIADKA & CICHOCKA (1990), TUZOVSKIJ (2007). Deutonymph: TUZOVSKIJ (1990, 2007).

Similar species: From the area treated, only females are known. Specimens reported from the Ukraine and Russia (SOKOLOW 1940a, TUZOVSKIJ 2000a) also have setae in the centre and at lateral margins of the genital flap surfaces, a scattered few in females, numerous and more densely arranged in males. *Parathyas magadanensis* (TUZOVSKIJ, 2000a, E Russia), another species with setated genital flap surfaces, differs in an even longer gnathosoma with a relatively longer rostrum and more slender palp. See also *P. barbigera*.

Habitat and Biology: Small standing water bodies. Life cycle unknown.

Distribution: Holarctic. In Europe recorded from Sweden, Ukraine and Russia.

3. *Parathyas colligera* (K. VIETS, 1923) (*8-15 c*)

1923a *Thyas rivalis colligera* K. VIETS, Arch. Hydrobiol., Suppl. 3: 199.

2007 *Thyas distinctus* TUZOVSKIJ, Acarina 15: 221 - DI SABATINO et al. (2009).

2009 *Parathyas colligera* (K. VIETS, 1923) - DI SABATINO et al. (2009).

Description: Frontal eye plate large (diameter 60-90 µm, similar to length of a lateral eye capsule), round or oval. Frontalia fused to an elongate platelet. Dorsocentralia 1 and 2 generally separate. Coxae with a few setae only, Cx-I with 3-4 at anterior tip and 1-2 at medial margin. Genital flap medial margin anteriorly straight, here bearing 5-6 fine setae, and posteriorly strongly concave, embracing Ac 2-3, with 7-10 long and strong setae. Acetabula small and rounded, Ac-3 located posterior to Ac-2. Excretory pore slit-shaped, with a large, crescent-shaped anterior sclerite, posterior sclerite small and roundish or absent. Gnathosoma with normally developed rostrum - as in *P. palustris* (*8-15 b*), neither bent ventrally, nor elongated. Palp setation: P-1, 2-3; P-2, 6; P-3, 4 (two shorter and two longer), P-4 with three hair-like distal setae, one dorsal, one medial and one ventral. Males: Idiosoma length/width 930-1150/750-900; genital flap length/width 165-210/ 65-85 µm; gnathosoma length 180-225 µm. Females: Idiosoma length/width 875-1300/700-1100; genital flap length/width 225-240/75-95 µm; pregenital sclerite not fused to anterior genital sclerite; gnathosoma length 215-280 µm.

Similar species: This species was initially considered a subspecies, later a synonym, of *P. palustris* (see DI SABATINO et al. 2009). A detailed description is given by TUZOVSKIJ (2007- under the name *Thyas distinctus*).

Preadult stages: Larva: TUZOVSKIJ (2007- under the name *Thyas distinctus*).

Habitat and Biology: Small stagnant water bodies such as ditches, terrestrializing zones in older, abandoned oxbow lakes and carr. In eutrophic to mesotrophic waters with presence of seepage. Life cycle unknown.

Distribution: Palaearctic. In Europe recorded only from northern Germany, The Netherlands and Russia.

4. *Parathyas dirempta* (KOENIKE, 1912) (*8-16 b, f*)

1912a *Thyas dirempta* KOENIKE, Zool. Anz. 40: 61.

Description: Frontal eye very small (diameter 30-35 µm, less than 50 % length of a lateral eye capsule); frontalia rarely fused, Dc-1+2 fused to a larger platelet, other dorsalia and lateralia small. Leg setation: III-L-4/5 with finely pennate setae at distal segment margins. Genital flaps anteriorly narrow, slightly oblique, posteriorly oblique, posterior extension completely surrounding the fused Ac-2+3 and here bearing 3-6 long setae, one or two setae located between Ac-2 and Ac-3. Excretory pore with anterior and posterior sclerites generally not fused to a ring. Gnathosoma very variable in shape (length/width ratio 1.5-1.9). Palp setation: P-1, 2-3; P-2, 8-10, all short and pennate; P-3, 3-4, 1-2 short and pennate, 2-3 finer and longer; P-4, 3 hair-like setae, all distal, in ventrolateral, ventromedial and dorsal position. Males: Idiosoma length/width 870-1200/800-1030; genital flap length/width 165-210/70-85; gnathosoma length 180-220 µm. Females: Idiosoma length/width 900-1700/800-1350; genital flap length/width 190-250/75-85; gnathosoma length 320 µm.

Similar species: Specimens with a stout or slender gnathosoma were formerly attributed to the subspecies, *P. d. bucculenta* (LUNDBLAD, 1925), *P. d. rostrata* (LUNDBLAD, 1925) respectively. As LUNDBLAD (1968) found them to grade continuously into each other, he synonymized both with the stem species. *Parathyas diremptellus* (TUZOVSKIJ, 1990) from Magadan (E Russia) differs in having shorter setae of Dgl-1 and a shorter gnathosomal rostrum.

Preadult stages: Larva: WAINSTEIN (1980), BIESIADKA & CICHOCKA (1990), TUZOVSKIJ (2007). Deutonymph: TUZOVSKIJ (1990, 2007).

Biology: Life cycle unknown. In The Netherlands found from the beginning of April until the end of September, with a peak in spring (Smit & Van der Hammen 2000).

Habitat: Small standing water bodies, also temporary vernal pools, coastal dune waters and ditches, eurythermous.

Distribution: Palaearctic. Recorded from Central, Western and Northern Europe.

5. *Parathyas inepta* (Lundblad, 1925) (*8-15 d-e*)

1925d *Thyas inepta* Lundblad, Ent. Tidskr. 46: 177.

1962 *Thyas pachystoma* [partim] Lundblad, Ark. Zool. (2) 14 (1): 37.

Description: Frontal eye plate small (diameter 35-50 µm, less than 50 % length lateral eye capsule). Setae of Dgl-1 smooth. On each side, frontalia separated by a wide interspace, small; Dc-1 not fused to Dc-2. Coxae, legs and genital field as described for *A. palustris*. Ac-2 located anterior to Ac-3. Excretory pore slit-shaped, with separate anterior and posterior sclerites or surrounded by a complete sclerotized ring. Gnathosoma stout, rostrum with large mouth opening set off against ventral margin by a subrectangular kink. Palp setation: P-1, 2 (rarely 1 or 3); P-2, 6-8; P-3, 5-6 (2-3 short, 3 longer), P-4 with 3 hair-like distal setae situated dorsally, medially and ventrally. Males: Idiosoma length/width 1000-1300/900-1100; genital flap length/width 190-220/70-85; gnathosoma length 205-225 µm. Females: Idiosoma length/width 1100-1700/900-1400; genital flap length/width 275-305/85-100 µm; pregenital sclerite fused with anterior genital sclerite; gnathosoma length 235-240 µm.

Preadult stages: Larva and deutonymph: Tuzovskij (2007).

Similar species: For a long time considered a subspecies, later a synonym of *P. pachystoma* (Lundblad 1962). However, the character state of separate frontalia, always found in combination with female pregenital sclerites separate from anterior sclerites, and *vice-versa*, allows the separation of adults of these two species. Further characters proposed for separation of adults by Tuzovskij (2007), who found diagnostic differences also in larvae, are unstable (extension of excretory pore sclerotization with infraspecific variability) or unclear (pennation of setae often not clearly visible - *P. inepta* appears to have Dgl-1 setae more slender and longer than *P. pachystoma*).

Habitat and Biology: Eurytopic, in many kinds of small water bodies, more frequent in standing than running water.

Distribution: Holarctic. Distribution in Europe poorly documented due to long-term unclear taxonomic status.

6. *Parathyas pachystoma* (Koenike, 1914) (*8-15 a*)

1914e *Thyas pachystoma* Koenike, Abh. naturw. Ver. Bremen 22: 383.

1920b *Thyas pachystoma paucispina* K. Viets, Arch. Hydrobiol. 12: 805 - Lundblad (1962).

1956a *Thyas pachystoma pannonica* Szalay, Acta Zool. Acad. Sci. Hung. 2: 274 - Lundblad (1962).

Description: Frontal eye plate small (diameter 35-55 µm, less than 50 % length of lateral eye capsule). On each side, frontalia fused to a roundish platelet; Dc-1/2 separate. Dgl-1 with pennate setae. Coxae, legs and genital field as described for *P. palustris*. Anterior margin of genital flaps oblique, anteromedial margin straight with about 10 fine setae, posteromedial margin oblique, with 12-14 setae increasing in length and with the tips embracing Ac-2+3, here bearing 3 long setae. Acetabula often elongated, Ac-3 posterior to Ac-2. Excretory pore slit-shaped, with separate anterior and posterior sclerites or surrounded by a complete sclerotized ring. Gnathosoma stout, large mouth opening on short rostrum set off against ventral margin by a subrectangular kink. Palp setation: P-1, 2; P-2, 8-11; P-3, 4-7 (2-5 short, 2 longer dorsodistally); P-4 with four hair-like distal setae, dorsally, medially and ventrally. Males: Idiosoma length/width 1000-1500/900-1100; genital flap length/width 205-280/90-100; gnathosoma length 195-265 µm. Females: Idiosoma length/width 1400-1800/1000-1400; genital flap length/width 280-330/75-90 µm; pregenital sclerite separate from anterior genital sclerite; gnathosoma length 280-325 µm.

Preadult stages: Larva: MÜNCHBERG (1935b, uncertain assignment), BIESIADKA & CICHOCKA (1990), TUZOVSKIJ (2007). Deutonymph: TUZOVSKIJ (1990, 2007).

Similar species: *Parathyas inepta* (with frontalia away from each other and female pregenital sclerite fused with anterior genital sclerite, see there for further discussion) was for a long time considered a subspecies, later a synonym, of *P. pachystoma*. A further species in need for better definition is *P. rarus* (TUZOVSKIJ, 2007), known from a single female collected in a mountain stream in Jaroslavl province (Russia). It was described for the partly sclerotized excretory pore, a character found frequently as an individual variation in Central European populations of *P. pachystoma*.

Habitat and Biology: In prealpine areas preferably in helocrenes at low elevation. Also, in semi-aquatic habitats (coastal dunes poor in lime, quagfens, swamps, temporary pools), occasionally in lowland streams. Adults in The Netherlands from the end of March onwards, with a peak in April/May (SMIT & VAN DER HAMMEN 2000). Life cycle unknown.

Distribution: Central and Northern Europe. No records from the Mediterranean or the Pyrenees.

7. *Parathyas palustris* (KOENIKE, 1912) (*8-5 b, 8-15 b*)

1912a *Thyas palustris* KOENIKE, Zool. Anz. 40: 62.

1912a *Thyas rivalis* KOENIKE, Zool. Anz. 40: 63 – LUNDBLAD (1962).

Description: Frontal eye plate large (diameter 50-70, similar to length of a lateral eye capsule). Generally frontalia fused to a large roundish platelet similar to size of a lateral eye capsule, but Dc-1/2 separate . In exceptional cases, frontalia may be separate on one or both sides, and / or Dc-1+2 may fuse. However, in the latter case the resulting platelet is elongate, not rounded as in species having these sclerites regularly fused. Coxae with sparse setation (Cx-I+II 13-15). Setae at distal margins of III/IV-L-4-5 pennate. Anterior and posterior margins of genital flaps oblique, their posterior extensions embracing Ac-2 and Ac-3 and here bearing a few setae. Ac-3 located posterior to Ac-2, their basal sclerites forming more or less elongated stalks. In both sexes pregenital sclerite round, not fused with anterior sclerite. Excretory pore with separate anterior and posterior sclerites or surrounded by a laterally narrowed sclerotized ring without extensions. Gnathosoma compact (length/width ratio 1.5), in lateral view with an equally concave ventrodistal margin. Palp robust, setation: P-1, 1; P-2, 6-7; P-3, 2-3, rather fine and long; P-4 with 3-4 fine distal setae arranged dorsally, ventrally and laterally. Males: Idiosoma length/width 870-1200/550-900; genital flap length/width 160-225/55-85 µm, with 19-22 medial setae; gnathosoma length 150-205 µm. Females: Idiosoma length/width 670-1350/500-840; genital flap length 155-230 µm, with 9-21 medial setae; gnathosoma length 160-205 µm.

Preadult stages: Larva and deutonymph: K. VIETS (1923a), BIESIADKA & CICHOCKA (1990), MARTIN (2003), TUZOVSKIJ (2007).

Habitat and Biology: Helocrenes, more rarely in swamps of riverine forests. In the Mediterranean area restricted to waters of montane forests, in Central, Western and Northern Europe also in lowlands. Larvae parasitic on Limoniidae (*Molophilus*, *Paradelphomyia*, *Pilaria*) and Chironomidae (*Parochlus*), prefered attachment site thorax. Parasitic larvae in Luxembourg found from July to September, mainly in August (MARTIN & STUR 2005, 2006, MARTIN et al. 2010).

Distribution: Holarctic (in Europe not recorded from the Iberian peninsula); North America.

8. *Parathyas thoracata* (PIERSIG, 1896) (*8-5 d, 8-14 a-b*)

1896a *Thyas thoracatus* PIERSIG, Zool. Anz. 19: 441.

1926b *Parathyas thoracata* LUNDBLAD, Ent. Tidskr. 47: 208.

1996a *Thyas thoracatus* TUZOVSKIJ, Acarina 15: 243.

Description: Frontal shield large (maximum diameter 140-250), transverse-oval (length/width ratio 0.7–1.0). Dorsalia and lateralia relatively large, increasing in diameter from anterior to posterior, Dc-1/2 not fused. Anterior edge of genital flaps truncate or oblique; Ac-2/3 on separate basal sclerites; Ac-3 posterior to Ac-2. Excretory pore with round or elongate sclerotized ring. Gnathosoma stout, mouth-opening large, on a short rostrum, in lateral view set off against ventral margin by a subrectangular kink. Palp setation: P-1, 1-2; P-2, 8-10; P-3, 7-9 (increasing in length from base to distal part); P-4, 2-4 in dorsal position (at least one of them strong) and 3 fine, hair-like distal setae, dorsally, medially and ventrally. Male: Idiosoma length 1100-1250; frontal shield length/width 180-220/200-250; genital flap length/width 220-230/70-75 µm; pregenital sclerites away from anterior sclerite; gnathosoma length 205-210 µm. Females: Idiosoma length 1400-1750; frontal shield length/width 180-200/190-250; genital flap length 250-280/70-90; pregenital sclerites and anterior sclerite fused; gnathosoma length 210-310 µm.

Preadult stages: Larva: WAINSTEIN (1980); TUZOVSKIJ (2007). Deutonymph: With two pairs of Ac, among the species of the area treated inconfusable due to the frontal sclerotization, but without the dorsal P-4 setae characteristic in adults (TUZOVSKIJ 1975, 1990, 2007). Life cycle unknown.

Habitat: Shallow vernal ponds and slowly flowing streams, euryoecious but very rare.

Distribution: Western Palaearctic. Scattered records from Maghreb to Northern Europe.

Genus *Parathyasella* K. VIETS, 1949

1949b *Parathyasella* K. VIETS, Abh. naturw. Ver. Bremen 32: 310.

Type species: *Thyasella (Parathyasella) frigida* LUNDBLAD, 1925b, by original designation.

Diagnosis: Idiosoma sclerites, as well as coxae and genital flaps well developed and clearly delineated. Frontal shield built up by frontal plate + frontalia + Dc-1. Pre- and postfrontalia on small dome-shaped projections. Posteromedial edge of Cx-I+II pointed, without a projection embracing gnathosomal bay. Coxae with few, but long setae. Anterior margins of genital flap embracing Ac-1 and bearing one to three setae at anterior tip. Posterior margin truncated, nearly straight, medial setae fine, also in the posterior part. Basal sclerites of Ac-2+3 at posterior margin of genital field separate. No sclerotized genitalia. Excretory pore with a sclerotized ring. Gnathosoma with short, broad rostrum; chelicerae stout, with strongly elongated air sacs. Palp short and stout, P-5 shortened, but with very long dorsodistal projection, P-5 long and slender (length = ventral length of P-4).

Remarks: *Parathyasella* was described as a subgenus of *Thyasella*, but elevated to genus rank by SMITH & COOK (1999). In addition to the only known European species, two species are described from North America. See also discussion of *Thyasella.*

Preadult stages: No descriptions available.

Parathyasella frigida (LUNDBLAD, 1925) (*8-17 c-d*)

1925b *Thyas frigida* LUNDBLAD, Ent. Tidskr. 46: 54.

1927c *Thyasella frigida* LUNDBLAD, Zool. Bidrag Uppsala 11: 231.

1949b *Thyasella (Parathyasella) frigida* K. VIETS, Abh. Naturw. Ver. Bremen 32: 310.

Description: Male (female unknown): Colour unknown. All sclerites well sclerotized and porose. Frontal eye visible as a distinct pore in frontal shield, but without pigmentation. Dc-1 fused to frontal shield. Excretory pore flanked by a pair of slender, crescent-shaped sclerites. Idiosoma length/width 1030/690; genital flap length 200 µm, with about 25-30 medial setae – no setae at posterior and lateral flap margins. LUNDBLAD (1927c) reported the presence of two paired bubble-shaped organs of uncertain identity, 75-100 µm in diameter, lying below the genital flaps. Gnathosoma length/width 257/161; chelicera length/height 243/86; air sac length 125 µm.

Habitat and Biology: A cold stream at 700 m asl. in the upper coniferous forest area (only one single specimen known).

Distribution: Northern Sweden.

Genus *Thyasella* K. VIETS, 1926

1926b *Thyasella* K. VIETS, Zool. Anz. 66: 147.

Type species: *Thyas mandibularis* LUNDBLAD, 1924b, by original designation.

Diagnosis: Idiosoma sclerites, as well as coxae and genital flaps weakly developed and faintly delineated. Frontal shield built up by frontal plate + frontalia, Dc-1 separate. Pre- and postfrontalia on small dome-shaped projections. Posteromedial edge of Cx-I+II pointed, without a projection embracing gnathosomal bay. Coxae with few, but long setae. Anterior margins of genital flap embracing Ac-1 and bearing a fine seta (or setal pore) at anterior tip. Posterior margin truncated, nearly straight, medial setae fine, also in the posterior part. Basal sclerites of Ac-2+3 at posterior margin of genital field, fused, but separate from genital flaps. No sclerotized genitalia. Excretory pore various. Gnathosoma with short, broad rostrum; chelicerae stout, with strongly elongated air sacs. Palp short and stout, P-4 shortened, but with very long dorsodistal projection, P-5 long and slender (length = ventral length of P-4).

Remarks: In addition to the European *T. mandibularis*, one species known from North America (SMITH & COOK 2009), *Thyasella* and *Parathyasella* are closely related genera, together forming a well-defined monophyletic clade rather distant from, and with uncertain taxonomic relations to, all other hydryphantid genera. *Thyas pustulosa* THOR, 1901a, a poorly documented *species incerta*, is most probably a junior synonym of one of the two species representing these genera in Europe (DI SABATINO et al. 2009).

Preadult stages: No descriptions available.

Thyasella mandibularis (LUNDBLAD, 1924) (*8-3 e, 8-17 a-b*)

1924b *Thyas mandibularis* LUNDBLAD, Ent.Tidskr. 45: 68.

1926f *Thyasella mandibularis* K. VIETS, Zool. Anz. 66: 147.

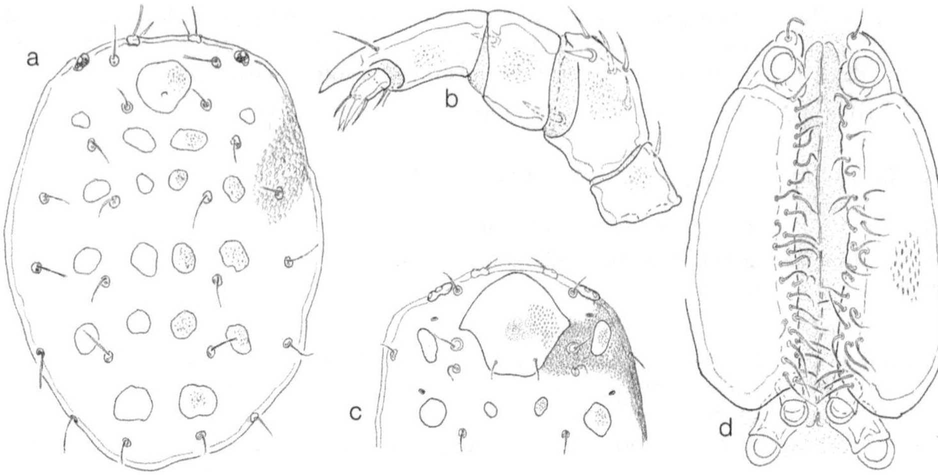

8-17: a-b, *Thyasella mandibularis* male; a, dorsal idiosoma; b, palp; c-d, *Parathyasella frigida* male; c, anterodorsal idiosoma; d, genital field.

1958a *Thyasella mandibularis torrenticola* SCHWOERBEL, Zool. Anz. 160: 197 - DI SABATINO et al. (2009).

Description: Colour yellow. Integument rather smooth, papillae flattened and separated from each other. All sclerites weakly hardened, with a fine granulation instead of being porose. Frontal shield rounded (length/width 110-115/110-115), area of weak sclerotization indicating position of frontal eye, Dc-1 separate from frontal shield. Legs slender, with few setae only. All Ac very small (diameter 25-30 µm). Excretory pore unsclerotized. Chelicera stout (length basal segment 120-125, claw 85, height 58-65, air sac 93-95 µm). Palp very stout, setation: P-1, 1; P-2, 5; P-3, 2 (one stout, the other more slender); P-4, 3 (fine, hair-like, in typical distal position). Males: Idiosoma length/width 700-870/450-620; genital flap length 180-195; gnathosoma length/width 195/130 µm. Females: Idiosoma length/width 700-800/480-520; genital flap length 230; gnathosoma length/width 180/130; length/height P-1, 40/48; P-2, 80/60; P-3, 45/62; P-4, 115/40; P-5, 45/15 µm.

Habitat and Biology: Springs (also in a temporary site: LUNDBLAD 1968) and interstitial habitats of first-order streams. Life cycle unknown.

Distribution: Southern Sweden, England, Germany (lowlands, Black Forest and Alps), a few scattered records only.

Genus *Thyasides* LUNDBLAD, 1926

1926b *Thyasides* LUNDBLAD, Ent. Tidskr. 47: 208.

Type species: *Thyas dentata* THOR, 1897c, by original designation.

Diagnosis: Lateral eyes encapsulated. Frontal eye with two pigment spots. Frontalia fused and generally attached to posterior border of frontal eye sclerite. This frontal shield anteriorly convex due to the projecting frontal eye sclerite, posteriorly concave or straight between the projecting frontalia (in rare cases frontal eye separate from frontalia and the latter lying as a transverse platelet posterior to frontal eye). Dc-1+2 fused. Cx-I, occasionally also Cx-II, with strong, tooth-like medial setae. III/IV-L-4-5 distal setae finely pennate. Genital flaps with numerous uniform, short setae both at medial margin and over the whole

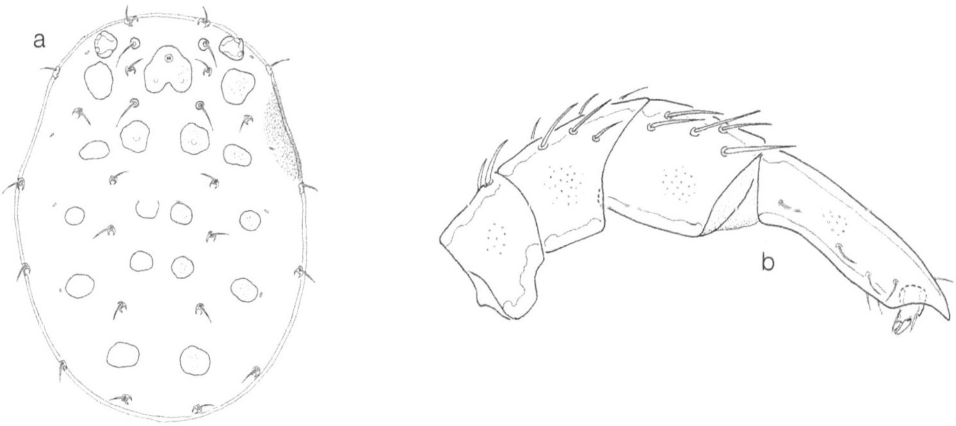

8-18: *Thyasides dentatus* female; a, dorsal idiosoma; b, palp.

flap surface. Excretory pore surrounded by a sclerotized ring bearing an anterior knob-shaped extension.

Remarks: A taxon well distinguished in the apomorphic shape of the frontal shield and setation of coxae, and the probably plesiomorphic intense setation of the genital flaps. Relationships to other hydryphantid genera unclear. In the study area only one species.

Thyasides dentatus (Thor, 1897) (*8-5 c, 8-18 a-b*)

1897c *Thyas dentata* Thor, Arch. Math. Naturvid. Kristiania 20: 9.

1926b *Thyasides dentatus* Lundblad, Ent. Tidskr. 47: 208.

Description: Dl-1 large, Dl-2 transverse oval. Anterior margin of genital flap transverse or slightly oblique, posterior margin inversely oblique (maximum length medially). Ac-3 located posteromedially to Ac-2 as in *Parathyas dirempta*, but no setae between Ac-2 and -3. Gnathosoma in ventral view slender (length/width ratio 1.7). Palp morphology without particular features, setation: P-1, 2-3; P-2, ca. 15; P-3, 4; P-4, 3 (fine hair-like, located distally). Males: Idiosoma length/width 1200/1000 µm. Pregenital sclerite away from anterior margin of genital field; genital flaps with very dense cover of setae. Females: Idiosoma length/width 1500-1750/1100-1400 µm. Pregenital sclerite close to anterior margin of genital field; genital flap length 300 µm, with lower number and density of setae than in males; gnathosoma length/width 310-360/180-210 µm (ratio 1.7).

Similar species: Only one other species described, *T. sphagnorum* Habeeb, 1958, from North America; diagnostic differences need revision.

Preadult stages: Larva: Lundblad (1913c), Biesiadka & Cichocka (1990). Deutonymph undescribed.

Biology: Larvae parasitic on Culicidae (*Ochlerotatus*): Lundblad (1927c). Preferred attachment site: thorax.

Habitat: Temporary vernal pools, bogs.

Distribution: Central, Northern and North-eastern Europe.

Genus *Thyopsis* PIERSIG, 1899

1899 *Thyopsis* PIERSIG, Zoologica 19: 378.

Type species: *Thyas cancellata* PROTZ, 1896b, by original designation.

Diagnosis: All dorsalia fused into a reticulate dorsal shield, leaving only small membranous areas surrounding the glandularia. Lateral eyes encapsulated, embraced or surrounded by extensions of the dorsal shield, frontal eye small, unpigmented, in the dorsal shield. Genital flaps equally enlarged from anterior to posterior edge, three pairs of Ac; Ac-2 and -3 lying nearly in a transverse line. Legs and palps relatively stout.

Remarks: The genus is highly distinct and still of uncertain relationships to other taxa in the Hydryphantidae.

Thyopsis cancellata (PROTZ, 1896) (*8-19 a-d*)

1896b *Thyas cancellata* PROTZ, Zool. Anz. 19: 408.

1983a *Thyopsis cancellatomima* BADER, Entomol. Basil. 8: 20 - GERECKE & SCHWOERBEL (1992).

Description: Dorsal shield in juveniles appearing like the "negative" of the typical hydryphantid dorsal plate pattern, with plate-interspaces sclerotized forming a large-meshed net. With increasing age, fine sclerotized trabeculae grow out from this primary reticulation to form a secondary reticulation filling all interspaces. In parallel, rings of similar secondary sclerotization are added to the dorsal plate margin, resulting in an age-dependent centrifugal extension of the plate. Cx-I with 8-16, Cx-II with 10-19 long setae located along suture lines, genital flaps triangular, with a highly variable number of setae (generally 30-60). Ventral margin of gnathosoma equally S-shaped, mouth opening diameter ca. 50 µm. Chelicera rather slender (basal segment/claw 2.7-3.2, length/height 3.8-4.8), claw short and slightly curved. Palp setation: P-1 generally 4; P-2, 6-9; P-3, 3 (one short and stout, one longer and finer); P-4, 3 (one strong medial peg seta at 2/3 segment length, two fine hair-like distoventrally), P-2/P-4 length ratio 0.6-0.7. Males: Idiosoma length/width 950-1220/750-1050; genital flap length 170-185 µm, anterior margin truncated,

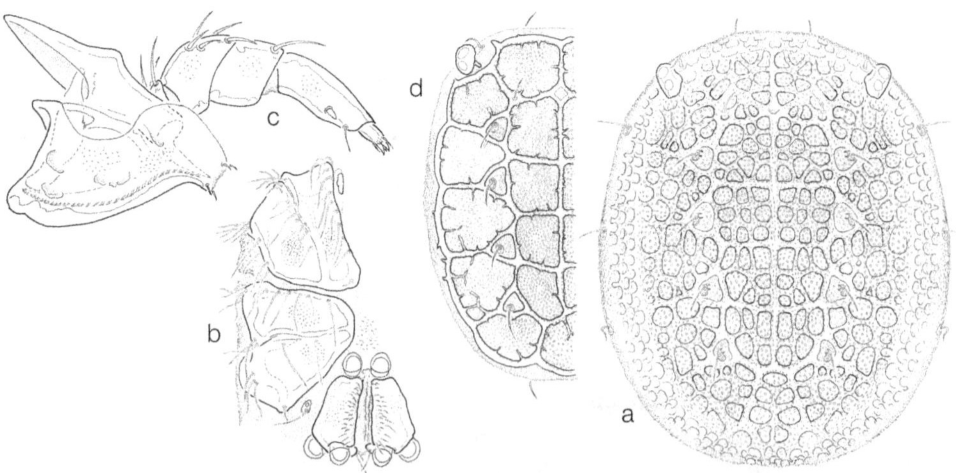

8-19: *Thyopsis cancellata* male (after GERECKE & SCHWOERBEL 1992); a, dorsal idiosoma; b, right coxae and genital field; c, gnathosoma and palp; d, dorsal idiosoma of a juvenile, partial view.

without a lateral seta. Gnathosoma ventral length 235-280; P-4 length/height 155-180/48-54 µm (ratio 3.3-3.6). Females: Idiosoma length/width 900-1450; genital flap length 160-200 µm, anterior margin oblique, with one seta located lateral to Ac-1. Gnathosoma ventral length 230-320; P-4 length/height 160-200/47-61 µm (ratio 3.2-3.5).

Similar species: *Thyopsis cancellata maderensis* LUNDBLAD, 1941 (Madeira, detailed description as *T. maderensis*: LUNDBLAD 1942c), differs in the dorsal shield being without secondary reticulation, but having three to four rows of well developed lateral meshes. *Thyopsis rothae* OEZKAN, 1982 (Turkey) differs in smaller Ac (diameter < 40 µm) (GERECKE & SCHWOERBEL 1992).

Preadult stages: Larva: MÜNCHBERG (1936b), MOTAŞ & TANASACHI (1944a), MARTIN (2000, 2006). Deutonymph: TUZOVSKIJ (1990).

Biology: Larva parasitic on representatives of many different genera of Limoniidae, Pediciidae and Tipulidae (MÜNCHBERG 1936b, MARTIN 2000, 2006). Attachment sites: Cervix and thorax (MARTIN 2000, MARTIN & STUR 2005).

Habitat: Springs, small streams, ditches, ponds,on most occasions found as single specimens only. GERECKE & SCHWOERBEL (1992) reported a large population from a helocrene at the immediate border of a prealpine lake.

Distribution: Central and Southwestern Europe, British Isles.

Genus *Trichothyas* K. VIETS, 1926

1926l *Trichothyas* K. VIETS, Zool. Anz. 69: 191.

Type species: *Thyas pennata* K. VIETS, 1913c, by original designation

Diagnosis: Frontal plate large, bearing the unpigmented frontal eye, formed by fusion of frontalia and Dc-1+2; Dc-3 and -4 each fused medially. Lateral eyes not in distinct capsules, attached to preocularia platelets. Coxal plate margins with finely pennate setae, in dense rows at medial margin Cx-I and suture Cx-I/II, more scattered elsewhere. Legs short and stout. Gnathosoma elongated and flattened, with nearly straight ventral margin. Palp rather slender, with few setae.

Remarks: Along with *Almuerzothyas* GOLDSCHMIDT & GERECKE, 2003, *Dacothyas* MOTAŞ, 1959, *Heterothyas* LUNDBLAD, 1941 and *Javathyas* K. VIETS, 1929, *Trichothyas* forms a probably monophyletic taxonomic unit characterized by the organization of lateral eyes and dorsal plates ("*Trichothyas*-like mites", COOK 1974). Within this clade, a trend to sexual dimorphism in the shape of I-L is observed. In addition to the monotypic nominate subgenus (Cameroon), *Trichothyas* includes two other monotypic subgenera, *Neothyas* LUNDBLAD, 1941c (Indonesia) and *Plesiothyas* K. VIETS, 1935c (Ivory Coast), *Kashmirothyas* LUNDBLAD, 1934a with two species (Himalayas, Turkey) and the more species-rich and wider distributed *Lundbladia* K. VIETS, 1929h. Only the latter is represented in the study area.

Subgenus *Lundbladia* K. VIETS, 1929

1929h *Lundbladia* K. VIETS, Zool. Anz. 86: 49.

Type species: *Lundbladia feuerborni* K. VIETS, 1929h, by original designation.

Diagnosis: Genital field with three pairs of Ac. Genital flaps rather straight or only slightly reduced in size, touching Ac-3 in both sexes. Males: Gonopore extending posteriorly beyond level of Ac-2.

Preadult stages and Biology: Larva: LUNDBLAD (1942c), MITCHELL (1953). Deutonymph: GERECKE (1996b). *Trichothyas muscicola* (MITCHELL, 1953) from N America parasitic on *Limonia* (Diptera: Limoniidae).

Remarks: In males of the N American *T. muscicola*, an equidistant location of Ac was observed (Ac-2 not close to Ac-1 and apart from Ac-3) and COOK (1974) considered this situation as diagnostic for the subgenus. However, this character state is not found in typical males neither in the type species, nor in the European species. The detection of heteromorphic males in *T. petrophila* (with male-like and female-like genital fields, coexisting in one population, see below) suggests that this phenomenon could be also found in the North American species. Representatives of *Lundbladia* are recorded from North and Central America, Africa, Asia and many parts of Europe.

Trichothyas (Lundbladia) petrophila (MICHAEL, 1895) (*8-20 a-d*)

1895a *Thyas petrophilus* MICHAEL, Proc. zool. Soc. 1895: 174.

1929h *Lundbladia petrophila* K. VIETS, Zool. Anz. 86: 49.

1935 *Trichothyas armata* WALTER, Arch. Hydrobiol. 28: 80 – K. VIETS (1951d).

1949b *Lundbladia plumifera* C. ANGELIER, Bull. Mus. Hist. natur. Paris 21 (3): 358 – K. VIETS (1951 d).

1956b *Trichothyas petrophila* K. VIETS, Die Milben des Süßwassers und des Meeres. Hydrachnellae et Halacaridae (Acari). Zweiter und dritter Teil: Katalog und Nomenklator. Fischer, Jena: 121.

Description: Colour deep red. Anterior margin of frontal shield convex, straight or in the centre slightly extended, length/width ratio 1.1-1.2, frontal eye near anterior margin. Dc-3 roundish or elongated (length/width ratio 0.9-1.9), Dc-4 transverse-oval (length/width ratio 0.7-0.9). Excretory pore plate elongated (length/width 1.5-2.1). Chelicera slender (length/height 6.0-7.6), with short claw (basal segment/claw length 2.8-3.1). Length ratio P-2/P-4 0.6-0.8. Males: Idiosoma length/width 860/510; length/width frontal shield 290/260, Dc-3, 117-128/103-138; Dc-4, 126-148/148-183 µm. Cx-I with 25-35 flattened, pennate setae, 21-30 of these in a dense row at the medial margin surrounding the gnathosomal bay. Genital flap length 176-190 µm, typically (in heteromorphic males) with subparallel lateral and medial margins, posteriorly not larger than Ac-3, but in homoiomorphic males shaped as in females (see below). Postgenitale elongated (length/width 31-54/25-41); gnathosoma length 265, chelicera length 245 µm. Females: Idiosoma length/width 1000-1200/600-820; length/width frontal shield 340-390/300-390, Dc-3, 140-210/10-200; Dc-4, 140-230/210-300 µm. Cx-I with 20-30 setae less flattened than in males, 15-20 of these at medial margin surrounding gnathosomal bay. Genital flap length 200-210 µm, equally enlarged from anterior to posterior margin, posterior margin distinctly larger than Ac-3 diameter. Postgenitale roundish or longish (length/width 48-76/36-67); gnathosoma length 270-300, chelicera length 250-310 µm. The original description is done very carefully and detailed and includes anatomy. An additional detailed account on the external morphology was published by LUNDBLAD (1933b).

Similar species: The subspecies *T. p. rutae* LUNDBLAD, 1941b (Madeira, Canary Islands) differs in the porosity of the frontal shield and dorsalia (marginally with distinctly finer pores than centrally) and relatively larger Ac. In males, the distal seta on I-L-3 is smaller and an additional small platelet is located anterior to V-2. *Trichothyas duplicata* K. VIETS, 1954c (Sardinia, Corsica) differs from *T. petrophila* among others in the extension of all dorsal and posteroventral plates (closely joined together, leaving only extremely narrow strips of membranous integument) and minor dimensions of appendages (GERECKE 1996b).

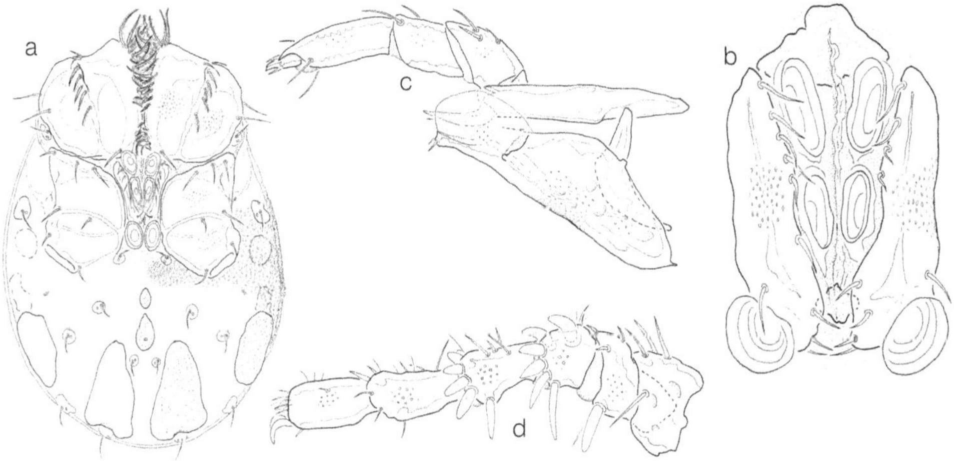

8-20: *Trichothyas petrophila*; a, male ventral idiosoma; b, female genital field; c, gnathosoma and palp; d, male I-L.

Preadult stages: Larva unknown, but desribed for the subspecies *T. p. rutae* by LUNDBLAD (1942c). Deutonymph: E. ANGELIER (1949b, "*Lundbladia plumifera*").

Habitat and Biology: Hygropetric areas of rheocrenes and cascades of low- and middle-order streams. Also found in an interstitial dig (GERECKE 1996b). Life cycle unknown. The coexistence of heteromorphic and homoiomorphic males within one population, implicating a complicated reproductive system, is a phenomenon well known in terrestrial mites, but unique in water mites. Some indications in the literature (e. g. LUNDBLAD 1956a) suggest that male dimorphism could be more frequent among *Trichothyas*-like mites (GERECKE 1996b).

Distribution: Mediterranean (here widely distributed, but on Corsica and Sardinia replaced by a sister species), England, Asia Minor. Always found close to the sea in England.

Genus *Vietsia* LUNDBLAD, 1926

1926b *Vietsia* LUNDBLAD, Ent. Tidskr. 47: 208.

Type species: *Thyas scutata* PROTZ, 1923, by original designation.

Diagnosis: Frontal shield composed of frontal eye sclerite and frontalia, median eye indistinct in posterior part of shield, unpigmented. Prae- and postocularia free in integument. Remaining dorsal and posteroventral idiosoma without sclerotization. Lateral eyes not encapsulated, lying on a pair of transverse platelets at anterior idiosoma margin. Coxae bearing only a few scattered setae. Genital field with three pairs of relatively small Ac in the membranous area between gonopore and genital flap, Ac-2 closer to Ac-3 than to Ac-1. Genital flaps with numerous setae in a row at medial margin and additional scattered setae on flap surface. Excretory pore unsclerotized. Palps and chelicerae stout.

Remarks: A distinct taxon with uncertain relationships to other Hydryphantidae, but arrangement of lateral eyes similar in *Panisellus* and the *Trichothyas*-like genera. Monotypic.

Vietsia scutata (PROTZ, 1923) (*8-3 d, 8-21 a-d*)

1923 *Thyas scutata* PROTZ, Zool. Anz. 15: 320.

1926b *Vietsia scutata* LUNDBLAD, Ent. Tidskr. 47: 208.

Description: Colour red. Maximum width of frontal shield in posterior part. Muscle attachments representing Dc-1 in touch with posterior edge of shield, but unsclerotized. Legs slender, with few, but rather long setae. Diameter Ac-1-2, 20-25, Ac-3, 35 µm. Gnathosoma with short rostrum bent ventrally. Palp stout, maximum height P-3 at proximal margin of segment. Palp setation: P-1, 1; P-2, 10-12; P-3, 2 (both stout and pennate); P-4, 3 fine (distally, in dorsal, ventral and medial position). Males: Genital flaps with convex lateral and undulating medial margin, densely covered by setae. Females: Idiosoma length/width 1500/1000-1100, frontal shield length/width 290/350, genital flap length 220 µm, with straighter lateral and medial margins, less densely covered by setae; gnathosoma length 240, chelicera basal segment 175, claw 148, height 78; length/height P-1, 50/70; P-2, 90/75; P-3, 70/70; P-4, 130/50; P-5 42/20 µm.

Preadult stages: Larva unknown. Deutonymph: K. VIETS (1936d).

Habitat and Biology: Shallow ponds and ditches, quagfens, flooded meadows. Often in semi-aquatic, terrestrializing biotopes. Life cycle unknown.

Distribution: Western, Central and northern Europe, British Isles. Very rare.

Genus *Zschokkea* KOENIKE, 1892

1892b *Zschokkea* KOENIKE, Zool. Anz. 15: 320.

Type species: *Zschokkea oblonga* KOENIKE, 1892b, by monotypy.

Diagnosis: Lateral eyes encapsulated. Idiosoma without sclerotized plates (but the typical muscle insertion pattern of Hydryphantidae visible as dots of weak papillosity – Dc-1+2 fused). Frontal eye with a small sclerotized ring, pigmented. Coxae with sparse setation. Genital flaps with fine setae, mostly at medial margins, some on the flap surfaces. All Ac

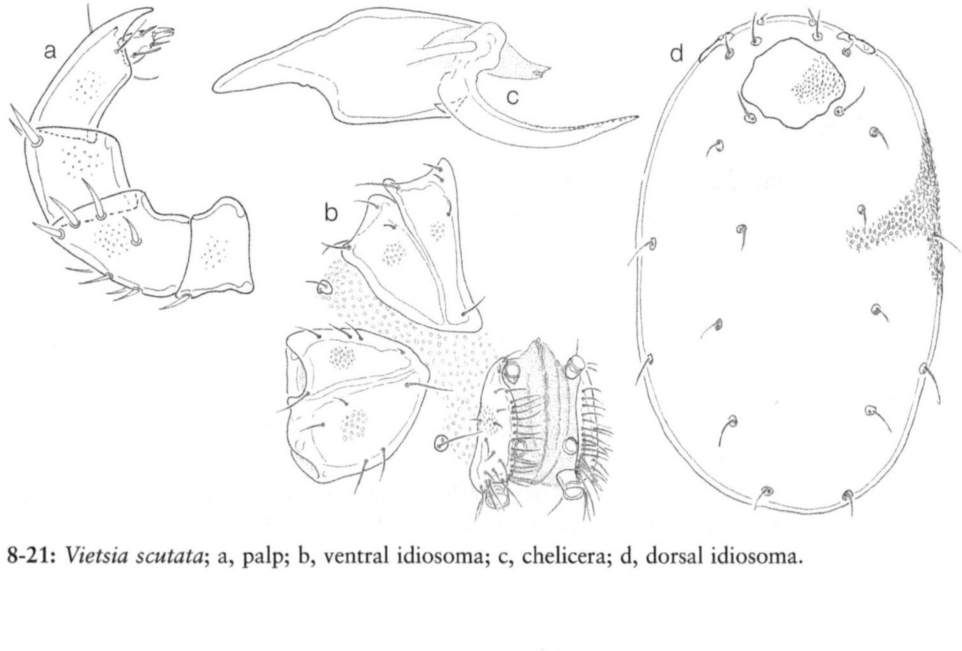

8-21: *Vietsia scutata*; a, palp; b, ventral idiosoma; c, chelicera; d, dorsal idiosoma.

completely included into genital flap margins, an extension of the flap separating Ac-2 and Ac-3. Excretory pore unsclerotized. Gnathosoma with very short rostrum (visible only in lateral view), mouth opening small. Chelicera rather slender and with a relatively short claw, directed ventrally, with very long air sacs. P-1 very short, P-3 highly variable in shape (length/height 0.9-1.5, as long as P-2 or longer).

Remarks: A highly distinct taxon with unclear relations to other Hydryphantidae, in addition to the European species, one further species known from North America (SMITH & COOK 2009).

Zschokkea oblonga KOENIKE, 1892 (*8-1 g, 8-22 a-c*)

1892b *Zschokkea oblonga* KOENIKE, Zool. Anz. 15: 320.

Description: Colour red. Integument papillae strong and flat, not bent posteriorly. Cx-I+II medially pointed, all coxae with only a few setae. Legs stout, setae in low numbers, but strong. Lateral margins of genital flaps equally convex, medial margins equally concave, Ac diameter ca. 20 µm. Chelicera basal segment length/height 160/65, claw length 65 µm. Palp robust, setation: P-1, 1; P-2, 7; P-3, 4; P-4, 3 (all distally, one dorsally, fine, one ventrolaterally and one ventromedially, basally thickened. Males: Genital flap setae numerous, rather long, posterior genital flap extension bearing 2-3 setae – no measurements published. Females: Idiosoma length 1000-1360; genital flap length 250 µm with setae shorter and fewer than in males, posterior genital flap extension bearing one seta; gnathosoma length/width 210/170 µm.

Preadult stages: Larva unknown. Deutonymph: TUZOVSKIJ (1990).

Habitat: Both in cold helocrene springs and summer-warm stagnant waters, e.g. moorland pools, quagfens and reedlands.

Distribution: Western, Central and Northern Europe. Farthest southern record: Retezat mountains, Romania (CÎMPEAN & GERECKE 2006).

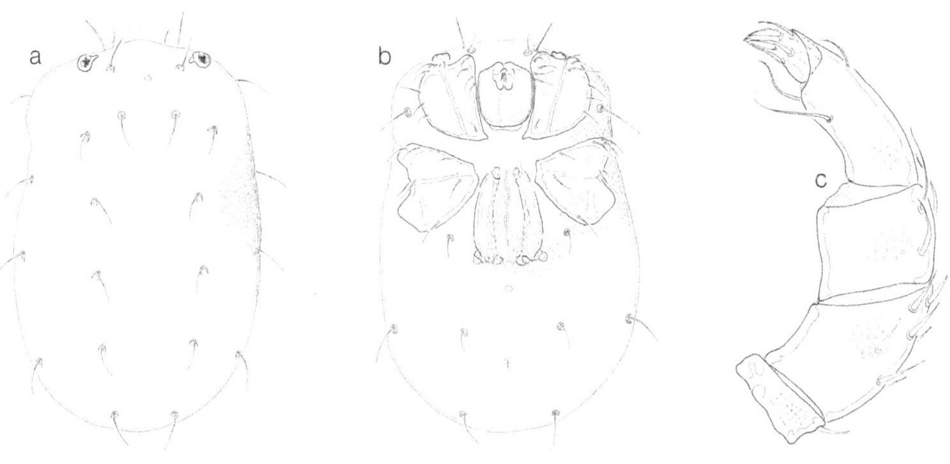

8-22: *Zschokkea oblonga* female; a, dorsal idiosoma; b, ventral idiosoma; c, palp

Subfamily Hydryphantinae PIERSIG, 1896

1896c Hydryphantinae PIERSIG, Sitzungsber. naturf. Ges. Leipzig 22: 48.

Diagnosis: Lateral eyes in capsules. Frontal plate including frontal eye, frontalia, pre- and postocularia, characteristic in shape due to a pair of more or less extending posterior extensions flanking the concave medioposterior margin. Posterior dorsal idiosoma generally without sclerites, rarely muscle attachments sclerotized as small dots or larger platelets. Legs of most species with swimming setae arranged in rows on the dorsal surface of II-L-4-5, III-L-3-5 and IV-L-3-5 (species lacking swimming setae not known from the area treated).

Discussion: A very distinct, surely monophyletic subfamily. The nominate genus *Hydryphantes,* includes species from all continents except Antarctica. Occasionally, misshapening of the frontal shield hampers recognition. In such cases, the paired posterior extension may be the only remnant of the frontal shield (DI SABATINO et al. 2009).

Genus *Georgella* KOENIKE, 1907

1907a *Georgella* KOENIKE, Abh. Naturw. Ver. Bremen 19: 129.

Type species: *Hydrodroma helvetica* HALLER, 1882a, by original designation.

Diagnosis: Coxae (*8-23 a*) with numerous long, fine setae on Cx-I-III, in lines away from the anterior/posterior plate margins, in dense groups at anterior/posterior corners, on Cx-IV scattered all over plate surface, but near posterior margin concentrated in a small group posterior to leg insertion, posterior margin Cx-IV with two concave sectors, the lateral one more extended, separated by a blunt projection. Arrangement of swimming setae rather homogeneous in all species (on II-L-4/5, III-L-3-5 and IV-L-3-5 on the posterior side, on IV-L-3-4 also on anterior side; range of numbers: II-L-4/5 5-20/20-30, III-L-3/4/5 5-10/15-25/25-35, IV-L-3 ant./post. 5/3-10, IV-L-4 ant./post. 5-20/25-35, IV-L-5 25-45). Genital field with numerous Ac, most of them arranged on the genital flaps in two groups (homologs of multiplied Ac-1 and Ac-3) separated by an indentation of the lateral flap margin. An additional pair of Ac free (homolog of Ac-2) in membranous integument generally present, rarely reduced. Acetabula disc-like, their basal sclerites not extended to stalks. Mouth parts stout, gnathosoma with short rostrum, palp with numerous setae (*8-23 c*).

Remark: This taxon has been variously treated as a genus on its own, or as a subgenus of *Hydryphantes*. TUZOVSKIJ (2007) demonstrated significant differences between the known larval instars in *Hydryphantes* and *Georgella* that suggest the treatment of both taxa as separate genera. This point of view also finds support in the autapomorphic character states concerning the morphology of the genital field, setation of coxae and swimming setae arrangement on the legs of adults. Since a former published association with coenagrionid Odonata (MÜNCHBERG 1936a) was based on misidentified larvae (TUZOVSKIJ 2007), larval host preference is unclear for all species of the genus. KARL VIETS (1930d) described an interesting dimorphism in the deutonymphal provisional genital organ, possibly indicating a preadult sexual differentiation.

Key to species

Excluded as *species incertae* (DI SABATINO et al. 2009): *Georgella berlesei* PIERSIG, 1901b, *G. imperfecta* BADER, 1955.

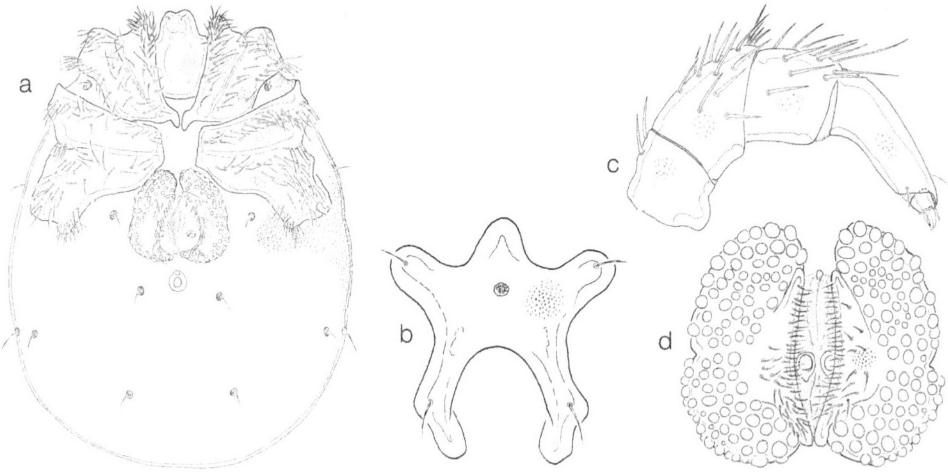

8-23: *Georgella*; a-c, *G. helvetica*; a, ventral idiosoma; b, frontal shield; c, palp; d, *G. koenikei*, genital field (after MAGLIO 1906).

1 Genital field with higher numbers of Ac (*8-23 d,* number per flap in males > 80, in females > 130). . . . . *Georgella koenikei* (page 44)

\- Genital field with lower numbers of Ac (*8-2 c, 8-23 a,* < 70 per flap in both sexes). . . . . *Georgella helvetica* (page 43)

1. *Georgella helvetica* (HALLER, 1882) (*8-2 c, 8-23 a-c*)

1882a *Hydrodroma helvetica* HALLER, Mitt. naturf. Ges. Bern 1881, 2: 49.

1903 *Hydryphantes apsteini* KOENIKE, Zool. Anz. 26: 535 – DI SABATINO et al. (2009).

Description: Colour red, with regularly-arranged darker spots. Idiosoma length/width in males 2000-2800/1600-2000, in females up to 3500/2500 µm. Frontal shield length/width in males 540-600/500-570, in females 650-720/600-660 µm (ratio 1.1-1.2), anterolateral extensions directed laterally or slightly bent anteriorly, but with their tips not reaching level of triangular projection of anterior margin. Frontal eye generally near base of this projection or in anterior third of central plate surface. Medial margin Cx-III+IV straight. Genital flaps on each side with 40-65 Ac and 30-40 fine setae. Chelicera ratio length/height 3.7-3.9, basal segment/claw 1.6-1.8. Palp seta numbers P-1, 2-6; P-2, 16-20; P-3, 10-18; P-4, 4; P-4 relatively short and stout (length ratio P-2/4, 0.7-0.9, length/height P-4, 2.4-2.6).

Similar species: *Georgella asteroidea* K. VIETS, 1930d (Spain) differs in generally smaller dimensions and the characteristic star-shaped frontal shield, with the central surface reduced, and the anterior projection narrow and long, similar to the two pairs of lateral extensions. *Georgella fimbriata* WALTER, 1925a (Algeria) is characterized by fewer Ac, the anterolateral extensions of the frontal shield more strongly bent anteriorly, reaching the level of the tip of the anteromedial projection, and medial margins of Cx-III+IV not straight, but Cx-IV slightly projecting over Cx-III.

Preadult stages: Larva unknown – description and host association given by MÜNCHBERG (1936a) based on misidentified specimens (TUZOVSKIJ 2007). Deutonymph: ÖZKAN (1981).

Habitat and Biology: Small stagnant water bodies (ditches, swamps, ponds). Life cycle unknown.

Distribution: Central, Eastern and South-eastern Europe, Turkey.

2. *Georgella koenikei* MAGLIO, 1906 (*8-23 d*)

1903 *Georgella berlesei* MAGLIO (nec PIERSIG, 1901), Rend. Istit. Lombard. Sci. Lett. Milano (2) 36: 294.

1906 *Georgella koenikei* MAGLIO, Zool. Anz. 30: 410.

1927 *Georgella intermedia* WALTER, in: WALTER & MOTAŞ, Trav. Labor. Piscicult. Univ. Grenoble 11: 80 – DI SABATINO et al. (2009).

1953 *Georgella intermedia* C. ANGELIER, Ann. Stat. Centr. Hydrobiol. appliquée, Fasc. hors série: 211.

1987 *Hydryphantes (Georgella) intermedia* K. O. VIETS, Abh. naturw. Ver. Hamburg 8: 361 (praeocc.).

1987 *Hydryphantes (Georgella) dauphinensis* K. O. VIETS, Abh. naturw. Ver. Hamburg 8: 361: DI SABATINO et al. (2009).

Description: Idiosoma length/width in males 2000-2800/1500-1900, in females up to 3700/2900 µm. Frontal shield total length/width in males 610/630, in females 750-780/650-700 µm, similar to *G. helvetica*, anterolateral extensions directed laterally or slightly bent anteriorly, antero- and posterolateral extensions slender or stouter depending on secondary sclerotization, anteromedial projection triangular, frontal eye near base of this projection, ratio total length/width 1.1. Genital flap length in males 480-530, in females 500-600 µm, laterally slightly indented at centre, on each side with more than 90 Ac in males, 130 Ac in females. Length ratio P-2/4, 0.87, length/height P-4, 3.0.

Habitat and Biology: Standing waters (ponds, swamps and ditches). Preadult stages and life cycle unknown.

Distribution: French Alps and Italian prealpine lowlands (Lombardy). All records from outside this area incorrect or questionable (DI SABATINO et al. 2009).

Genus *Hydryphantes* KOCH, 1841

1841 *Hydryphantes* KOCH, Deutschlands Crustaceen, Myriapoden und Arachniden 37: 18.

1985 *Hydrobaderia* ÖZKAN, Entomologica Basiliensia 10: 19 – DI SABATINO et al. (2009).

Type species: *Hydryphantes ruber* GEER, 1778 (sub nom. *H.hilaris* KOCH, 1841) by monotypy.

Diagnosis: Genital field with well developed flaps, bearing three to numerous Ac. In polyacetabulate species all Ac located on the flaps, except for Ac-1 (at anterior plate margin, occasionally doubled) and Ac-2 (between gonopore and medial flap margin). Genital flaps without lateral indentations.

Remark: Of the three subgenera, *Papilloporus* WALTER, 1935 (recorded from tropical Asia, Africa and America) is absent from the study area.

Biology: Early records concerning larval parasitism of *Hydryphantes*-species on Culicidae probably refer to misidentified euthyadine species (SMITH & OLIVER 1976). Feeding by deutonymphs and adults on insect eggs and a capacity to survive drought conditions in moist sediments observed in North American species (WIGGINS et al. 1980).

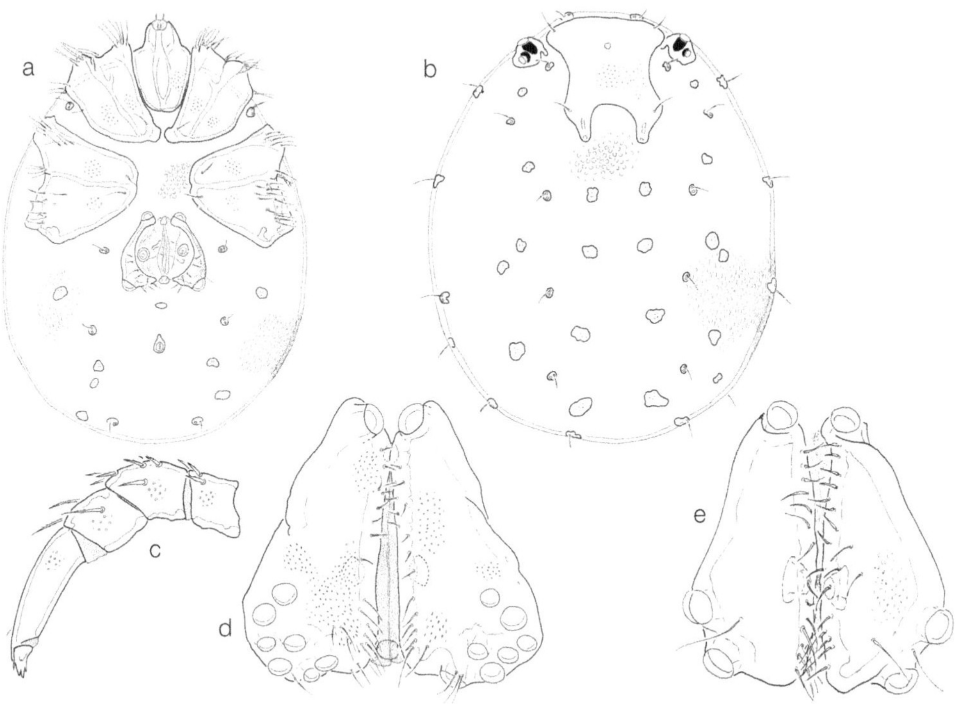

8-24: *Hydryphantes*; a-c, *H. armentarius*; a, ventral idiosoma; b, dorsal idiosoma; c, palp; d, *H. flexuosus* genital field; e, *H. octoporus* genital field.

Key to subgenera and species

Species recorded from the study area but excluded as *species incertae* (Di Sabatino et al. 2009): *Hydryphantes (Hydryphantes) bayeri* Písařovic, 1896; *H. (H.) clypeatus* Thor, 1899c; *H. (H.) frici* Thon, 1899a; *H. (H.) peroviensis* Udalzow, 1907b.

1 Genital field with more than three pairs of Ac (*8-24 d-e* – one anterior, one free in the centre between gonopore and genital flaps, two or more at posterior genital flap margin). (Subgenus *Polyhydryphantes*). 2

- Genital field with three pairs of Ac (*8-2 b*, *8-24 a*). (Subgenus *Hydryphantes s. str.*). . 3

2 (1) Genital field with four to six pairs of Ac (*8-24 e* – two to four pairs at posterior genital flap margin); III/IV-L-4/5 each with at most 15 swimming setae per segment. . . . . *Hydryphantes octoporus* (page 55)

- Genital field with more than six pairs of Ac (*8-24 d* – four or more pairs at posterior and lateral genital flap margin); III/IV-L-4/5 each with more than 15 swimming setae per segment. . . . . *Hydryphantes flexuosus* (page 54)

3 (1) Legs with very few, relatively short swimming setae (restricted to III-L-4/5 and IV-L-4/5, maximum number per segment 5, *8-25 e*); all dorsalia and ventralia sclerotized as small platelets, in size not exceeding frontal eye capsules (*8-24 a*). . *Hydryphantes armentarius* (page 48)

- Legs with swimming setae in higher numbers and longer (also on II-L, III-L-4/5 and IV-L-4/5 with more than 5 setae per segment, *8-27 e*); dorsalia and ventralia often absent (if muscle attachments sclerotized, <u>or</u> small, similar to glandularia, <u>or</u> larger than frontal eye capsules). 4

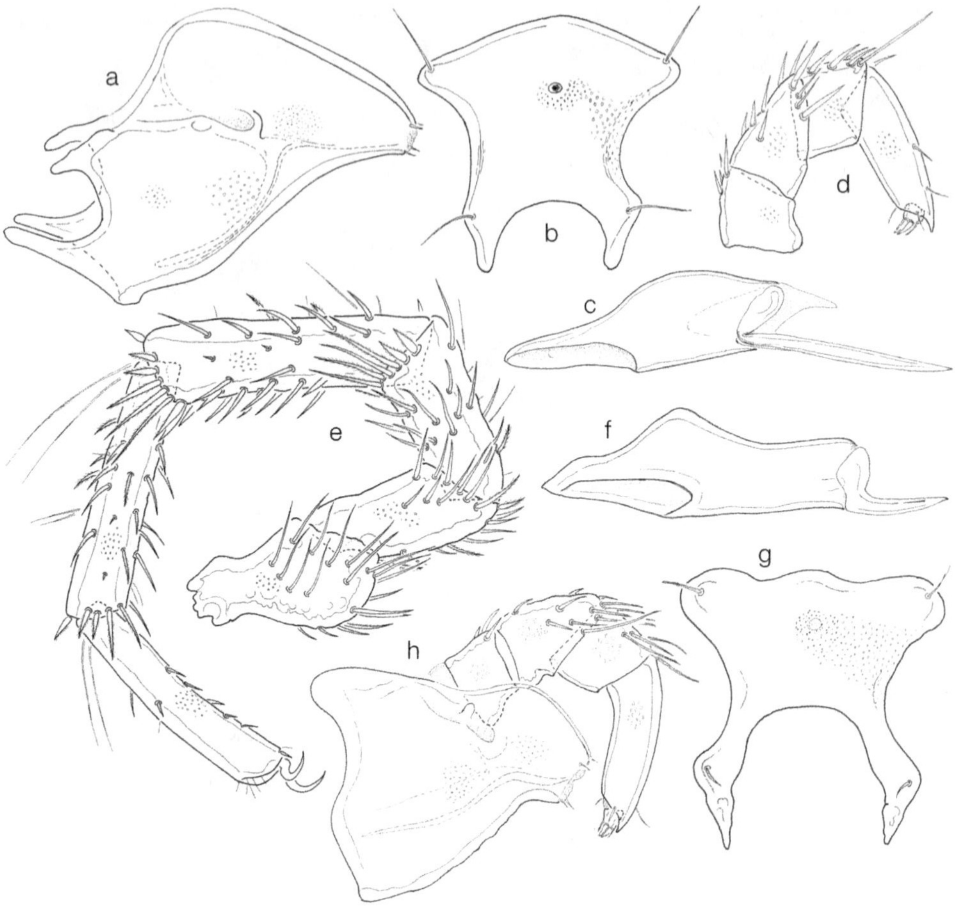

8-25: *Hydryphantes*; a-d, *H. fontinalis*; a, gnathosoma; b, frontal shield; c, chelicera; d, palp; e, *H. armentarius* IV-L; f-h, *H. placationis*; f, chelicera; g, frontal shield; h, gnathosoma and palp.

4 (3) Cheliceral claw modified, stylet-like, very fine and long (nearly as long as basal segment), with an acute-angled basal kink (*8-25 c*); swimming setae reduced in number (III-L-4/5, 6; IV-L-4/5, 7-8). . . . . *Hydryphantes fontinalis* (page 51)

\- Cheliceral claw more robust and shorter (occasionally straight, but at least 30 % shorter than basal segment), without a basal kink (*8-25 f*, *8-27 c*); swimming setae in higher numbers (III-L-4/5, 7-20, IV-L-4/5, 9-30). 5

5 (4) Posterior extensions of frontal shield long (*8-25 g*, *8-26 a* - total length/medial length 1.6-2.1). 6

\- Posterior extensions of frontal shield short (e. g. *8-27 b*, *d* – total length/medial length 1.1-1.5). 7

6 (5) Frontal shield anteriorly hardly projecting, enlarged (*8-25 g* - anterior width > medial length); cheliceral claw very short (*8-25 f* – basal segment/claw > 2.8); P-4 long and slender (*8-25 h* - length > 200 µm, length/width > 3.0). . . . . *Hydryphantes placationis* (page 52)

- Frontal shield anteriorly projecting in an obtuse angle, more slender (*8-26 a* - anterior width < medial length); cheliceral claw relatively longer (*8- 26 c* – basal segment/claw < 2.8); P-4 short and stout (*8- 26 b* – length < 200 µm, length/width < 3.0). . . . . *Hydryphantes dispar* (page 50)

7 (5) Dorsal and posteroventral muscle attachments sclerotized as symmetrically-arranged plates, some of them distinctly larger than eye capsules. . . . *Hydryphantes parmulatus* (page 51)

- Dorsal and posteroventral muscle attachments unsclerotized or developed as small sclerotized spots, in diameter not larger than glandularia. 8

8 (7) Palp very stout, in particular P-4 (*8-26 g* – length/height < 2.3). . . . . *Hydryphantes crassipalpis* (page 50)

- Palp more slender, P-4 length/height > 2.8 (*8-27 a, f, h*). 9

9 (8) Chelicera with long and straight claw (*8-27 c* – basal segment/claw ratio < 1.8); gnathosomal rostrum long (*8-27 a*); body large in size (frontal shield length/width > 520/440, Ac-3 diameter > 75, P-4 length > 270 µm). . . . . *Hydryphantes tenuipalpis* (page 54)

- Chelicera with shorter, slightly curved claw (*8-27 i* - basal segment/claw ratio > 1.8); gnathosomal rostrum various in shape (*8-27 f, h*); body minor in size (frontal shield length/width < 520/440, Ac-3 diameter < 75, P-4 length < 270 µm). 10

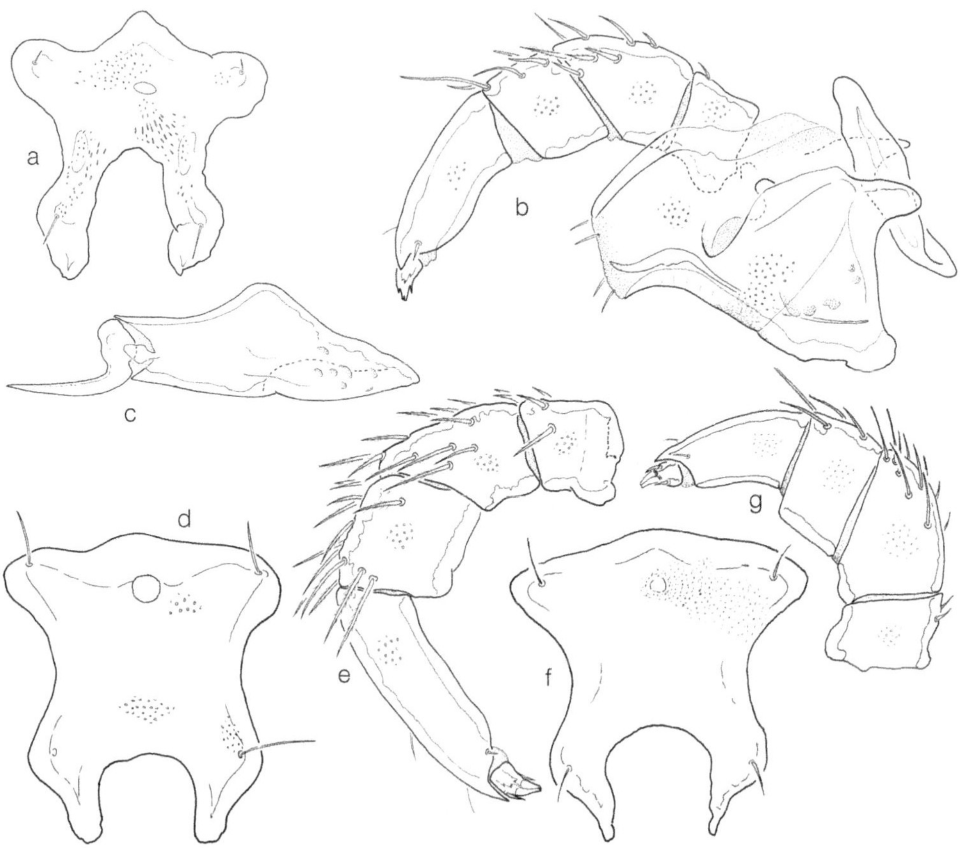

8-26: *Hydryphantes*; a-c, *H. dispar;* a, frontal shield; b, gnathosoma and palp; c, chelicera; d-e, *H. parmulatus*; d, frontal shield; e, palp; f-g, *H. crassipalpis*; f, frontal shield; g, palp.

10 (9) Gnathosomal rostrum longer (*8-27 f*); occasionally idiosomal muscle insertions sclerotized as small dots. . . . . *Hydryphantes ruber* (page 53)

\- Gnathosomal rostrum shorter (*8-27 h*); idiosomal muscle insertions unsclerotized (*8-27 g*). . . . . *Hydryphantes planus* (page 52)

Subgenus *Hydryphantes* Koch, 1841

1841 *Hydryphantes* Koch, Deutschlands Crustaceen, Myriapoden und Arachniden 37: 18.

1985 *Hydrobaderia* Özkan, Entomologica Basiliensia 10: 19 – Di Sabatino et al. (2009).

Type species: *Hydryphantes ruber* Geer, 1778 (sub nom. *H. hilaris* Koch, 1841) by monotypy.

Diagnosis: Shape of frontal shield and swimming setation of legs species-specifically very diverse. Genital field with three pairs of Ac; Ac-3 stalked.

Remark: Possibly a paraphyletic clade that gave rise to *Polyhydryphantes*- and *Papilloporus*-like species at one or several times.

Biology: Following observations made on North American populations of *H. ruber* and *H. ramosus* Daday, 1905 (syn.: *H. tenuabilis* Marshall, 1926), spermatophore transfer indirect, eggs are laid in masses of 30-200 and larvae are able to jump from the water surface film to reach hosts. Pre-parasitic larvae of *H. ramosus* attacking hydrometrids able to transfer from one host instar to the next, and start to engorge only after the hosts adult moult. Transformation to the protonymphal stage on the host (Smith & Oliver 1986 and bibliography cited therein).

1. *Hydryphantes (Hydryphantes) armentarius* Gerecke, 1996 (*8-24 a-c, 8-25 e*)

1996b *Hydryphantes armentarius* Gerecke, Arch. Hydrobiol., Suppl. 77: 363.

1985 *Hydryphantes pyrenaicus* nom. nud. E. Angelier, Annls. Limnol. 21: 6 – Di Sabatino et al. (2009).

Description: Frontal shield with equally convex anterior margin, posteriorly slightly narrowed, posterior extensions short and stout; measurements similar in both sexes (total length/medial length/maximum width 240-390/190-260/230-350. Swimming setae: II-L-4/5, 0; III-L-4, 1; III-L-5, 1 (both reduced in size); IV-L-3, 0-1; IV-L-4, 1-5; IV-L-5, 1-4. Gnathosoma dorsoventrally high, rostrum short, mouth opening small, cheliceral claws short and slightly curved. Palp rather slender, P-3 as long as high, P-4 variable in size (length/height 3.1-3.7). Males: Idiosoma length/width 840-1100/560-850; genital flap length ca. 200, Ac-1-3 diameter 30-40, 25-30, 35-55; gnathosoma length 170-250 µm. Females: Idiosoma length/width 1000-1700/730-1300; length genital flap 250; gnathosoma 270 µm.

Similar species: Other Western Palaearctic species with poorly developed or completely reduced swimming setation are the East Anatolian *H. baderi* Özkan, 1982 and *H. yalvaci* Özkan, 1982 (both with a characteristic blade-like cheliceral claw as described below for *H. fontinalis*) and *H. handschini* Walter, 1924c from North Africa and Sicily (differing in a frontal shield with very long posterior extensions and a slender gnathosoma with a very long rostrum) (Gerecke 1996b).

Habitat and Biology: Crenobiont. Generally in weakly flowing rheohelocrenes exposed to sunlight. Hardly sensitive to eutrophication; also in farmland fountains and temporary pools; at 700-1700 m asl. Life cycle unknown.

Distribution: Central Mediterranean (Tyrrhenian Islands, Central Italy, Balkan peninsula, Pyrenees – here reaching the Southwestern border of the area treated).

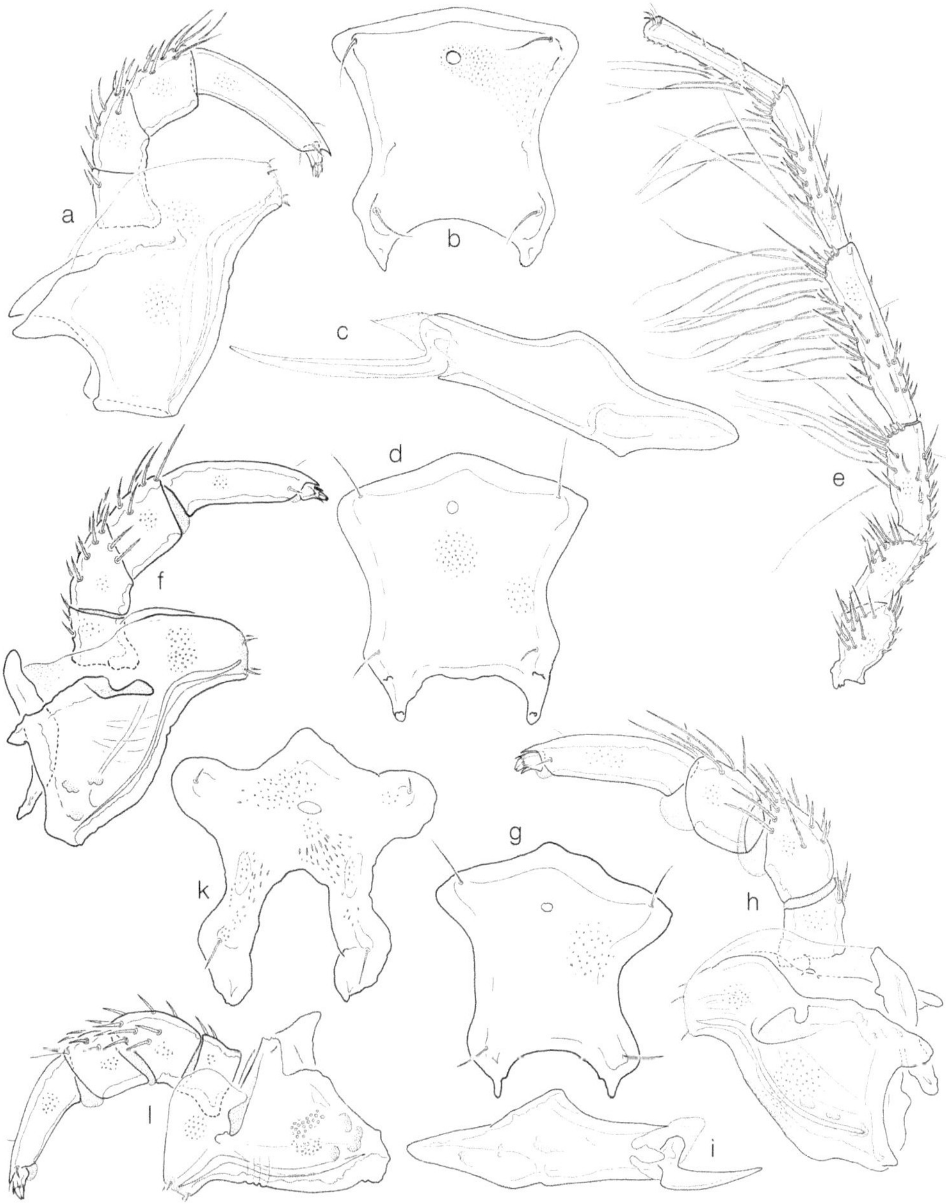

8-27: *Hydryphantes*; a-c, *H. tenuipalpis*; a, gnathosoma and palp; b, frontal shield; c, chelicera; d-f, *H. ruber*; d, frontal shield; e, IV-L; f, gnathosoma and palp; g-i, *H. planus*; g, frontal shield; h, gnathosoma and palp; i. chelicera; k-l, *H. flexuosus*; k, frontal shield; l, gnathosoma and palp.

2. *Hydryphantes (Hydryphantes) crassipalpis* KOENIKE, 1914 (*8-26 f-g*)

1914e *Hydryphantes crassipalpis* KOENIKE, Abh. naturw. Ver. Bremen 22 (2): 387.

1923 *Hydryphantes crassipalpis lacustris* HALBERT, J. Linn. Soc. London, Zool. 35: 389 – ÖZKAN (1988).

1985 *Hydrobaderia ilicaensis* ÖZKAN, Entomologica Basiliensia 10: 19 – DI SABATINO et al. (2009).

Description: Frontal shield compact, pentagonal in shape due to a bluntly pointed medioanterior projection. Gnathosoma variable in shape with equally S-shaped ventral margin, rostrum of medium size or elongated and cheliceral claws small, slightly curved. Palp extremely robust, P-3 distinctly shorter than high, P-4 only about twice as long as high. Males: Idiosoma length/width 1450/1200; frontal shield length/width 470-540/455-475; genital flap length 285-300, Ac-1-3 diameter 45/35-45/60; gnathosoma length 325; P-1-5 length/height 65/100, 145/115, 70/115, 135/80, 35/22 µm. Females: Idiosoma length/width 1500-2400/1200-1700; frontal shield length/width 435-555/360-525; genital flap length 245-300, Ac-1-3 diameter 40-55/30-40/50-70; gnathosoma length 290-425; P-1-5 length/height 60-80/85-110, 130-160/100-130, 65-80/110-130, 110-150/70-85, 30-40/25-30 µm. For a thorough analysis of variability see ÖZKAN (1988).

Preadult stages: Larva: WAINSTEIN (1980), BIESIADKA & CICHOCKA (1990), ÖZKAN (1988). Deutonymph: ÖZKAN (1988); TUZOVSKIJ (1990).

Habitat and Biology: Standing waters, also in temporary pools and ditches. Life cycle unknown. Adults appear in The Netherlands from April until October, with a peak in spring (SMIT & VAN DER HAMMEN 2000).

Distribution: Palaearctic; widespread in Europe, but not recorded from the Mediterranean area.

3. *Hydryphantes (Hydryphantes) dispar* (SCHAUB, 1888) (*8-26 a-c*)

1888 *Hydrodroma dispar* SCHAUB, S. B. Akad. Wiss., Wien, math.-naturw. Cl. 97: 107.

1901c *Hydryphantes intermedius* DADAY, In: G. HORVATH, Dritte asiatische Forschungsreise des Grafen Zichy 2, zool. Ergebn.: 423 – DI SABATINO et al. (2009).

1908a *Hydryphantes abnormis* KOENIKE, Abh. naturw. Ver. Bremen 19: 248 – LUNDBLAD (1962).

1925a *Hydryphantes algeriensis* WALTER, Bull. Soc. Hist. Natur. Afrique Nord 16: 206 – DI SABATINO et al. (2009).

Description: Frontal shield slender, with a bluntly pointed medioanterior projection and prominent anterolateral and posterior extensions. Swimming setae: II-L-4/5, 2-4/7-13; III-L-3/4/5, 2-3/9-13/15-17. Gnathosomal rostrum short, mouth opening large. Cheliceral claws small, slightly curved. Palp robust, P-3 shorter than high, P-4 length/height 2.4-2.6; palp setation P-1-4: 1-3, 10-11, 2-4, 3. Males: Idiosoma length/width 1300-1500/1100-1200; frontal shield total length/medial length/width 390-450/220-240/380-410; genital flap length 230-275, Ac-1-3 diameter 40-50/35-40/50-55; gnathosoma length 210-300; P-1-5 length/height 60-70/85-95, 115-135/95-105, 65-80/95-120, 150-170/60-75, 30-35/20-25 µm. Females: Idiosoma length/width 1700-1800/1300; frontal shield total length/medial length/width 400-450/250/450-460; genital flap length 240, Ac-1-3 diameter 45/30-40/50-55; gnathosoma length 270-290; P-1-5 length/height 70/100, 140/120, 80/105, 195/75, 45/20 µm.

Preadult stages: Larva: SPARING (1959). Deutonymph: TUZOVSKIJ (1990).

Habitat and Biology: Standing waters, also in temporary pools and ditches. Larvae parasitic on Ephydridae (*Hydrellia*) (SPARING 1959).

Distribution: Palaearctic. Recorded from most parts of Europe, but rare in the Mediterranean area (Di Sabatino et al. 2009).

4. *Hydryphantes (Hydryphantes) fontinalis* Sokolow, 1936 (*8-25 a-d*)

1936b *Hydryphantes fontinalis* Sokolow, Arch. Hydrobiol. 30: 471.

Description: Frontal shield enlarged at the level of preocular setae, anterior margin equally convex, frontal eye pigmented. Swimming setae: II-L-4/5, 0-3, 2-5; III-L-3/4/5, 0-1, 5-6, 5-7; IV-L-3/4/5, 0-2, 4-8, 6-7. Ventral margin of gnathosoma weakly curved, rostrum not distinctly set off, mouth opening small. Cheliceral claw fine, long and straight. Palp robust, P-1 and P-3 subrectangular, P-4 maximum height at segment base. Setation P-1-P-3: 4/15/6. Males: Idiosoma length/width 800/600; frontal shield total length/medial length/width 400/300/430 (a probably male specimen from Russia total length/width 330/350); genital flap length 230, Ac-1-3 diameter 50/45/55; gnathosoma length 325; chelicera basal segment/claw/height 300/270/110; P-1-5 length/height 85/95, 132/100, 90/92, 200/62, 42/25 µm. Females: Idiosoma length ca. 1000; frontal shield length/width 480/420; genital flap length 290 µm.

Similar species: *Hydryphantes fontinalis* and two species described from Anatolia, *H. baderi* Özkan, 1982 and *H. yalvaci* Özkan, 1982, represent a highly distinct species group characterized by long, blade-like cheliceral claws and the reduction of swimming setae. *Hydryphantes baderi* differs in a relatively longer cheliceral claw (basal segment/claw ratio 0.8) and slightly higher swimming seta numbers, *H. yalvaci* in complete reduction of swimming setae. Both Turkish species are known from a single female only; minor differences in the shape of chelicerae and swimming setation could be the result of sex-specific characters (Gerecke 1996b).

Habitat and Biology: Crenobiont, preferably in helocrenes. Preadult stages and life cycle unknown.

Distribution: Only recorded from a few sites in NE Russia (near St. Petersburg) and one spring in the Italian Alps (Sokolow 1936b, Crema et al. 1996).

5. *Hydryphantes (Hydryphantes) parmulatus* Koenike, 1912 (*8-26 d-e*)

1912a *Hydryphantes parmulatus* Koenike, Zool. Anz. 40: 67.

1926c *Hydryphantes peltatus* Walter, Bull. Soc. Sci. natur. Maroc 6: 141 – Gerecke (1996b).

1989 *Hydryphantes algeriensis* Bader (nec Walter 1926c), Bijdr. Dierk. 39: 33 – Gerecke (1996b).

Description: Frontal shield compact, anterior margin slightly undulating and projecting in the centre, lateral and posterior extensions small. Muscle insertion sclerites on dorsum and posterior venter large (maximum size ca. 200 µm in Dc-5 and V-4), regularly arranged in the pattern typical for many Euthyadinae. Swimming setae: II-L-4/5, 2, 10; III-L-3-5, 3, 15, 20; IV-L-3-5, 5, 15, 15. Ventral margin of gnathosoma equally S-shaped, rostrum and mouth opening of medium size, and cheliceral claws slender, slightly curved, basal segment/claw 1.9. Palp slender, P-3 as long as high, P-4 length/height 3.5; setation P-1-3, 4, 20, 12. Males: Idiosoma length/width 1200/900; frontal shield total length/medial length/width 370/260/340; genital flap length 250, Ac-1-3 diameter 45, 30, 50; gnathosoma length 310; chelicera basal segment/claw length/height 300/120/110; P-1-5 length/height 72/80, 126/80, 90/94, 200/63, 41/16 µm. Females: Idiosoma length 1500; frontal shield width 430 µm.

Habitat and Biology: Permanent and temporary ponds, in Central Europe restricted to lowlands. Life cycle unknown. Adults in The Netherlands appearing from mid-April until mid-May (SMIT & VAN DER HAMMEN 2000).

Distribution: Central and Western Europe, Mediterranean area.

6. *Hydryphantes (Hydryphantes) placationis* THON, 1899 (*8-25 f-h*)

1899a *Hydryphantes placationis* THON, Zool. Anz. 22: 101.

Description: Frontal shield large and medially shortened, with long posterior extensions. Legs with numerous swimming setae: II-L-4/5, 5, 18; III-L-3-5, 5-8, 18-24, 18-20; IV-L-3-5, 11-13, 25, 28. Ventral margin of gnathosoma strongly curved, rostrum short and distinctly set off. Cheliceral claws very small, slightly curved. Palp rather slender, length/height P-2, 0.9; P-3, 0.8-0.9; P-4, 3.1-3.4. Setae numbers: P-1, 4-5; P-2, 15-17; P-3, 3-4. Males: Idiosoma length/width 1400/1000; frontal shield total length/medial length/width 440/210/440; genital flap length 310, Ac-1-3 diameter 40/40/70; gnathosoma length 260; chelicera basal segment/claw length/height 330/110/120; length/height P-1, 65/75; P-2, 135/110; P-3, 80/95; P-4, 205/65; P-5, 38/20 µm. Females: Idiosoma length/width 1500-2000/1100-1600; frontal shield total length/medial length/width 400/250/460; genital flap length 290, Ac-1-3 diameter 50/55/70; gnathosoma length 300; chelicera basal segment/claw length/height 360/120/140; length/height P-1, 85/95; P-2, 155/130; P-3, 100/118; P-4, 240/70; P-5, 42/23 µm.

Preadult stages: Larva unknown. Deutonymph: TUZOVSKIJ (1990).

Habitat and Biology: Stagnant waters.

Distribution: Europe, widespread except for northern Fennoscandia and Iceland.

7. *Hydryphantes (Hydryphantes) planus* THON, 1899 (*8-27 g-i*)

1899i *Hydryphantes planus* THON, Rozpr. Česke Akad. Praze (s. 2) 8: 26.

1919a *Hydryphantes bayeri nonundulatus* K. VIETS, Abh. naturw. Ver. Bremen 24: 13 – LUNDBLAD (1962), DI SABATINO et al. (2009).

1964b *Hydryphantes nonundulatus* LÁSKA, Biológia, Casopis Slovenskej Akademie Vied 19: 926 – DI SABATINO et al. (2009).

Description: Frontal shield compact, anterior margin obtuse-angled, lateral and posterior margins often concave, but occasionally straight, lateral and posterior extensions rather short. Posterodorsal and posteroventral muscle attachments unsclerotized. Swimming setae: II-L-4/5, 6, 9-13; III-L-3-5, 3, 7-8, 10-19; IV-L-3-5, 3-4, 10, 10-15. Palp rather slender, P-1 subquadratic, length/height P-2, ca. 1.2; P-3, 0.8-1.0; P-4, 3.2-3.8. Setae numbers: P-1, 5; P-2, 20; P-3, 3-4; P-4, 6. Males: Idiosoma length/width 1100-1300/900-1100; genital flap length 240-260, Ac-1-3 diameter 40, 40, 60; gnathosoma length 260-300 µm. Females: Idiosoma length/width 1100-1700/9000-1400; frontal shield total length/medial length/width 400-500/240-390/310-460; genital flap length 200-270, Ac-1-3 diameter 25-50, 25-35, 40-70; gnathosoma length 265-275 µm, chelicera basal segment/claw 1.9-2.6; length/height P-1, 65-80/75-85; P-2, 115-137/95-110; P-3, 72-90/92-95; P-4, 175-225/55-60; P-5, 38-40/18-20 µm.

Similar species: *Hydryphantes planus* is similar to *H. ruber*. See there for a discussion.

Preadult stages: Larva: WAINSTEIN (1980), BIESIADKA & CICHOCKA (1990). Deutonymph: TUZOVSKIJ (1990).

Habitat and Biology: Stagnant waters, pools of streams in riverine forests. Life cycle unknown.

Distribution: Western, Central and Northern Europe. Exact distribution unclear due to frequent confusion with other taxa.

8. *Hydryphantes (Hydryphantes) ruber* (Geer, 1778) (*8-2 a-b, 8-27 d-f*)

1778 *Acarus aquaticus ruber* Geer, Mémoires pour servir à l'histoire des Insectes, Stockholm: 141.

1899a *Hydryphantes hellichi* Thon, Zool. Anz., 22 (581): 101 – Lundblad (1962).

1919a *Hydryphantes hellichi auriculatus* K. Viets, Abh. Naturw. Ver. Bremen 24: 11 – Lundblad (1962).

1899a *Hydryphantes ruber prolongata* Thon, Zool. Anz., 22 (581): 101 – Lundblad (1962).

1922a *Hydryphantes spinipes* Walter, Rev. Suisse Zool. 29 (7): 276 – Di Sabatino et al. (2009).

Description: Frontal shield compact, anterior margin obtuse-angled, lateral and posterior margins concave or straight – in specimens with very short projections, the shield may have a subquadratic or elongated rectangular appearance. Posterodorsal and posteroventral muscle attachments often sclerotized as small dots. Swimming setae: II-L-4/5, 0-5, 7-16; III-L-3-5, 3-5, 8-15, 9-15; IV-L-3-5, 3-8, 10-20, 10-20. Gnathosoma with equally S-shaped ventral margin and rather long, slender rostrum. Cheliceral claws small, slightly curved. Palp rather slender, similar to *H. planus*. Setae numbers: P-1, 2-5; P-2, 9-16; P-3, 3-7. Males: Idiosoma length/width 1000-1800/800-1500; frontal shield total length/ medial length/width 300-500/220-420/290-520; genital flap length 200-300, Ac-1-3 diameter 37-50/25-40/40-60; gnathosoma length 210-430 µm; chelicera basal segment/claw ratio 1.9-2.3; length/height P-1, 50-90/60-100; P-2, 100-150/75-120; P-3, 65-110/65-110; P-4, 165-260/40-80; P-5, 30-40/15-20 µm. Females: Idiosoma length/width 1800-2500/ 1200-2000; frontal shield total length/medial length/width 330-470/250-360/340-470; genital flap length 210-250, Ac-1-3 diameter 35-50, 30-40, 30-70; gnathosoma length 250-460; length/height P-1, 50-85/65-100; P-2, 100-125/75-110; P-3, 70-100/70-100; P-4, 185-220/40-75; P-5, 35-45/15-20 µm.

Similar species: Extensive variability studies of Besseling (1941) and Lundblad (1962) show a rather wide concept of the morphological range of this species, including also specimens with slightly, or strongly, elongated frontal shields (former subspecies *H. ruber prolongatus* Thon, 1899a and *H. ruber hellichi* Thon, 1899a). The relatively long gnathosomal rostrum was the only stable character found in large populations, allowing a differentiation between *H. ruber* and *H. planus* (with a shorter rostrum). However, both taxa may grade into each other and sometimes species attribution is difficult. The taxonomic complex *H. ruber/planus* merits particular attention in future taxonomic research.

Preadult stages: Larva: Numerous descriptions, e. g., Piersig (1896-1899), Lundblad (1927c), Prasad & Cook (1972), Wainstein (1980), Biesiadka & Cichocka (1990). Deutonymph: Tuzovskij (1990).

Biology: Larvae parasitic on culicid midges (Piersig 1896-1899, Koenike 1909c – uncertain association). Deutonymphs and adults feeding on insect eggs and able to survive conditions of drought if a minimum of moisture in sediments is guaranteed.

Habitat: Many kinds of temporary and permanent stagnant waters, also in lakes and limnocrenes.

Distribution: Holarctic (? – North American records questionable, partly as subspecies *H. r. mozleyi* Marshall, 1929c and *H. r. novatus* K. Viets, 1949b). Widespread in Europe.

9. *Hydryphantes (Hydryphantes) tenuipalpis* Thon, 1899 *(8-27 a-c)*

1899a *Hydryphantes ruber tenuipalpis* Thon, Zool. Anz. 22: 101.

1962 *Hydryphantes tenuipalpis* Lundblad, Ark. Zool. (ser. 2) 14: 42.

Description: Frontal shield compact, often subquadratic, with obtuse-angled anterior, and concave lateral and posterior margins, extensions slightly projecting. Swimming setae: II-L-4-5, 3-4, 10-20; III-L-3-5, 2-6, 12-20, 16-24; IV-L-3-5, 3-4, 16-30, 18-26. Ventral margin of gnathosoma equally S-shaped, rostrum relatively long, mouth opening small. Cheliceral claws relatively long and straight (basal segment/claw 1.5-1.7). Palp slender, P-1 subquadratic, P-2 and generally also P-3 longer than high, P-4 length/height 3.6-4.3. Setae numbers: P-1, 3-5; P-2, 12-18; P-3, 4-9. Males: Idiosoma length/width 1200-1800/1000-1400; frontal shield total length/medial length/width 600/480/600; genital flap length 320, Ac-1-3 diameter 50, 40, 90; gnathosoma length 440; length/height P-1, 105/115; P-2, 150/130; P-3, 125/118; P-4, 305/85; P-5, 45/20 µm. Females: Idiosoma length/width 1600-2200/1200-1600; frontal shield total length/medial length/width 530-640/450-540/550-700; genital flap length 290-340, Ac-1-3 diameter 45-70, 35-45, 80-100; gnathosoma length 380-450; length/height P-1, 95-110/105-135; P-2, 140-180/100-150; P-3, 120-140/100-135; P-4, 290-325/70-90; P-5, 45-55/20-25 µm.

Similar species: All other species with an elongated cheliceral claw belong to the *fontinalis* – species group (see above), characterized by a relatively longer and more blade-like cheliceral claw and reduced swimming setation.

Preadult stages: Larva and deutonymph: Tuzovskij (2008).

Habitat and Biology: Stagnant waters. Life cycle unknown.

Distribution: Central, Northern and Eastern Europe, Turkey.

Subgenus *Polyhydryphantes* K. Viets, 1926

1926l *Polyhydryphantes* K. Viets, Zool. Anz. 69: 191.

1927d *Octohydryphantes* Lundblad, Arch. Hydrobiol. 18: 409 – Lundblad (1962).

Type species: *Hydryphantes flexuosus* (Koenike, 1885a), by original designation.

Diagnosis: Frontal shield rather slender, with convex anterior margin and well developed anterolateral and posterior projections embracing a deep, reversed U-shaped posterior bay. Legs always with numerous swimming setae. Genital field with four to many pairs of Ac, in species with numerous Ac their basal sclerites not developed to stalks. Surface of Ac not papillate.

Remark: In the closely related tropical genus *Papilloporus*, the surface of Ac is strawberry-like papillate, and the frontal shield is H-shaped, its central part being reduced to a transverse bar bearing the frontal eye.

10. *Hydryphantes (Polyhydryphantes) flexuosus* (Koenike, 1885) *(8-24 d, 8-27 k-l)*

1885a *Hydrodroma flexuosa* Koenike, Abh. naturw. Ver. Bremen 9: 222.

1896a *Hydryphantes flexuosus* Koenike, Forschber. Biol. Stat. Plön 4: 226.

1930d *Hydryphantes vastipalpis* K. Viets, Arch. Hydrobiol. 21: 234 – Lundblad (1962), Di Sabatino et al. (2009).

1959a *Hydryphantes longicornis* Jankowskaja, Trudy Zool. Inst., Akad. Nauk SSSR 26: 361 – Lundblad (1962).

Description: Frontal shield with bluntly pointed medioanterior projection and long posterior extensions, rather slender (length/width 1.9-2.1). Swimming setae: II-L-4/5, 5-14/8-20; III-L-3-5, 4-8, 15-30, 20-40; IV-L-3-5, 2-4, 15-30, 20-45. Gnathosoma ventral margin equally S-shaped, rostrum of medium size; cheliceral claws small, slightly curved. Palp robust, P-3 shorter than high, P-4 less than three times longer than high. Males: Idiosoma length/width 1600-1800/1200-1400; frontal shield total length/medial length/width 500-600/230-280/500 µm; IV-L-4-6 with short, peg-like ventral setae and strong claws; genital flap length 250-280, Ac number 17-19; genital flap setae number 36-44; gnathosoma length 300-370; P-1-5 length/height 80-90/100-130, 160-175/135-145, 90-100/115-120, 220-250/80-85, 45-55/25-30 µm. Females: Idiosoma length/width 2200/1800; frontal shield total length/medial length/width 490-520/230-280/500 µm; IV-L-4-6 with longer and pointed ventral setae and finer claws; genital flap length 310-360 µm, Ac number 18-22; genital flap setae number 80-100; gnathosoma length 300-410; P-1-5 length/height 90-105/125-130, 160-180/160-170, 105/145, 220-160/90, 50-55/25-30 µm.

Preadult stages: Larva: IMAMURA (1954b), BIESIADKA & CICHOCKA (1990). Deutonymph: TUZOVSKIJ (1990).

Similar species: See discussion of *H. octoporus*.

Habitat and Biology: Temporary and permanent stagnant waters, with a preference for high electrolyte concentrations. Often in brackish coastal habitats. Life cycle unknown.

Distribution: Palaearctic. Widespread in Europe, except for the northernmost areas.

12. *Hydryphantes (Polyhydryphantes) octoporus* KOENIKE, 1896 (*8-24 e*)

1896b *Hydryphantes octoporus* KOENIKE, Zool. Anz. 19: 358.

1900a *Hydryphantes flexuosus thoni* PIERSIG, Annuaire Mus. zool. Acad. Sci. 4: 486.

1900b *Hydryphantes thoni* PIERSIG, Zool. Anz. 23: 209.

Description: Frontal shield with bluntly pointed medioanterior projection and long posterior extensions, stout (length/width 1.3-1.9). Leg swimming setae: II-L-4/5, 0, 7-10; III-L-3-5, 0-3, 7-10, 8-13; IV-L-3-5, 3-4, 8-13, 9-15. Posterior legs without sexual dimorphism. Ventral margin of gnathosoma equally S-shaped, rostrum of medium size; cheliceral claws small, slightly curved. Palp robust, P-3 shorter than high, P-4 about three times longer than high. Males: Idiosoma length/width 900-1500/750-1300; frontal shield total length/medial length/width 270-350/160-260/210-320; genital flap length 185-290 µm, Ac number 8; genital flap setae number 18-20; gnathosoma length 210-280; P-1-5 length/height 50-70/60-80, 70-120/70-105, 70-80/65-95, 135-180/40-60, 30-40/15-20 µm. Females: Idiosoma length/width 1100-2200/900-1600; frontal shield total length/medial length/width 310-390/180-250/290-410; genital flap length 220-240 µm, Ac number 8-10; genital flap setae number 35-40; gnathosoma length 250-290; P-1-5 length/height 50-65/70-90, 100-115/95-110, 65-80/90-100, 150-160/50-70, 35/20-22 µm.

Preadult stages: Larva: TUZOVSKIJ (2007).

Similar species: LUNDBLAD (1962) doubted the taxonomic value of Ac numbers not only at subgenus- but also at species level. In fact, if larger individual numbers are studied an overlap of *H. octoporus* and *H. flexuosus* is possible. However, in addition to the other characters described here, differences in size also allow for a clear separation of the two species, with *H. octoporus* being distinctly smaller in dimensions of idiosoma sclerites and appendages.

Habitat and Biology: Temporary and permanent stagnant waters, with a preference for high electrolyte concentrations. Often in brackish coastal habitats. Larvae parasitic on damselflies (ZAWAL & DYATLOVA 2008).

Distribution: Palaearctic. Widespread in Europe, except for the northernmost areas. One record from Zaire (LUNDBLAD 1949).

Subfamily Protziinae KOENIKE, 1909

1909c Protziinae KOENIKE, Acarina, In: Brauer, A: Die Süsswasserfauna Deutschlands: 23.

Diagnosis: All species red or orange in colour. Lateral eyes encapsulated. Frontal eye absent or (rarely) visible as a small ring not surrounded by sclerotization. Dorsal and posteroventral idiosoma surface without sclerites in all taxa known from the area treated (small sclerites in some species of the Neotropical genus *Neocalonyx*). Legs without swimming setae in all taxa known from the area treated (swimming setae present in the monotypic subgenus *Paracalonyx* of the genus *Neocalonyx* known from Argentina). Leg claws simple or with lateral clawlets. Genital flaps well developed or reduced in various degrees, occasionally their site only indicated by a row of setae surrounding the membranous gonopore area. Genital field with more than three, generally numerous pairs of Ac. Acetabula in most species stalked, with stalks longer in posterior, shorter in anterior genital field, never fused with genital flaps. Excretory pore smooth.

Remark: In addition to the genus *Protzia*, represented in considerable diversity in the area treated, the subfamily includes the neotropical genus *Neocalonyx* WALTER, 1919a, characterized by genital flap sclerites (or their vestiges) located medially to the majority of the Ac. These and other considerable differences (in some species of *Neocalonyx* absence of the dorsodistal projection of P-4, presence of swimming setae) suggest an early separation of the two clades. In view of unclear relationships to certain genera of Euthyadinae a polyphyletic origin of Protziinae is possible (COOK 1974).

Genus *Partnunia* PIERSIG, 1896

1896c *Partnunia* PIERSIG, Sitzungsber. Naturf. Ges. Leipzig 22/23: 52.

Type species: *Thyas angusta* KOENIKE, 1893b, by original designation.

Diagnosis: Frontal eye present or absent. Dorsal and posteroventral idiosoma without muscle attachment sclerites. Legs without swimming setae, leg claws simple, without lateral clawlets. Genital flaps well developed and porose, flanking and posteriorly embracing a row of stalked Ac. Integument flanking gonopore papillate in females, smooth in males.

Remarks: All species rather homogeneous in external morphology and habitat preference (confined to cold montane springs and spring brooks) (GERECKE 1993, 1996b). The genus has a holarctic distribution, but so far no species described from the Nearctic (records based on material determined to genus level only, COOK 1974).

Preadult stages: Larva and deutonymph: IMAMURA (1950a) and bibliography cited in the species section. Deutonymph with three pairs of Ac.

Key to species

1 Palp slender, P-1 elongated (length/height 0.7-1.0), P-3 ventral margin convex, P-4 maximum height at the centre of the segment; posterior Ac rounded; male without tent-like structures

surrounding the gonopore; ejaculatory complex long (> 180 µm), with anterior and posterior keels well developed. . . . . *Partnunia steinmanni* (page 58)

- Palp stout, P-1 shortened (length/height 0.5-0.7), P-3 ventral margin straight; P-4 maximum height at the base of the segment; posterior Ac elongated; males often with tent-like sclerotized structures surrounding the gonopore; ejaculatory complex short (< 150 µm), with anterior keel knob-shaped, posterior keel completely reduced. 2

2 (1) Males (females not distinguishable): posterior Ac longer (> 30 µm); ejaculatory complex stout, with more robust distal arms almost in right-angled position. . . . . *Partnunia angusta* (page 58)

- Males: posterior Ac shorter (< 30 µm); ejaculatory complex slender, with narrow distal arms in acute-angled position. . . . . *Partnunia tabernaculata* (page 59)

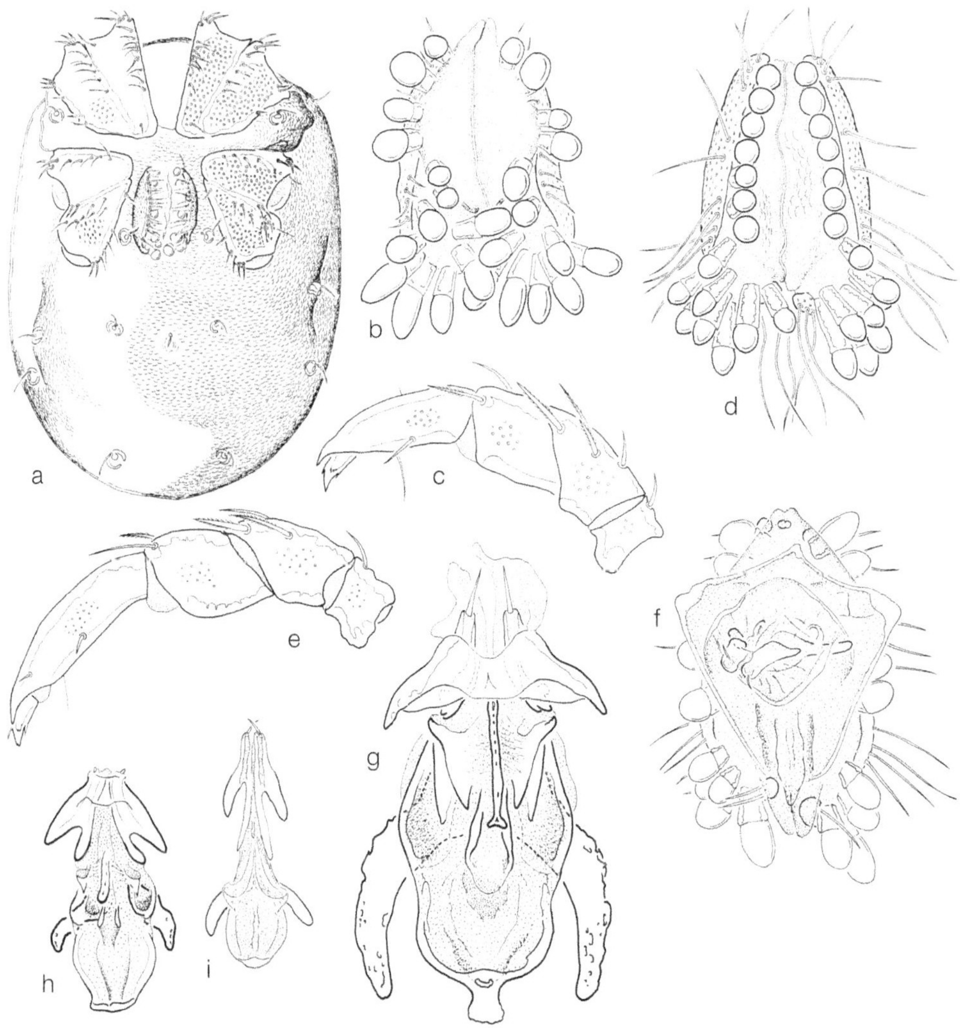

8-28: *Partnunia* (from Gerecke 1993 and Gerecke & Smit 1993, modified); a, ventral idiosoma; b-c, *P. angusta*; b, genital field; c, palp; d-e, *P. steinmanni*; d, genital field; e, palp; f, *P. tabernaculata*, genital field; g-i, ejaculatory complex, anterior view; g, *P. steinmanni*; h, *P. angusta*; i, *P. tabernaculata*.

1. *Partnunia angusta* (KOENIKE, 1893) (*8-28 b-c, h*)

1893b *Thyas angusta* KOENIKE, Zool. Anz. 16: 94.

Description: Idiosoma flattened and with parallel lateral margins. Genital flaps with rather fine setae. Length of posterior Ac stalks 20-50 µm (ratio stalk/Ac length 0.5-1.3). Palp stout, length/height ratio P-1, 0.5-0.7; P-2, 1.4-1.7; P-3, 0.8-1.0; P-4, 3.1-3.7. Males: Idiosoma length/width 700-850/450-550 µm; genital field with a trabecular structure as described for *P. tabernaculata*, but its parts more weakly sclerotized; gonopore length 170-180 µm, Ac number 23-26; ejaculatory complex length 115-130 µm, proximal arms not longer than distal arms, anterior keel knob-shaped, posterior keel reduced. Females: Idiosoma length/width 950-1100/600-700, gonopore length 160-230 µm, Ac number 20-30.

Habitat and Biology: Rheocrenes and low order streams, hyporheophilous. Preadult stages and life cycle unknown.

Distribution: Alps (Austria, Italy, Germany).

2. *Partnunia steinmanni* WALTER, 1906 (*8-1 h, 8-28 d-e, g*)

1906b *Partnunia steinmanni* WALTER, Zool. Anz. 30: 570.

1956a *Partnunia angusta* LUNDBLAD (*nec* KOENIKE, 1893b) Ark. Zool. 10: 45 – GERECKE (1993).

1975 *Partnunia raetica* BADER, Ergebn. wiss. Unters. schweizer. Nationalpark 14: 70 – GERECKE (1993).

Description: Idiosoma not flattened, with rounded lateral margins. Genital flaps with long robust setae. Length of posterior Ac stalks 27-55 µm (ratio stalk/Ac length 1.2-2.7). Palp slender, length/height ratio P-1, 0.7-1.0; P-2, 1.4-1.8; P-3, 0.9-1.4; P-4, 3.6-4.5. Males: Idiosoma length/width 680-1000/450-700, gonopore length 160-210 µm, Ac number 23-37; ejaculatory complex length 240-260 µm, proximal arms much longer than distal arms, keels normally developed. Females: Idiosoma length/width 740-1340/500-1000, genital flap length 220-260 µm, Ac number 28-55.

Similar species: Several species similar in shape of Ac (roundish, not elongated) are recorded from mountain ranges in the Mediterranean area (GERECKE 1993, 1996b). They all differ from *P. steinmanni* in a more stout palp (P-1 shortened, maximum height P-4 at segment base). These species differ from each other in the shape of the male ejaculatory complex which is generally smaller in size than in *P. steinmanni*.

Preadult stages: Larva: MARTIN (2003, 2006). Deutonymph characterized by a round Ac-3 and more slender palp.

Biology: Larvae with a wide host range: Plecoptera, Trichoptera, Diptera (Chironomidae, Empididae, Limoniidae, Pediciidae; attachment site: thorax) (GERECKE & MARTIN 2006; MARTIN et al. 2010).

Habitat: Crenobiont. Preferably in shaded rheo- and rheohelocrenes rich in gravel and mosses. In lower mountain ranges a typical species of forest springs, often associated with stygophilous amphipods of the genus *Niphargus* (GERECKE et al. 2005).

Distribution: Alps, Tatra, Central and Western European lower mountain ranges. Records from the Carpathians uncertain, records from the Pyrenees probably refer to *P. tabernaculata* (GERECKE 1993, GERECKE & SMIT 1994).

3. *Partnunia tabernaculata* GERECKE & SMIT, 1994 (*8-28 f, i*)

1994 *Partnunia tabernaculata* GERECKE & SMIT, Spixiana 17: 201.

Description: Idiosoma flattened, with parallel lateral margins. Genital flaps with rather fine setae. Length of posterior Ac stalks 22-35 µm (ratio stalk/Ac length 0.7-1.1). Palp stout, length/height ratio P-1, 0.5-0.7; P-2, 1.4-1.7; P-3, 0.8-1.1; P-4, 3.1-3.8. Males: Idiosoma length/width 720-780/480-490 µm; genital field with interlocking bars when extended expanding the gonopore as a tent-like structure, gonopore length 160-170 µm, Ac number 20; ejaculatory complex length 138-145 µm, proximal arms as long as distal arms, equally curved, distal arms in an acute-angled position, anterior keel knob-shaped, posterior keel reduced. Females: Idiosoma length/width 950-1130/580-730, gonopore length 210-230 µm, Ac number 28-40.

Habitat and Biology: Spring near a lake above the tree line at about 1900 m asl. Life cycle unknown.

Distribution: Type locality in Spain the only ascertained locality, but records under *P. steinmanni* from the French Pyrenees (see GERECKE 1993) possibly refer to this species.

Genus *Protzia* PIERSIG, 1896

1896c *Protzia* PIERSIG, Sitzungsber. Naturf. Ges. Leipzig 22/23: 52.

1905b *Sporadoporus* WOLCOTT, Stud. Zool. Univ. Nebraska. 31: 191 - K. VIETS (1956).

1907a *Calonyx* WALTER, Zool. Anz. 31: 298 – LUNDBLAD (1969).

1966 *Diplocalonyx* IMAMURA, Results Kyoto Univ. Sci. Exp. Karakorum to Hindukush 1955 (8): 190 - GERECKE (1996a).

Type species: *Protzia invalvaris* PIERSIG, 1898b, by original designation.

Diagnosis: Colour red or orange due to carotenoids. Frontal eye rudimentary and unpigmented or absent. Dorsal and posteroventral idiosoma without muscle attachment sclerites. Legs without swimming setae. Leg claws with lateral clawlets, often arranged in the shape of a fan. Genital flaps either developed as porose longish sclerites, or as narrow, lightly sclerotized bars, or completely absent. Acetabula generally on stalks, at least the posterior ones.

Remarks: Probably a monophyletic clade distributed mostly in the Northern Hemisphere, including diversified evolutionary lines. These cannot be characterized by the degree of development of the genital flaps, as was suggested by the former subdivision in the subgenera *Protzia* s. str., *Calonyx* and *Diplocalonyx*. Instead, differences in the shape of mouthparts, coxae, leg claws and Ac allow for separation of several species groups of still uncertain phylogenetic relationships (*squamosa-*, *invalvaris-*, *lata-*, *rotunda-*, *eximia-*, *octopora*-group: GERECKE 1996a). Of these, the latter is not recorded from the area treated. It includes only *P. octopora* LUNDBLAD, 1954b (Iberian peninsula, Italy), among others characterized by a low Ac number of only 4 pairs.

Biology: In addition to the details given below in species discussions, host ranges based on North American material not determined to species level also include the chironomid subfamily Diamesinae and the trichopteran families Brachycentridae, Lepidostomatidae and Philopotamidae (SMITH & OLIVER 1986).

Key to species

Excluded as *species incertae* (GERECKE 1996a): *Protzia invalvaris barsica* SZALAY, 1936; *Protzia invalvaris multispina* MIGOT, 1926a; *Protzia larrieui* MOTAŞ & SOAREC, 1939a - for a discussion of the also excluded *P. walteri*, see under *P. invalvaris*, similar species.

1 Leg claws with few and diverging lateral clawlets (laterally 2-3, medially 3-4), inserting far distally and short (*8-29 e*); genital setae differentiated (anteriorly flattened, posteriorly rounded, inserting on a pair of rather strong sclerotized bars forming a tongue-shaped centromedial projection; posterior Ac roundish, distinctly shorter than their stalks (*8-29 d*); palp slender (*8-29 c*), P-4 with subparallel dorsal and ventral margins or maximum height in the centre. . . . . *Protzia squamosa* (page 68)

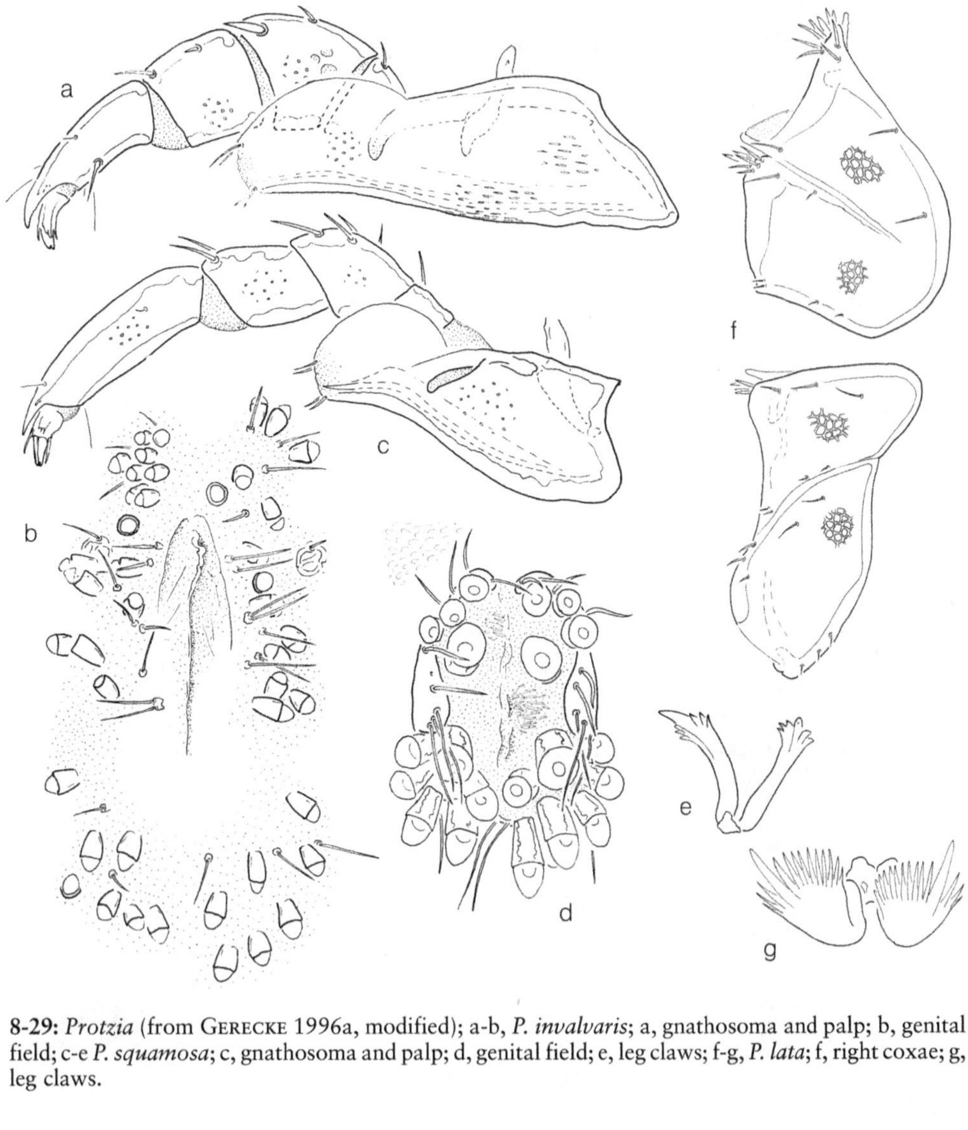

8-29: *Protzia* (from GERECKE 1996a, modified); a-b, *P. invalvaris*; a, gnathosoma and palp; b, genital field; c-e *P. squamosa*; c, gnathosoma and palp; d, genital field; e, leg claws; f-g, *P. lata*; f, right coxae; g, leg claws.

- Leg claws with higher numbers of lateral clawlets (if only 4 medial clawlets present, these inserting more basally and relatively longer, *8-30 f*); all genital setae rounded, inserting on sclerotized bars or in the membranous integument; posterior Ac often as long as, or longer than, their stalks (*8-30 a-d*); palp more compact (*8-29 a*), P-4 maximum height basally. 2

2 (1) Cx-III with a nose-shaped medial projection extending beyond medial margin of Cx-IV (*8-29 f*); genital setae on a sclerotized bar; genital field with less than 18 pairs of Ac. 3

- Cx-III medially not extending beyond medial margin of Cx-IV (*8-30 e*); genital field sclerotization and Ac number various. 4

3 (2) Posterior Ac roundish, shorter than wide, on very short stalks (*8-30 a*); cheliceral claw relatively long (basal segment/claw 2.4-2.9). . . . . . *Protzia lata* (page 66)

- Posterior Ac longer than wide, on long stalks (*8-30 b*); cheliceral claw relatively shorter (basal segment/claw ca. 3.0). . . . . . *Protzia brevipalpis* (page 63)

4 (2) Genital setae on a sclerotized bar flanking the whole genital field, rarely with one or two small spaces; less than 13 pairs of Ac; cheliceral claw short. 5

- Genital setae generally in the membranous integument lateral to the genital field, rarely one to three of the anterior setae on a very weakly sclerotized strip; Ac number and shape of cheliceral various. 6

5 (4) Posterior Ac little longer than wide (length/width 1.1-1.5), on short stalks (< 35 µm, *8-30 d*). Leg claws with 3-4 lateral and 5-6 medial clawlets (*8-30 f*); chelicera slender (total length/height ca. 5.0). . . . . . *Protzia rotunda* (page 67)

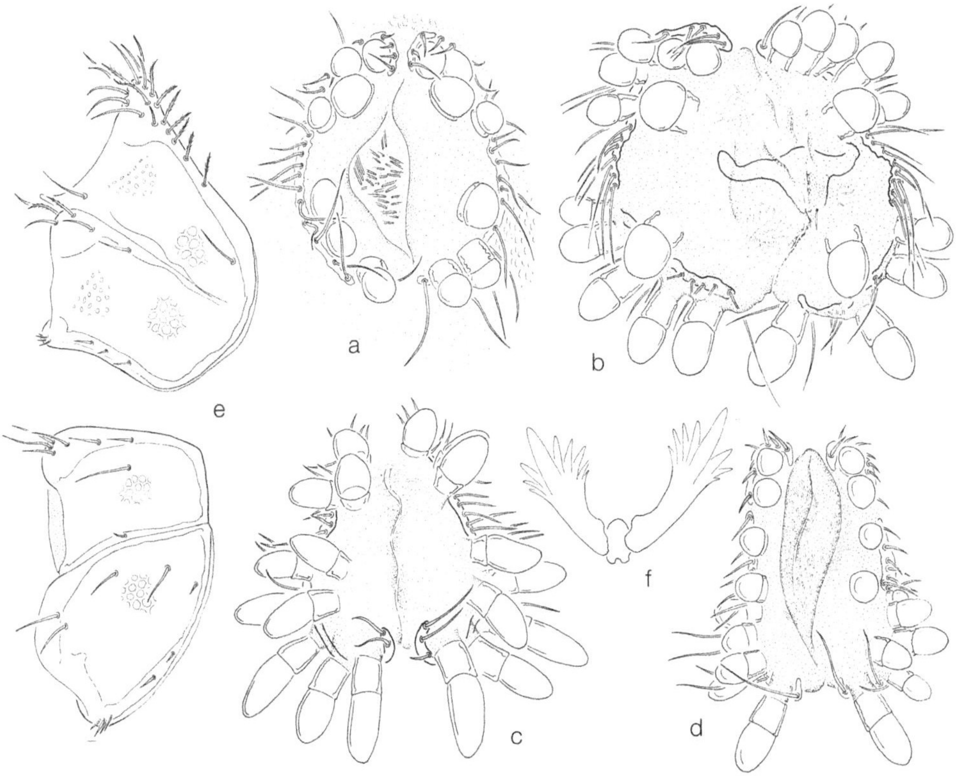

8-30: *Protzia* (from Gerecke 1996a, modified); a-d, genital fields; a, *P. lata*; b, *P. brevipalpis*; c, *P. halberti*; d, *P. rotunda*; e-f *P. rotunda*; e, right coxae; f, leg claws.

-	Posterior Ac distinctly longer than wide (length/width 1.3-2.1), on long stalks (> 35 µm, *8-30 c*). Leg claws with 4-5 lateral and 7-9 medial clawlets; chelicera more robust (total length/height 4.0-4.5). . . . . *Protzia halberti* (page 65)
6 (4)	Genital field with a high number (> 22 pairs) of Ac, all roundish or a little longer than wide, on basally narrowed stalks; genital setae partly scattered between Ac (*8-29 b*). 7
-	Genital field with a lower number (< 22 pairs) of Ac which are anteriorly often shorter than posteriorly, their stalks not narrowed basally; all genital setae arranged in a line surrounding the Ac (*8-31 e*). 8
7 (6)	Gnathosoma flattened dorso-ventrally, chelicera more slender (length/height > 5.5); P-4 in females more slender than in males (length/height P-4 in males < 3.0, in females > 3.0), ventrodistally with a basally thickened seta (more developed in males than in females); Cx-IV with 1-2 lateral setae anterior to insertion of IV-L. . . *Protzia invalvaris* (page 65)
-	Gnathosoma more robust, chelicera stouter (length/height < 5.5); P-4 stout in both sexes (length/height < 3.0); Cx-IV with 3-5 lateral setae anterior to insertion of IV-L. . . . . *Protzia pachygnatha* (page 66)

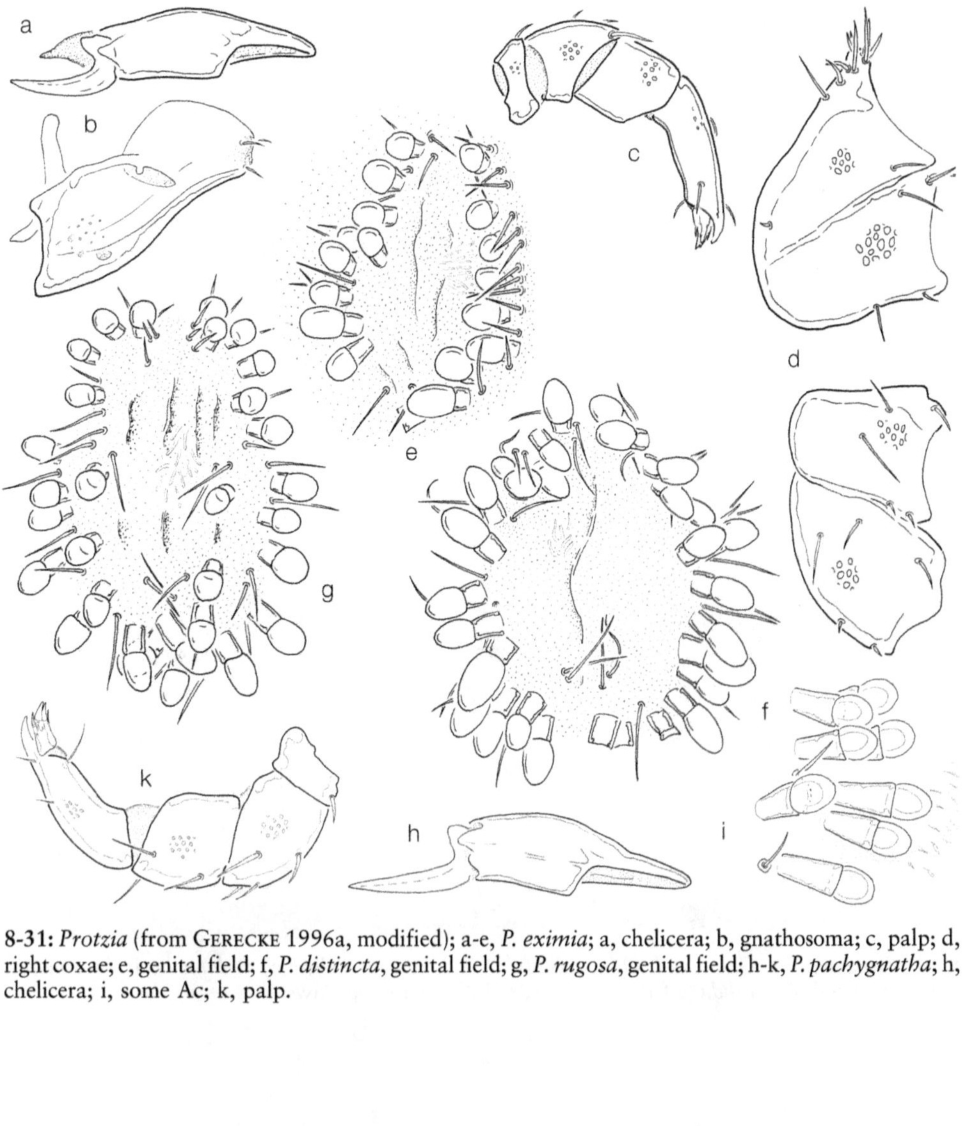

8-31: *Protzia* (from GERECKE 1996a, modified); a-e, *P. eximia*; a, chelicera; b, gnathosoma; c, palp; d, right coxae; e, genital field; f, *P. distincta*, genital field; g, *P. rugosa*, genital field; h-k, *P. pachygnatha*; h, chelicera; i, some Ac; k, palp.

8 (6) Posterior Ac globular or slightly elongated (length/width 1.2-1.3), their stalks longer than Ac diameter; Ac often in higher number (12-20 pairs) (*8-31 g*). . . *Protzia rugosa* (page 68)

- Posterior Ac more elongated (length/width ca. 1.5), their stalks shorter than Ac diameter; Ac often in lower number (8-13, exceptionally 20 pairs) (*8-31 e-f*). 9

9 (8) P-3 (length/height < 1.0) and often also P-4 (length/height < 3.6) more stout; all anterior Ac shorter than the posterior ones; male ejaculatory complex longer (> 130 µm). . . . . *Protzia eximia* (page 64)

- P-3 (length/height > 1.0) and often also P-4 (length/height 3.4-4.0) more slender; some of the anterior Ac may be as long as the posterior ones; male ejaculatory complex shorter (< 130 µm). . . . . *Protzia distincta* (page 63)

1. *Protzia brevipalpis* MAGLIO, 1909 (*8-30 b*)

1909 *Protzia brevipalpis* MAGLIO, Atti Soc. It. Sci Nat. 48: 258.

Description: Coxae with relatively large pores, only a few setae on coxal surface, groups of short and strong setae at anterolateral tips of Cx-I and Cx-II; Cx-III medially extending beyond the medial margin of Cx-IV. Leg claws with 4 lateral and 8-15 medial clawlets (specimens from French Alps: 8-10, specimens from NE Italy 14-15). Genital field with genital flaps reduced to narrow, but continuous strips, posterior Ac elongated (length > 45 µm, length/width 1.3-2.0). Ventral margin of gnathosoma straight or slightly S-shaped. Chelicera robust, but claw small (basal segment/claw ratio 3.0, length/height 4.2-4.4). Palp stout, P-4 with strongly converging dorsal and ventral margins, ventral margin proximally slightly concave. Males: Idiosoma length/width 1400/900 µm; genital field with 9-14 pairs of Ac and 25-30 pairs of setae; ejaculatory complex length 270, gnathosoma length 280 µm. Females: Idiosoma length/width 1200-1600/800-1200 µm; genital field with 10-20 pairs of Ac and about 20 pairs of setae; gnathosoma length 310 µm.

Preadult stages: Larva: MOTAŞ (1929).

Habitat and Biology: Rhithrobiont. Low- and middle-order streams, occasionally in rheocrenes, at 100-1100 m asl.. Life cycle unknown.

Distribution: Southern Alps (France, Italy). Records from Corsica refer to *P. lata* (GERECKE 1996b).

2. *Protzia distincta* WALTER, 1922 (*8-31 f*)

1922a *Protzia distincta* WALTER, Rev. Suisse Zool. 29: 235.

1922a *Protzia alpina* WALTER, Rev. Suisse Zool. 29: 235 – GERECKE (1999).

1944 *Protzia macrognatha* WALTER, Int. Rev. Hydrobiol. Hydrogr. 43: 287 – DI SABATINO et al. (2009).

1994 *Protzia raetica* BADER, Ergebn. wiss. Unters. schweizer. Nationalpark 16: 248 – GERECKE (1999).

1996a *Protzia eximia distincta* GERECKE, Arch Hydrobiol., Suppl. 77: 447 – GERECKE (1999).

1996a *Protzia eximia macrognatha* GERECKE, Arch. Hydrobiol., Suppl. 77: 328 – DI SABATINO et al. (2009).

Description: Coxae with large pores, Cx-I medial margin straight, Cx-I and -II with few setae at anterior margins, one or no lateral setae on C-IV. Leg claws with 5-6 lateral and 8-10 medial clawlets. Genital field with about 10-20 pairs of Ac and 14-23 pairs of setae; all posterior Ac (length/width 1.2-1.8), but also some of the anterior ones elongated. Gnathosoma dorso-ventrally elevated and with short rostrum. Cheliceral claw robust (basal segment /claw ratio 1.8-2.1). Palp relatively slender (L/H ratio P-3, 1.0-1.2; P-4, 3.4-4.0); P-3 ventral margin convex; length ratio P-2/P-4, 0.55-0.60. Males: Idiosoma length/width 700-900/500-700; gnathosoma length 165-200 µm; gonopore field with fine villi; Females: Idiosoma length/width 700-1100/500-800; gnathosoma length 190-220 µm; gonopore field smooth.

Similar species: This species includes alpine populations of the *eximia*-group with a morphological variability range overlapping with *P. eximia*. Attribution of single specimens may be problematic (Gerecke 1999). Exceptionally large-sized specimens (length Cx-I+II > 205, chelicera > 310, palp total length > 380 µm) with elevated Ac numbers up to 20 pairs, previously published under the name *P. macrognatha* (Di Sabatino et al. 2009).

Preadult stages: Larva: Martin (2003). Deutonymph with three pairs of Ac.

Biology: Larvae parasitic on Diptera (Chironomidae: *Corynoneura, Heleniella, Thienemanniella*) and Trichoptera (Gerecke & Martin 2006; Martin et al. 2010); attachment site: thorax. Adults feeding on water mite eggs (Martin 2005).

Habitat: Crenophilous. Rheocrenic springs and first order streams on mineral substrata, preferably in woodlands.

Distribution: Alps.

3. *Protzia eximia* (Protz, 1896) (*8-1 f, 8-31 a-e*)

1896a *Thyas eximia* Protz, Zool. Anz. 19: 25.

1896c *Protzia eximia* Piersig, Zool. Anz. 20: 52.

Description: Coxae with large pores, Cx-I medial margin straight, Cx-I and -II with few setae at anterior margins, with one or without lateral setae on C-IV. Leg claws with 4-5 lateral and 7-9 medial clawlets. Genital field with about 8-13 pairs of Ac and 15-20 pairs of setae; anterior Ac generally roundish, posterior ones elongated (length/width ca. 1.5). Gnathosoma dorso-ventrally elevated and with short rostrum. Cheliceral claw robust (basal segment /claw ratio 1.8). Palp relatively stout (L/H ratio P-3, 0.8-1.0; P-4, 3.1-3.6); P-3 ventral margin convex; length ratio P-2/P-4, 0.6. Males: Idiosoma length/width 600-800/450-600; gnathosoma length 160-200 µm; gonopore field with fine villi; Females: Idiosoma length/width 700-1100/500-800; gnathosoma length 190-220 µm; gonopore field smooth.

Similar species: *Protzia eximia* gives the name to a rather homogeneous species group including, in the area treated, several doubtful species. *Protzia felix* Gerecke, 1996b (Central Mediterranean) is distinguished by rounded posterior Ac.

Preadult stages: Larva: Jones (1967), Böttger (1972c), Ullrich (1976), Biesiadka & Cichocka (1984), Martin (2000, 2006). Deutonymph with three pairs of Ac.

Biology: Larvae parasitic on Diptera (Empididae, Psychodidae, Simuliidae, the chironomid subfamilies Tanypodinae, Orthocladiinae and Chironominae) and Trichoptera (Apataniidae, Glossosomatidae, Hydropsychidae, Limnephilidae Polycentropodidae, Psychomyiidae, Rhyacophilidae) (Jones 1967, Böttger 1972c, Ullrich 1978, Martin 2000). Adults feeding on water mite eggs (Martin 2005).

Habitat: Low- and middle-order streams, occasionally in rheocrenes at low altitude. In areas on limestone bedrock more frequent than in silicaceous areas (Schwoerbel 1959).

Distribution: Holarctic?

4. *Protzia halberti* (WALTER, 1920) (*8-30 c*)

1920 *Calonyx halberti* WALTER, Festschr. Zschokke 24: 9.

1922a *Calonyx rotundus* WALTER, Rev. Suisse Zool. 29: 235, *nec C. rotundus* WALTER, 1908 – GERECKE (1999).

1955e *Calonyx bayeri* LÁSKA, Biológia 10: 420 – GERECKE (1996a).

Description: Coxae with small pores, Cx-I anteriorly projecting, with a group of strong, pennate setae; medial margin slightly concave. Cx-III medially not projecting beyond medial margin of Cx-IV, Cx-IV laterally with a group of strong setae; leg claws strong, with 4-5 lateral and 7-9 medial clawlets. Genital field with 15-18 pairs of setae and 8-10 pairs of Ac, posterior Ac very elongated (length 40-60 µm, length/width 1.3-1.7), stalk length 25-40 µm. Ventral margin of gnathosoma nearly straight. Chelicera length/height 4.1-4.5, with small claw (length basal segment /claw 2.5-2.8). Palp stout, P-2 with two strong medial setae, P-3 ventral margin convex, P-4 maximum height basally, length/height P-3, 0.7-0.8; P-4, 2.9-3.2. Males: Idiosoma length/width 950-1100/700-800; gnathosoma length 230-275 µm. Females: Idiosoma length/width 850-1500/550-1100; gnathosoma length 270-315 µm.

Similar species: *Protzia halberti* has been confused for a long time with the predominantly Mediterranean *P. rotunda* (see there).

Preadult stages: Larva unknown. Deutonymph with two pairs of Ac, the posterior pair generally elongated.

Habitat and Biology: Middle-order streams.

Distribution: British Isles, Central and Eastern Europe (GERECKE 1996a).

5. *Protzia invalvaris* PIERSIG, 1898 (*8-1 i, 8-29 a-b*)

1898b *Protzia invalvaris* PIERSIG, Zool. Anz. 21: 452.

1939a *Protzia walteri pyrenaica* K. VIETS, Zool. Anz. 125: 3 – GERECKE (1996a).

1936 *Protzia sculptopetiolata* SZALAY, Zool. Anz. 115: 51 – GERECKE (1996a).

Description: Idiosoma often anteriorly narrowed. Coxae with large, often elongated pores. Cx-I anteriorly pointed, here bearing a tuft of ca. 10 pectinate setae; medial margin distally straight, gradually passing into the convex posteromedial edge of Cx-I+II; Cx-IV laterally with 1-2 setae. Leg segments 5 and 6 strongly enlarged, segment 5 forming a sheath embracing base of terminal segment. III/IV-L-3-5 with long, slightly flattened dorsal setae. Leg claws with 3-5 lateral and 8-10 medial clawlets. Genital field without sclerites, with 30-40 (20-55? – see below) pairs of small, homeomorphic Ac (round, on long stalks, stalk/Ac length ratio 1.5-3.5) and 10-20 pairs of setae scattered between Ac; gonopore field smooth in both sexes. Gnathosoma dorso-ventrally flattened and long, with nearly straight ventral margin. Chelicera slender, with maximum height of basal segment in anterior quarter, basal segment /claw 2.5-3.0. Palp with sexual dimorphism, very stout, with few setae; P-4 with a basally thickened seta ventrodistally on the medial surface. Males: Idiosoma length/width 900-1200/650-900; ejaculatory complex length 180-200; gnathosoma length 290-330 µm; P-4 length/height 2.8-2.9, mediodistal seta basally strongly swollen, near a small denticle away from distal margin. Females: Idiosoma length/width 1000-1300/700-1000; gnathosoma length 300-360 µm; P-4 length/height 3.3-3.7, mediodistal seta less swollen than in male and closer to distal segment margin.

Similar species: Possibly a species complex. The name *P. walteri* K. VIETS, 1930d has been introduced for populations with higher Ac numbers (40-55 pairs) from the Spanish

Pyrenees. In contrast, lower numbers (20-35 pairs) were recorded from *P. invalvaris*-populations in South Italy (GERECKE 1996b). In both cases no further significant differences were found, and the question about genetic differentiation is open.

Preadult stages: Larva unknown. Deutonymph with more than three Ac.

Habitat and Biology: Rhithrobiont. Middle-order streams. Lifecycle unknown. Adults feeding on water mite eggs (MARTIN 2005).

Distribution: Palaearctic. In Europe recorded everywhere except for northern latitudes (e. g. British Isles, Fennoscandia).

6. *Protzia lata* (WALTER, 1906) (*8-1 d, 8-29 f-g, 8-30 a*)

1906b *Partnunia lata* WALTER, Zool. Anz. 30: 571.

1907a *Calonyx latus* WALTER, Zool. Anz. 31: 298.

1996a *Protzia lata* (WALTER, 1906) – GERECKE, Arch. Hydrobiol., Suppl. 77: 297.

Description: Coxae with large pores; most setae of Cx-I+II concentrated to anterior tips, slightly flattened; III/IV-L-3-5 with long, slightly flattened dorsal setae. Leg claws with 4-5 lateral and 10-16 medial clawlets. Genital field with 15-25 pairs of Ac flanked by narrow sclerotized strips which in the distal third are a little bowed towards the centre, each bearing 15-30 setae; posterior Ac large and flat (length/width 0.6-0.9) on short stalks (maximum length 25 µm). Gnathosoma with S-shaped ventral margin, in dorsal view rostrum distinctly set off from base. Chelicera length/height 4.4-5.1, with short claw (basal segment /claw length ratio 2.5-3.0). Palp stout, length/height P-3, 0.7-0.9; P-4, 2.8-3.3, length ratio P-2/P-4, 0.6-0.7. Males: Idiosoma length/width 1100-1500/700-1100; genital field length 200-270 µm, gonopore field with fine villi; gnathosoma length 235-275 µm. Females: Idiosoma length/width 1200-2000/850-1400; genital field length 250-300 µm, gonopore field smooth; gnathosoma length 300-340 µm.

Similar species: Three southern European species agree with *P. lata* in having rounded posterior Ac, not elongated: *Protzia pullinidentis* GERECKE, 1999 (Spain) differs in the leg claws with only 4-5 medial clawlets, *P. pycnica* GERECKE, 1996b (Corsica, Sardinia) and *P. faber* GERECKE, 1999 (Gomera) both in the reduced number of Ac (6-14 pairs). *Protzia annularis* LUNDBLAD, 1954b (Spain, Portugal, with Ac in a circular configuration) is a little documented species and possibly a synonym of *P. lata*.

Preadult stages: Larva unknown. Deutonymph with two pairs of Ac.

Habitat and Biology: Rhithrobiont. In low- and middle-order streams with well developed hyporheic zone where probably preadult stages develop. Life cycle unknown.

Distribution: Southern Europe and North Africa; the *locus typicus* in the Engadin (Switzerland) is the only certain record from the area treated. Records under this name from Corsica (E. ANGELIER 1959) refer to *P. pycnica* (GERECKE 1996b).

7. *Protzia pachygnatha* K. VIETS, 1930 (*8-31 h-k*)

1930d *Protzia pachygnatha* K. VIETS, Arch. Hydrobiol. 21: 214.

1958a *Protzia subalpina* SCHWOERBEL, Zool. Anz. 160: 193 – GERECKE (1999).

Description: Coxae with large, slightly elongated pores, Cx-I+II medially equally rounded, forming an obtuse posteromedial angle; Cx-IV with 3-5 lateral setae. Terminal leg segments not particularly enlarged; leg claws with 4 lateral and 10-13 medial clawlets. Genital field with 30-55 pairs of homeomorphic, rounded Ac (length/width 0.7-1.1).

Gnathosoma dorso-ventrally elevated; chelicera length/height 4.3-5.3, basal segment /claw 2.3-2.6. Palp stout, P-3 length/height 1.0-1.1, P-4 with sexual dimorphism in shape, mediodistal fine seta basally not thickened. Males: Idiosoma length/width 950/700 µm; 30-35 pairs of Ac; gnathosoma length 215-235 µm; P-4 shortened in populations from the Pyrenees (length/height 2.8-3.1), but not in Black Forest, Germany (3.4). Females: Idiosoma length/width 1300/1000 µm; 35-55 pairs of Ac; gnathosoma length 250-290 µm; P-4 more slender than in Pyrenean males (length/height 3.1-3.7).

Similar species: *Protzia curvipalpis* LUNDBLAD, 1956a (Portugal), similar in setation of Cx-IV, relatively stout chelicera and high gnathosoma, differs in a lower Ac number (20-22 pairs) and P-4 basally curved (GERECKE 1996a).

Preadult stages: Larva unknown. Deutonymph with more than three pairs of Ac.

Habitat and Biology: Low order streams and springs. Life cycle unknown.

Distribution: Pyrenees, one population recorded from S Germany (Black Forest: SCHWOERBEL 1958a).

8. *Protzia rotunda* WALTER, 1908 (*8-30 d-f*)

1908b *Protzia rotunda* WALTER, Arch. Hydrobiol. 4: 9.

1919a *Calonyx rotundus* WALTER, Rev. Suisse Zool. 27: 27.

1940a *Calonyx cabardinicus* SOKOLOW, In: SERNOW & STACKELBERG: Faune de l'URSS. Arachnides 5 (2): 96 – PEŠIĆ (2005).

Description: Coxae laterally with fine, in the centre with large pores; Cx-I anterior tip projecting and densely covered by strong setae. III/IV-L-3-5 with 3-5 long, slightly flattened dorsal setae. Leg claws with strong principal claw, 3-4 lateral and 4-6 strongly diverging medial clawlets. Genital field with 8-10 pairs of Ac on each side flanked by a rather strong sclerotized bar bearing 12-18 fine setae, these setae shorter in the anterior part, but elongated in the centre, where the sclerite is enlarged and extends between the Ac, and in the posterior part. Posterior Ac roundish or slightly elongated (length/width 1.1-1.5), on short stalks (length stalk/Ac 0.6-0.9). Gnathosoma dorso-ventrally flattened, with weakly S-shaped ventral margin; chelicera slender (length/height 4.7-5.3, with small, straight claw (basal segment /claw 2.3-2.6). Palp stout, P-2 with a strong, pennate medial seta, ventral margin P-3 convex, P-4 length/height 2.9-3.3. Males: Idiosoma length/width 640-860/450-580 µm; gonopore field with fine villi; ejaculatory complex length 150-180; gnathosoma length 170-230 µm. Females: Idiosoma length/width 1100-1400/760-950 µm; gonopore field smooth; gnathosoma length 220-280 µm.

Similar species: Often confused with *P. halberti* (see there). From Southern Italy populations with increased numbers of leg clawlets (7-13, generally 9-10 medial clawlets) are recorded (GERECKE 1996b). *Protzia musculosa* GERECKE, 1996b from Corsica and Sardinia differs in flattened Ac (length/height 0.6-1,1) on short stalks (length 13-26 µm) and more slender chelicerae and P-4.

Preadult stages: Larva unknown. Deutonymph with two pairs of Ac.

Habitat and Biology: Low- and middle-order streams. Life cycle unknown.

Distribution: Southern Europe, Asia Minor, Caucasus. Records published from Central and Eastern Europe probably refer to *P. halberti* (GERECKE 1996a).

9. *Protzia rugosa* WALTER, 1918 (*8-31 g*)

1918 *Protzia rugosa* WALTER, in STEINMANN & SURBECK: Die Wirkung organischer Verunreinigungen auf die Fauna schweizerischer fließender Gewässer. Bern: 438.

1996a *Protzia eximia rugosa* GERECKE, Arch. Hydrobiol., Suppl. 77: 317.

Description: Coxae with small pores, Cx-I+II medial margin straight, merging into the equally rounded posterior margin without an angle; anterior tip of Cx-I pointed, with few setae; leg claws with 4-5 lateral and 6-7 medial clawlets. Genital field without sclerites, 12-20 pairs of setae surrounding 12-20 pairs of Ac; posterior Ac slightly elongated, often globular (length/width 1.2-1.6), on relatively long stalks (length stalk/Ac 0.7-1.1, length stalk/width Ac > 1.0). Gnathosoma variable in shape, often slender (length/width 1.4-1.9). Chelicera length/height 4.4-5.1, with strong claws (basal segment /claw 1.7-2.0). Palp relatively stout (L/H ratio P-3, 0.9-1.0; P-4, 3.3-3.7); P-3 ventral margin convex; length ratio P-2/P-4, 0.6-0.7. Males: Idiosoma length/width 750-900/500-550 µm; gonopore field with fine villi; gnathosoma length 180-210 µm. Females: Idiosoma length/width 550-1100/450-700 µm; gonopore field smooth; gnathosoma length 180-240 µm.

Similar species: *Protzia caucasica* SOKOLOW, 1927c (Russia), similar in the slender gnathosoma, differs in lower numbers of Ac, with posterior Ac more elongated and on shorter stalks (length stalk/Ac 0.6) (GERECKE 1996a).

Preadult stages: Larva undescribed. Deutonymph with three pairs of Ac.

Habitat and Biology: Middle-order streams. Life cycle unknown.

Distribution: Central and Southeastern Europe.

10. *Protzia squamosa* WALTER, 1908 (*8-29 c-e*)

1908b *Protzia squamosa* WALTER, Arch. Hydrobiol. 4: 6.

1919a *Calonyx squamosus* WALTER, Rev. Suisse Zool. 27: 27.

1922a *Protzia multipora* WALTER, Rev. Suisse Zool. 29: 254 – DI SABATINO et al. (2009).

1996a *Protzia squamosa squamosa* WALTER, 1908 – GERECKE, Arch. Hydrobiol., Suppl. 77: 280.

Description: Coxae with fine pores and few setae; medial margin Cx-I straight, posterior margin equally rounded, suture Cx-I/II reaching medial plate margin; leg claws with 2-3 lateral and 3-4 medial clawlets. Genital field with strongly developed sclerotized bars bearing 10-18 setae, in the central part with a tongue-shaped extension separating a group of anterior, short-stalked Ac from the posterior longer stalked Ac. Total Ac number 9-17 pairs, two pairs of Ac larger than the remaining, forming a square in the centre of genital field. Posterior Ac rounded or slightly pointed, shorter than their stalks. Gnathosoma dorso-ventrally flat, with weakly S-shaped ventral margin. Chelicera slender, with relatively short claw (basal segment /claw 2.0-2.6). Palp with relatively long P-1, P-3 with weakly convex dorsal and ventral margins, P-4 maximum height near centre of segment. Males: Idiosoma length/width 600-1000/450-675 µm; gonopore area with fine villi; ejaculatory complex length 210-220 µm. Females: Idiosoma length/width 800-1250/450-800 µm; gonopore field smooth.

Similar species: The Mediterranean subspecies *Protzia squamosa paucipora* (K. VIETS, 1955) differs in a lower number of Ac. Specimens with more than 15 pairs of Ac previously attributed to a separate species, *P. multipora* (synonymy: DI SABATINO et al. 2009)

Preadult stages: Larva undescribed. Deutonymph with three pairs of Ac (K. VIETS 1922d).

Habitat and Biology: Crenobiont. Preferably in rheohelocrenes at low and middle elevations (GERECKE & MARTIN 2006).

Distribution: Central and Eastern Europe (records from Pyrenees and High Atlas possibly refer to *P. s. paucipora*, GERECKE 1996a).

Subfamily Pseudohydryphantinae K. VIETS, 1926

1926l Pseudohydryphantinae K. VIETS, Zool. Anz. 69: 196.

Diagnosis: Lateral eyes encapsulated. Frontal eye in a sclerotized ring. Dorsal and posteroventral idiosoma without muscle attachment sclerites (but in a genus recorded from outside the area treated, dorsolateral glandular platelets narrow and elongated). III-L and IV-L with swimming setae. Leg claws simple. Genital field with three pairs of Ac flanked by paired, movable flaps. Anterior margin of flaps bearing several setae and extending beyond anterior margin of Ac-1. P-2 without ventromedial setae.

Remarks: In addition to *Pseudohydryphantes*, with 17 species recorded from the Holarctic and Australian Region, the subfamily includes one further genus, *Cyclohydryphantes* LUNDBLAD, 1941c, only known from Australia (COOK 1974, HARVEY 1998).

Genus *Pseudohydryphantes* K. VIETS, 1907

1907b *Pseudohydryphantes* K. VIETS, Abh. Naturw. Ver. Bremen 19: 142.

Type species: *Pseudohydryphantes parvulus* K. VIETS, 1907b, by original designation.

Diagnosis: Integument papillae may be rounded or with fine corona-like distal extensions. Glandular platelets rounded (lunate in species from outside the area covered), not on expanding sclerites. Excretory pore smooth.

Remarks: Palaearctic species differ from all other representatives of the genus in the formation of the integument papillae.

Preadult stages: Larva: TUZOVSKIJ (2001g). Deutonymph: HARVEY (1987).

Pseudohydryphantes parvulus K. VIETS, 1907 (*8-1 k, 8-32 a-b*)

1907b *Pseudohydryphantes parvulus* K. VIETS, Abh. Naturw. Ver. Bremen 19: 142.

Description: Idiosoma colour iridescent (light refraction in the integument papillae). Coxae with numerous fine long setae, particularly dense at lateral margins and along sutures. Cx-I+II posterior margin projecting and pointed, medially nearly in touch, but not fused, forming a gnathosomal bay. Cx-III+IV subtriangular, medial margin formed by Cx-III only, posterior margin slightly convex. Legs rather slender, swimming setae strong and long, restricted to the distal third of segments, numbers on posterior surface: III-L-3-5, 2, 4, 7; IV-L-3-5, 2, 5, 3; in addition, IV-L-4/5 each with 1-2 anterodistally. Leg claws with two fine dorsal clawlets, one in the centre, the other distal. Genital flaps strong, anteriorly extending beyond Ac-1, with 15-20 pairs of very fine setae; Ac-1-2 longish, Ac-3 globular. Gnathosoma stout, with small mouth opening. Chelicera length/height 3.8-4.5, with strong claws (basal segment /claw 1.7-1.8). Palp slender, length ratio P-2/P-4 0.6. Males: Idiosoma length/width 650/560; genital flap length 125 µm, all Ac small, Ac-2-3 more rounded than in females; Ac-2 away from Ac-1 (interspace larger than Ac diameter); gnathosoma length 170 µm. Females: Idiosoma length/width 800/600; genital flap length 160 µm; Ac-1-3 with small interspaces, length 50-45-40; gnathosoma length 230 µm.

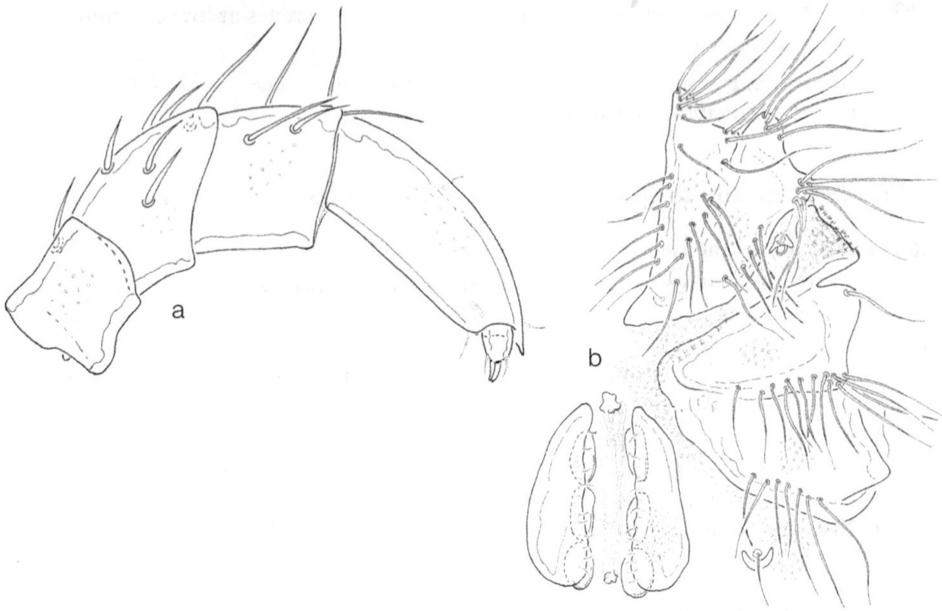

8-32: *Pseudohydryphantes parvulus* female; a, palp; b, left coxae and genital field.

Similar species: *Pseudohydryphantes rivalis* TUZOVSKIJ, 2001g (Russia), the only other species recorded from Europe, differs in a less strongly developed cheliceral claw (basal segment/claw ratio 2.0). With regard to the number of setae on P-3, older descriptions of *P. parvulus* were incomplete: *Pseudohydryphantes parvulus* and *P. rivalis* agree in having four or more setae on this segment.

Habitat and Biology: The few published records of this species all refer to single specimens found accidentally in standing or slowly running waters. Preadult stages and life cycle unknown.

Distribution: Europe; scattered records from Germany, Italy, Sweden, Ireland, Latvia and Russia.

Subfamily Tartarothyadinae K. VIETS, 1934

1934a Tartarothyasinae K. VIETS, Zool. Anz. 105: 133.

Diagnosis: Lateral eyes reduced, their lenses lying below the integument, median eye absent. Idiosoma slightly elongated, with parallel lateral margins, without dorsal and posteroventral muscle insertion sclerites. Legs without swimming setae, leg claws simple. Genital field with large genital flaps flanking three pairs of Ac and anteriorly extending beyond the distal margin of Ac-1. Chelicerae with relatively large claws.

Remarks: Monotypic subfamily including rather uniform species recorded from the Western Palaearctic, Nearctic and Indo-Australian regions.

Genus *Tartarothyas* K. VIETS, 1934

1934a *Tartarothyas* K. VIETS, Zool. Anz. 105: 133.

Type species: *Tartarothyas micrommata* K. VIETS, 1934a, by monotypy.

Diagnosis: Only genus, with the characters of the subfamily.

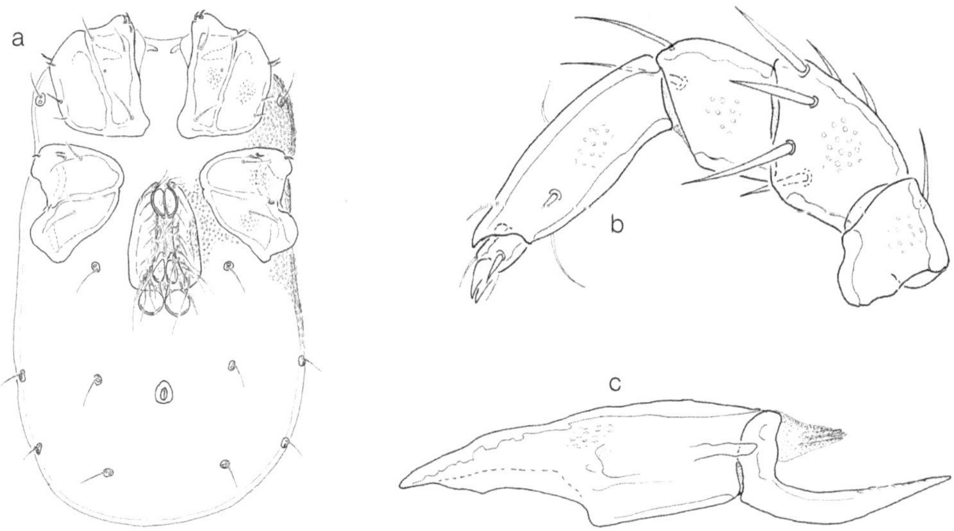

8-33: *Tartarothyas romanica*; a, male venter; b, palp; c, chelicera.

Tartarothyas romanica HUSIATINSCHI, 1937 (*8-1 l-m, 8-33 a-c*)

1937b *Tartarothyas romanica* HUSIATINSCHI, Zool. Anz. 117: 206.

1989 *Tartarothyas raetica* BADER, Entomologica Basiliensia 13: 7 – GERECKE (1996b).

1989 *Tartarothyas suecica* BADER, Entomologica Basiliensia 13: 13 – GERECKE (1996b).

Description: Colour yellow to brownish. Integument papillae flattened. Lateral eyes visible at anterior margin as two pairs of small pigment dots. Coxae with triangular, quadrangular or polygon-shaped pores, bearing very few, short and flattened setae at medial and anterior margins of Cx-I+II. Cx-I+II medioposteriorly rounded, with slightly projecting, medially-directed porose extension. Suture line Cx-I/II almost parallel to main idiosoma axis, ending in the straight posterior margin of Cx-I+II. Medial margin of Cx-III+IV formed by the rounded Cx-III, Cx-IV medially pointed and with slightly convex posterior margin. Legs stout, with rather strong setation at distal margins on segments 3-4 of all legs. Leg claws fine, weakly curved, without clawlets. Genital flaps anteriorly extending beyond Ac-1, with 12-20 pairs of very fine medial setae; Ac-1-2 elongated, often drop-shaped, their basal sclerites forming pointed extensions (posteriorly on Ac-1, anteriorly on Ac-2), Ac-3 globular on a short stalk. Gnathosoma stout, with straight ventral margin and small mouth opening. Chelicera slender (length/height 4.8-5.5), with very strong claws (basal segment/claw 1.5-1.7). Palp slender, setae numbers P-1-4, 1, 7-10, 2-3, 6 (4 of them fine, hair-like, dorsodistally); P-4 ventral margin proximally slightly concave, distally convex (maximum height in distal part), dorsodistal extension often set off from the segment like a peg-seta; length ratio P-2/P-4 ca. 0.6. Males: Idiosoma length/width 900-1100/500-750; Cx-I+II length 280-300; genital flap length 210-230, Ac-1-3 length 58-74, 49-58, 36-52 µm. Ac-2 away from Ac-1 (interspace similar to Ac diameter); ejaculatory complex length 180-200, gnathosoma length 210-230 µm. Females: Idiosoma length/width 950-1250/650-850; Cx-I+II length 260-320; genital flap length 220-280; egg diameter ca. 190 µm; genital setae less densely arranged, Ac-1-3 more elongated than in males, separated by equal, small interspaces; gnathosoma length 210-260 µm.

Similar species: *Tartarothyas micrommata* K. VIETS, 1934a (Montenegro, Romania, Israel), the only other Palaearctic species (syn.: *Vietsthyas fonticola* MOTAŞ & TANASACHI 1957) differs in smaller dimensions (idiosoma length/width ♂ < 800/550, ♀ < 1000/600, Cx-I+II length ♂ < 190, ♀ < 200 µm), lower genital setae numbers (with slight overlap: ♂ 23-31, ♀ 17-28) and in males Ac-2 closer to Ac-1.

Preadult stages: Larva: MARTIN (2003). Deutonymph with two pairs of Ac (K. VIETS 1934a, 1934c).

Biology: Larvae parasitic on chironomid diptera (*Micropsectra* - MARTIN et al. 2010), attachment site: abdomen.

Habitat: Crenobiont. Preferably in helocrenes with weak flow, also in flooded grassland vegetation, but sensitive to eutrophication (GERECKE & MARTIN 2006).

Distribution: Many parts of Europe including S England, but not recorded from France and the Iberian peninsula; rare.

Subfamily Wandesiinae SCHWOERBEL, 1961

1961b Wandesiinae SCHWOERBEL, Arch. Hydrobiol. Suppl. 25: 345.

1990 Wandesiidae SCHWOERBEL, 1961 – TUZOVSKIJ, Nauka: 62.

Diagnosis: Idiosoma extremely elongated, postgenital area at least as long as, generally longer than, pregenital idiosoma. Apart from coxal plates and sclerites in the genital field (and in one extra-European taxon a small, postocularia-bearing sclerite) idiosoma surface membranous. Lateral eyes not encapsulated, with more or less reduced lenses, often lacking pigmentation. Three to numerous pairs of Ac flanking the gonopore, genital flaps present or reduced. Legs without swimming setae, claws simple or with a dorsal clawlet. Palp chelate or with a dorsodistal seta on P-4.

Remarks: A well-defined clade, but very distant from the remaining Hydryphantidae, and therefore proposed by TUZOVSKIJ (1990) to be ranked at superfamily level, i. e. Wandesioidea. Based on larval morphology, similarities are found in *Ignacarus*, at present placed in the heterogeneous Euthyadinae (MORENO et al. 2008). A large-scale investigation on all clades now united in Hydryphantoidea is indicated. From the study area only species of the nominate genus *Wandesia* are recorded.

Genus *Wandesia* SCHECHTEL, 1912

1912 *Wandesia* SCHECHTEL, Bull. Acad. Sci. Cracowie Ser. B, Sci. Natur.: 463.

Type species: *Wandesia thori* SCHECHTEL, 1912, by original designation.

Diagnosis: Frontal idiosoma without sclerotization. Suture line Cx-I/II medially incomplete. Legs slender, I-L-6 and II-L-6 ventrodistally with two stout peg-like setae surrounded by numerous fine hair-like setae. Acetabula varying from three pairs to numerous, stalked or disk-shaped, not covered by flanking genital flaps. Palp without a mediodistal seta on P-4.

Discussion: The only other known genus in this subfamily, *Euwandesia* ANDRÉ & NAUDO, 1962, with a small frontal sclerite bearing postocularia, Ac partly covered by flap-like folds of the flanking integument, and a heavy mediodistal seta on P-4, is known from South America and New Zealand (COOK 1974).

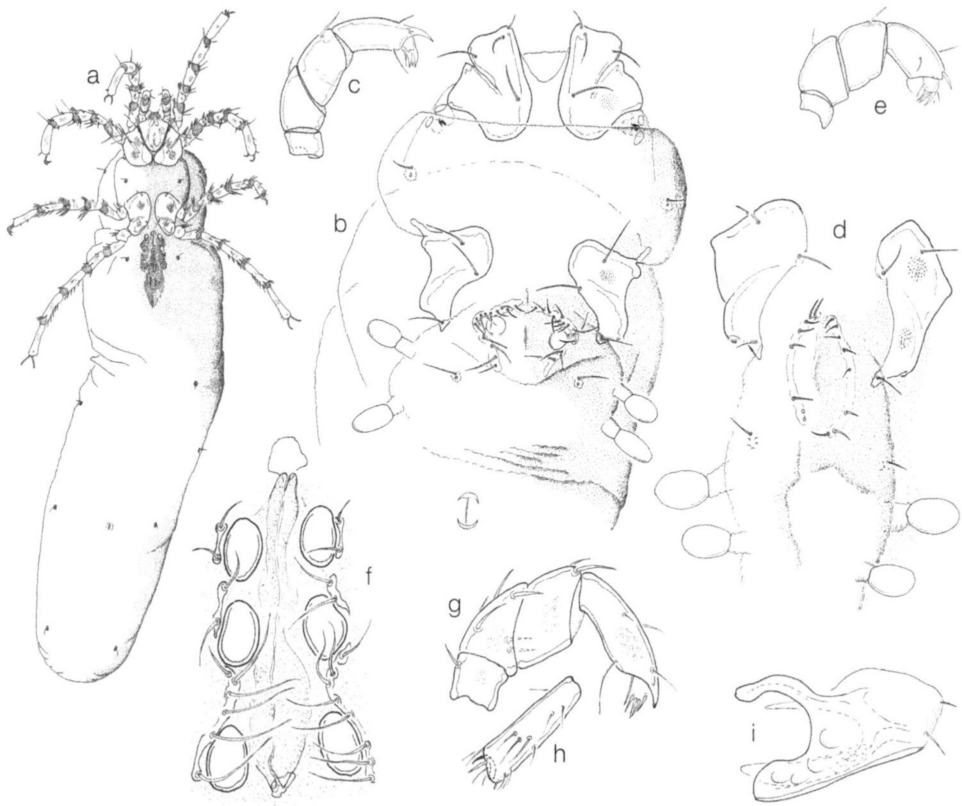

8-34: *Wandesia*; a, *W. lychnobia* ventral idiosoma; b-c, *W. thori*, male; b, coxal and genital field; c, palp; d-e, *W. racovitzai*, male; d, Cx-III/IV and genital field; e, palp; f-i, *W. somnambula*, male; f, genital field; g, palp; h, I-L-6; i, gnathosoma, lateral view (a, DI SABATINO et al. 2000; b-e, GERECKE 1999).

Key to subgenera and species

1 Acetabula disc-shaped, not stalked (subgenus *Pseudowandesia*). . . . . *Wandesia somnambula* (page 74)

\- Acetabula elongated, stalked (subgenus *Wandesia* s.str.). 2

2 (1) P-4 relatively long and slender, with maximum height in centre of segment (L/H > 2.6, P-2/P-4 ratio 0.5-0.6); posterior Ac > 40 µm; in males three pairs of Ac. . . . . *Wandesia thori* (page 75)

\- P-4 shorter and thickened, with maximum height near distal end of segment (L/H < 2.6, P-2/P-4 ratio 0.7); posterior Ac length < 40 µm; in males two pairs of Ac. . . . . *Wandesia racovitzai* (page 75)

Subgenus *Pseudowandesia* HABEEB, 1958

1958 *Pseudowandesia* HABEEB, Leafl. Acadian Biol. 17: 2.

Type species: *Wandesia stygophila* SZALAY, 1944a, by original designation.

Diagnosis: Acetabula disc-shaped, unstalked, arranged in three pairs rather close to each other, flanked by genital setae on weakly sclerotized bars.

Discussion: In the shape of the Ac, *Pseudowandesia* differs clearly from the nominate subgenus. A further subgenus, *Partnuniella* K. VIETS, 1938k from North and South America is closely related, primarily differing in a higher number of Ac. As Ac numbers obviously vary intraspecifically, separation of these two taxa is uncertain (COOK 1974). Only one species recorded from the area treated.

1. *Wandesia (Pseudowandesia) somnambula* GERECKE, 1996 (*8-34 f-i*)

1996b *Wandesia somnambula* GERECKE, Arch. Hydrobiol., Suppl. 77: 452.

Description: Colour red. Lateral eyes unpigmented. Cx-I+II subrectangular, with weakly concave medial margin merging with a right angle, into the straight posterior margin. Cx-III forming a blunt, slightly acute mediodistal angle. Legs at anterior margin of the claw furrow with one slender, slightly bifurcate, rigid seta. Genital setae on two fine sclerotized strips divided by a narrow membranous interruption near Ac-2; Ac-2 slightly closer to Ac-1, separated from Ac-3 by a projection of the setae-bearing sclerotized strip. Gnathosoma in lateral view dorsoventrally high, P-4 with nearly straight ventral, and slightly convex dorsal margins; ventral seta long, located in distal third. Males: Idiosoma length/width 2500-3400/500-700; length/width Cx-I+II 150-170/95-120, Cx-III+IV laterally 180-200/100-110; genital field length 150-170 µm, on each side 14-15 setae; Ac-1-3 length 28-35/35-42/37-45; ejaculatory complex length 160; gnathosoma length 140-160 µm; P-4 stout (length/height 2.5). Females: Idiosoma length/width 2400-2900/500-600; length/width Cx-I+II, 160-180/120, Cx-III+IV 170/100; genital field length 150 µm, on each side with 10-12 setae; eggs oval, length 150; gnathosoma length 100-140 µm; P-4 more slender than in male (length/height 3.5).

Similar species: *Wandesia stygophila* SZALAY, 1944a (Romania, unsufficiently documented) differs in the shape of Cx-I+II (medial margin anteriorly strongly concave, posteriorly forming a sector where the opposite plates meet with parallel margins) and the presence of an additional pointed seta at the claw furrow, on the posterior margin. *Wandesia propinqua* WALTER, 1947 (Romania) is similar in these two aspects, but differs in smaller dimensions and a lower number of genital setae (DI SABATINO et al. 2009). *Wandesia helvetica* WALTER, 1947, decribed from a deutonymph, and synonymized with *W. stygophila* by LÁSKA (1957), should be considered a *species incerta*.

Habitat and Biology: Hyporheobiont. In gravel- and sand substrata of middle-order streams. Preadults stages and life cycle unknown.

Distribution: Romania, Bosnia, S Germany.

Subgenus *Wandesia* SCHECHTEL, 1912

1912 *Wandesia* SCHECHTEL, Bull. Acad. Sci. Cracowie Ser. B, Sci. Natur.: 463.

1961b *Allowandesia* SCHWOERBEL, Arch. Hydrobiol., Suppl. 25: 345 – GERECKE (1999).

Type species: *Wandesia thori* SCHECHTEL, 1912, by original designation.

Diagnosis: Acetabula flanking the gonopore as long as wide or longer than wide, separated from each other, placed in the membranous integument on rather long, narrow stalks. A few genital setae on weakly sclerotized bars or in the membranous integument.

Preadult stages: Larva: SCHECHTEL (1912). Deutonymph: TUZOVSKIJ (1990).

Discussion: Earlier, differences in Ac shape between *Wandesia* s. str. and *Pseudowandesia* were considered characteristic for different developmental stages (SZALAY 1944a). Instead, the shape of these organs is a stable character expressed both in deutonymphs and adults, indicating that the subgenera represent distinct phylogenetic units. The subgenus

Mesowandesia André & Naudo, 1962 from North America, differing only in the presence of an additional pair of stalked Ac is closely related to *Wandesia* s. str.. Larvae of extra-European species of the genus were found parasitic on Plecoptera of the families Capniidae, Chloroperlidae, Gripopterygidae, Nemouridae, Perlodidae and Taeniopterygidae (Mullen 1979, Mullen & Vercammen-Grandjean 1980).

2. *Wandesia (Wandesia) thori* Schechtel, 1912 (*8-34 b-c*)

1912 *Wandesia thori* Schechtel, Bull. Acad. Sci. Cracowie Ser. B, Sci. Natur.: 463.

Description: Colour red. Lateral eyes unpigmented. Cx-I+II with straight medial margins, broadly rounded medioposteriorly, posterolaterally slightly concave. Cx-III medially rounded, Cx-IV hardly developed, subtriangular. Numerous slender, slightly curved rigid setae surrounding the claw furrows of I/II-L-6 (3 anteriorly, 4-5 posteriorly and 2-4 ventrally). Genital setae on two fine sclerotized strips embracing Ac-1, divided by a narrow membranous interruption; Ac-2 and -3 more elongated and located posterolaterally. Gnathosoma in lateral view dorsoventrally flattened, in ventral view with broadly rounded posterior margin. Palp slender, P-4 with nearly straight ventral, and slightly convex dorsal margins; ventral seta short, located in distal third. Males: Idiosoma length/width 1200-1600/400-500; length Cx-I+II, 110-120, Cx-III+IV laterally 90-110 µm; genital field with 10-14 pairs of setae; Ac-3 length/width 40-60/20-30; ejaculatory complex length 120; gnathosoma length 120-145 µm; P-4 length/height 3.3-3.4. Females: Idiosoma length/width 1300-2000/400-600; length Cx-I+II, 125-135, Cx-III+IV, 110/120 µm; genital field with 3-4 pairs of fine setae inserting in the membranous integument around Ac-1; Ac-3 length/width 45-50/20-30; gnathosoma length 140-160 µm; P-4 slightly more stout than in male (length/height 2.8).

Similar species: *Wandesia lucifuga* Gerecke, 1999 (S Italy, Maghreb) differs in shorter Ac-3 (length/width 1.2-1.7), longer P-2 (length P-2/P-4 0.7-0.8) and less pronounced sexual dimorphism in the shape of P-4.

Preadult stages: Larva and deutonymph: Schechtel (1912).

Biology: Larvae probably parasitic on Empididae (Diptera) (Gerecke & Martin 2006).

Habitat: Hyporheobiont (Schwoerbel 1961a, b), in montane areas crenophilous, preferably in mossy rheocrenes above the treeline (Gerecke & Martin 2006).

Distribution: Black Forest, Alps, Tatra and northern Apennines, Spain (?) (Gerecke & Martin 2006).

3. *Wandesia (Wandesia) racovitzai* Gledhill, 1970 (*8-34 d-e*)

1970 *Wandesia racovitzai* Gledhill, Acad. R. S. Roumaine, Livre du centenaire Emile G. Racovitza 1868-1968: 231.

Description: Colour red. Lateral eyes unpigmented. Cx-I+II with slightly concave medial margins, broadly rounded medioposteriorly, posterolaterally slightly concave. Cx-III medially rounded. Sutures Cx-I/II and Cx-III/IV medially incomplete. I/II-L-6 ventral and posterior margin of claw furrow each with 2-3 slightly curved, rigid setae. Distal segments of II-IV-L thickened. Gnathosoma compact, in ventral view with broadly rounded posterior margin. Palp stocky, P-4 with diverging ventral and dorsal margins, maximum height distally (length/height 2.3-2.4); ventral seta short, located in distal third. P-2/4 length ratio 0.7. Males: Idiosoma length/width 1600-1700/350-450; length Cx-I+II 110, Cx-III+IV laterally 80-100 µm; a pair of fine sclerotized strips embracing gonopore, each bearing 6-9 fine setae. Ac-1 reduced, Ac-2 and -3 elongated, located posterolaterally; posterior Ac length/width 30/20; ejaculatory complex length 90-95; gnathosoma length 140 µm;

P-4 length/height 2.3. Females: Idiosoma length/width 1600-2200/350-500; length Cx-I+II, 115, Cx-III+IV laterally 80-100 µm; genital field with 3 pairs of fine setae inserting in the membranous integument around Ac-1; Ac-1-3 inserting on a transverse line on the level of the posterior margin of gonopore, length/width Ac-3, 30-40/18-21; gnathosoma length 140 µm; palp slightly more slender than in male.

Habitat and Biology: Hyporheobiont; in gravel and sand substrata of middle-order streams. Preadult stages and life cycle unknown.

Distribution: British Isles, S Germany. Only known from two streams (GERECKE 1999).

Superfamily Lebertioidea THOR, 1900

1900b Lebertiidae THOR, Nyt. Mag. Naturv. 38: 2.

Diagnosis: Lateral eyes lateral in position, not on a common eye plate. Genital field with Ac (with rare exceptions) in two parallel rows in the soft integument flanking the gonopore. A pair of window-shutter-like genital flaps generally well developed, in closed position covering gonopore and Ac. Palps neither chelate nor subchelate. Chelicera two-segmented. Larvae aquatic, adapted for efficient swimming.

Remarks: This superfamily is surely a paraphyletic unit, defined exclusively by plesiomorphies. Based on larval, deutonymphal and adult character states, TUZOVSKIJ (1987) proposed the synonymy of Lebertioidea with Arrenuroidea and Hygrobatoidea (see vol. III of this book), uniting all included families in the single superfamily Hygrobatoidea. This point of view also found support from PANESAR (2004) and GOLDSCHMIDT (2004b) who described taxa representing transitional situations between Lebertioidea and Hygrobatoidea, with Ac merging to, and more or less fusing with, the medial genital flap margins. Based on the classic system, in the extreme case of the neotropical *Rheolimnesia*, males fit the diagnosis given here for Lebertioidea, but females show a typical hygrobatoid genital field (GOLDSCHMIDT 2004b). Obviously, shifting of Ac to the genital flap surface (which then became immovable and are called "genital plates") has happened several times in evolution, and also development of swimming behaviour in larvae occurred in several clades independently (TUZOVSKIJ et al. 2001). However, with regard to the Western Palaearctic fauna, the traditional system is still useful for keeping order among those numerous families of "higher Hydrachnidia", and phylogenetically convincing alternatives have not yet been elaborated. A new system requires additional morphological data, mainly concerning the little known preadult stages, and new taxonomic approaches including molecular techniques. All known families are present in the study area except for the monotypic Stygotoniidae (New Zealand: COOK 1992).

Key to families

1 All leg insertions concentrated in the anterior part of the idiosoma and directed anteriorly; coxal plate extending far onto dorsal surface (*8-35 a-b*). . . . . Oxidae (page 145)

\- Leg insertions not concentrated in the anterior part of the idiosoma, directed laterally; coxal plate generally not extending onto dorsal surface. 2

2 (1) P-4 shortened, P-5 in dorsal view foot-like enlarged and with nail-like claws (*8-35 c-d*); genital field with three pairs of Ac (*8-57 b, e*); dorsal and posteroventral idiosoma bearing paired plates not fusing to compact shields (*8-57 a-b*); male genital flaps fused to a penis tube (*8-57 c-d*). . . . . Rutripalpidae (page 152)

\- P-4 longer, P-5 not foot-like, with normal, hook-like claws (*8-36 b, e, 8-37 h*); number of Ac and idiosoma sclerotization various; male genital flaps not fused to a penis tube. . . . 3

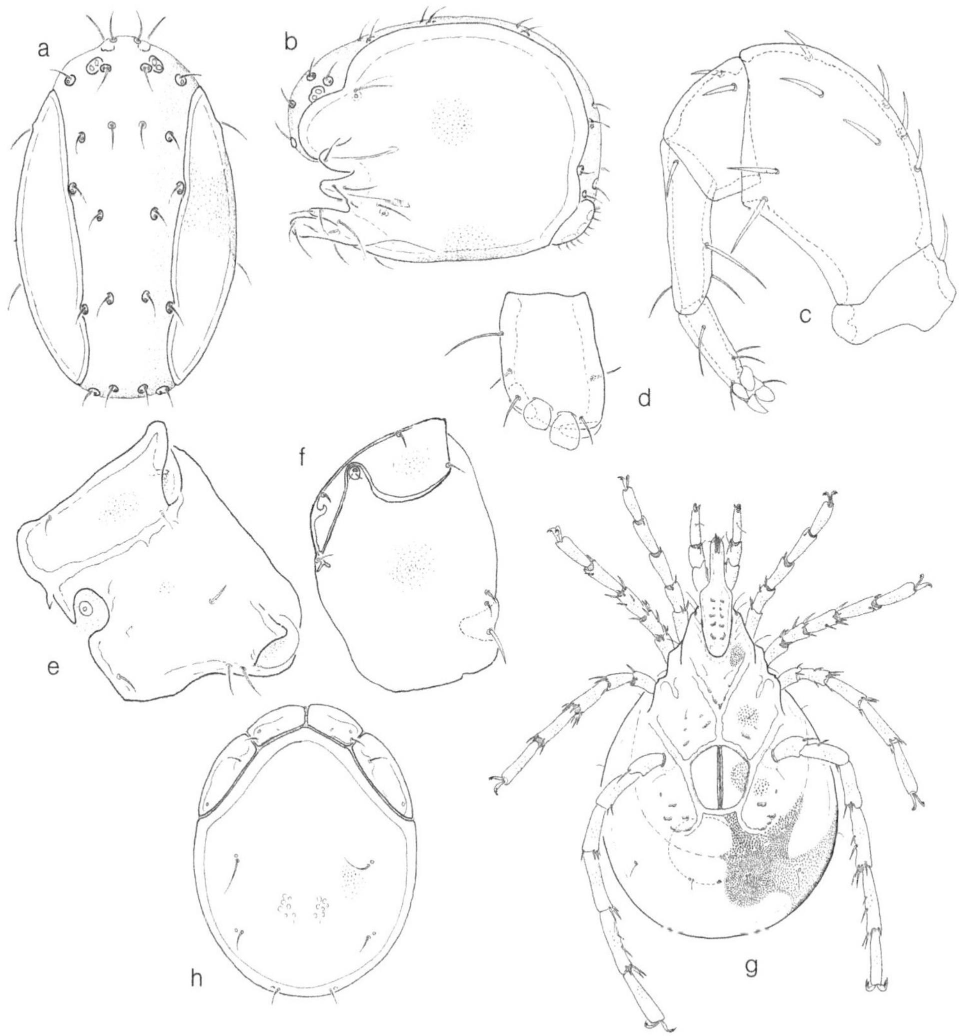

8-35: a-b, *Oxus* sp., idiosoma; a, dorsal view; b, lateral view; c-d, *Rutripalpus limicola*; c, palp lateral view; d, P-5 anterior view (from GERECKE & TUZOVSKIJ 2001); e, *Teutonia cometes*, Cx-III+IV; f, *Limnolegeria longiseta*, Cx-III+IV; g, *Torrenticola anomala* female, ventral view (from DI SABATINO et al. 2000); h, *Torrenticola laskai* male, dorsal shield (from DI SABATINO et al. 2009).

3 (2) A pair of glandularia on Cx-IV, located or near medial margin of plate and surrounded by a conspicuous membranous area (*8-35 e*), or on an anterior protrusion indenting the posterior edge of Cx-III (*8-35 f*); legs with swimming setae. . . . . Teutoniidae (page 173)

\- If glandularia present on Cx-IV, located neither in a membranous area in the plate centre, nor on an anterior protrusion (*8-37 f*); legs with or without swimming setae. 4

4 (3) Idiosoma sclerotized both dorsally and ventrally; dorsum typically with a large central shield and one or two pairs of anterior plates (*8-35 h*) – if anterior plates fused to main shield, sutures remain visible; a Y-shaped line extending from anterior genital field margin to tips of Cx-I (*8-35 g*); six pairs of Ac. . . . . Torrenticolidae (page 177)

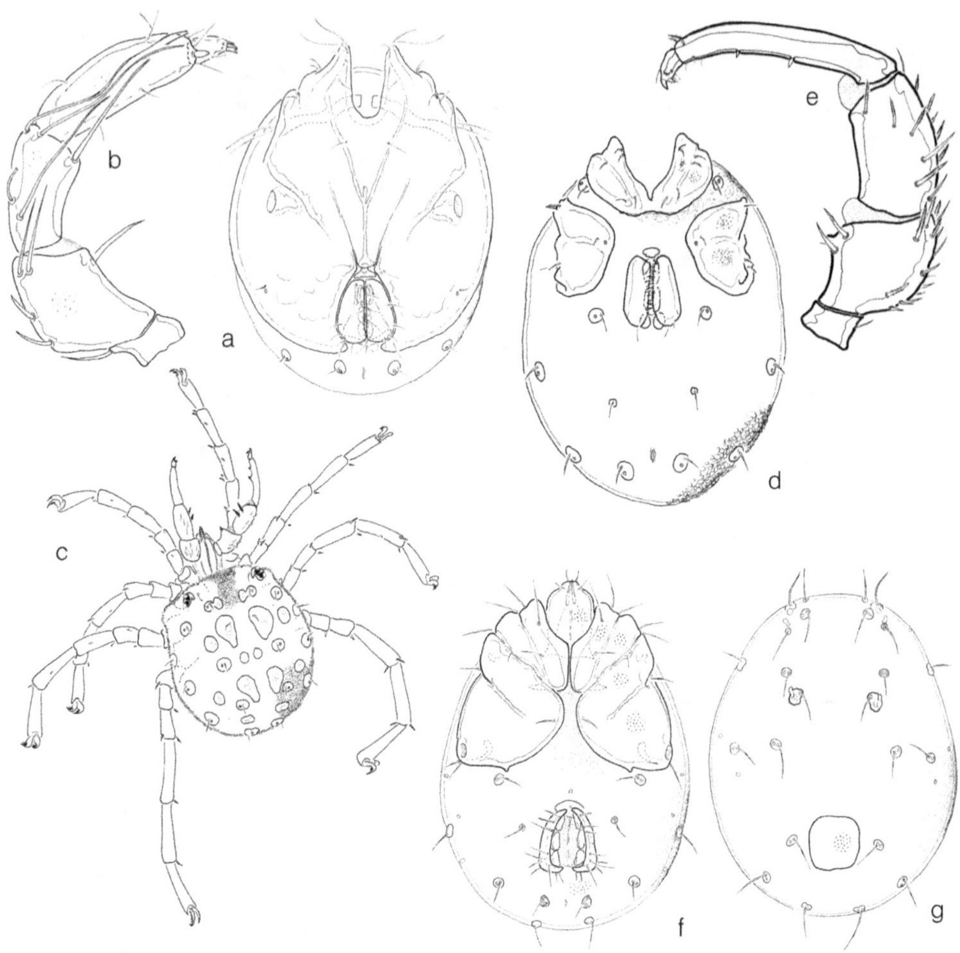

8-36: a, *Lebertia holsatica* male, ventral idiosoma; b, *Lebertia castalia* male palp; c-e, *Sperchon* male; c, *S. setiger*, dorsal view; d-e, *S. mutilus*; d, ventral idiosoma; e, palp; f-g, *Nilotonia* sp. male; f, ventral idiosoma; g, dorsum (a-b from GERECKE 2009, f-g after GERECKE 1992).

-	Idiosoma with soft integument or various sclerotizations; if a dorsal shield present, ventral shield without a Y-shaped suture line (e. g., *8-37 a-b, e, 8-38 a, c*); normally three (in one exceptional, soft-bodied species six) pairs of Ac.	5
5 (4)	Coxae fused, forming a characteristic coxal shield with suture lines Cx-II/III medially obsolete, a Y-shaped suture extending from anterior genital field margin to tips of Cx-I (*8-36 a*); P-2 with one (rarely 2-3) long ventrodistal seta(e), P-3 with 5-6 (rarely more) long medial setae (*8-36 b*); dorsum in all species from the treated area without sclerotizations	Lebertiidae (page 90)
-	A membranous area (*8-36 d*) or at least a complete suture line (*8-38 a*) separating Cx-II/III on each side; if coxae fused, not bearing a Y-shaped line between genital field and tips of Cx-I; palp not with the chaetotaxy described for Lebertiidae (*8-36 e, 8-38 e*); dorsum soft or sclerotized to various degree	6

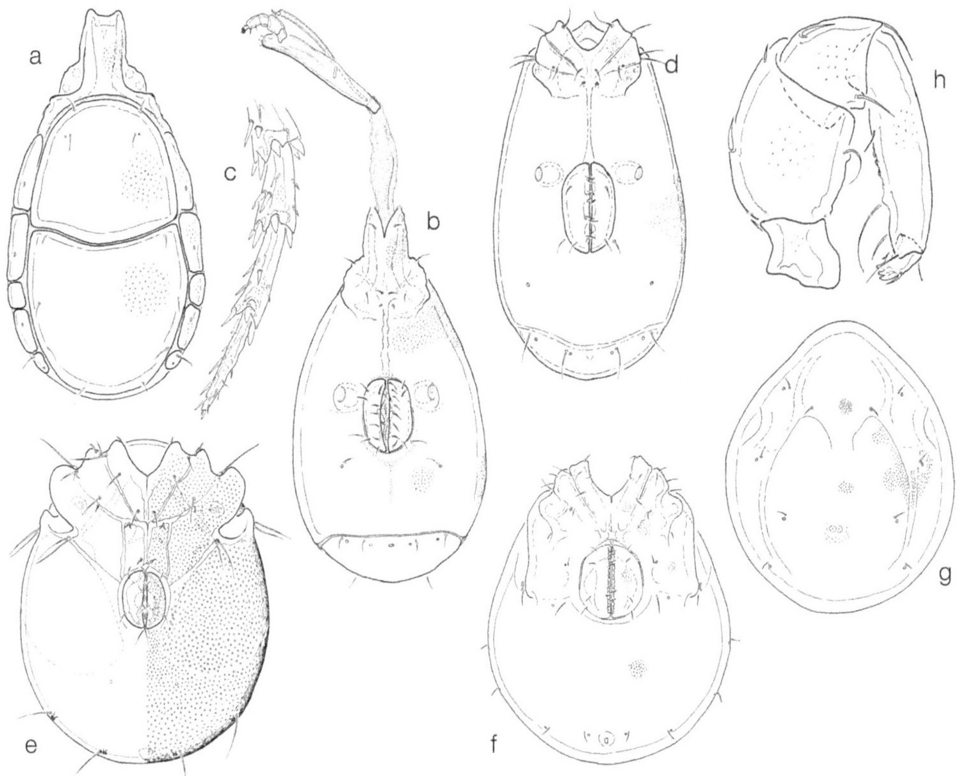

8-37: a-c: *Psammotorrenticola gracilis* female; a, dorsum; b, venter and gnathosoma; c, IV-L; d, *Stygomamersopsis viedmai* male, venter (after Cook 1974); e, *Utaxatax ventriplax* male, venter (from Gerecke 1990); f-h, *Sigthoria nilotica* male (from Gerecke 2004b); f, venter; g, dorsum; h, palp.

6 (5) Lateral eyes in rounded capsules (*8-36 c*); posterior margin of Cx-II and anterior margin of Cx-III in most cases separated by a strip of soft integument (*8-36 d*); dorsum soft or sclerotized to various degree, but posteroventral area never with a continuous shield (*8-36 d*); IV-L with claws. . . . . Sperchontidae (page 154)

\- Lateral eyes not in capsules (but in sclerotized species occasionally on platelets lying between dorsal and ventral shields or fused with one of the shields); posterior margin of Cx-II and anterior margin of Cx-III well matched, fused (*8-37 b, e, 8-38 a*) or separated by a very narrow membranous strip (*8-36 f*); dorsum never completely soft, in species from the treated area with three platelets (*8-36 g*) or a dorsal shield (*8-37 a, 8-38 d*); posteroventral area may be sclerotized (*8-37 b, 8-38 a*); IV-L with or without claws. . . . Anisitsiellidae (page 79)

Family Anisitsiellidae Koenike, 1910

1910b Anisitsiellidae Koenike, Abh. naturw. Ver. Bremen 20: 128.

2004 Limnesiidae *partim* Panesar, Bonn. Zool. Monogr. 52: 49.

Diagnosis: Integument completely smooth to completely sclerotized with ventral and dorsal shields separated by a narrow membranous furrow. Lateral eyes free under the integument, rarely on longish platelets (which may fuse to the neighbouring shields) or reduced. Coxae in four groups or fused to various degrees, occasionally integrated into a continuous ventral shield. Genital field with movable flaps that may cover the gonopore completely, generally three (in some taxa up to 20) pairs of Ac flanking the gonopore (in a few extra-European taxa fused to the genital flap margin). Palp five-segmented, P-2 typically with a ventral seta.

Remarks: The phylogenetic relationships of the genera assembled in this family are still open to many questions, and COOK (1974) designated it as a "dumping ground" for those members of the higher Lebertioidea which do not fit into the other families. Moreover, evidence on larval morphology in some genera included here (CRAMER & SMITH 1993, GERECKE & SMITH 1993) and morphological features in adults of recently described taxa (PANESAR 2004) blurred the limits between certain members of this family and basic Limnesiidae. As a consequence, PANESAR (2004) proposed the synonymization of Anisitsiellidae with Limnesiidae, where former anisitsiellids should be placed in three subfamilies, Anisitsiellinae KOENIKE, 1910b (including most of the former anisitsiellid genera), Bandakiinae PANESAR, 2004 (*Bandakia* and *Utaxatax*, probably a monophyletic unit) and Bharatoniinae PANESAR, 2004 (with *Bharatonia* COOK, 1967 and *Shivatonia* PANESAR, 2004, only known from India and the Far East). However, in many genera not one species has been described at the larval and/or deutonymphal stage, and newly detected taxa continue to open new fields for discussion rather than contributing convincing ideas for a more natural system. As said above with regard to superfamilies, lower-level taxonomy also needs elaboration of new data, and application of new techniques, before systematics can be improved in a phylogenetic sense. PANESAR's subfamiliar division, with four genera, three of them present in the European fauna, remaining without placement as "*incertae sedis*", is not satisfactory. Therefore, Anisitsiellidae is retained here *sensu* COOK, (1974), as an unique taxonomic unit above genus level.

Key to genera

1 European species with three small dorsal platelets and a small postgenital sclerite, the remaining dorsal and posteroventral idiosoma soft (*8-36 f-g*). . . . . *Nilotonia* (page 83)

\- Idiosoma with large shields completely covering venter (*8-37 b, d*) and dorsum (*8-37 a, g, 8-38 d*). 2

2 (1) IV-L lacking claws (occasionally, a pair of seta-like nails may be interpreted as remnants, *8-37 c*); dorsal shield divided into two parts (*8-37 a*). 3

\- IV-L with normal claws; dorsal shield not divided (*8-38 d*). 4

3 (2) Anterior part of Cx-I forming a funnel-shaped camerostome, gnathosoma at the tip of a protrusible membranous tube (*8-37 a-b*). . . . . . . *Psammotorrenticola* E. ANGELIER, 1951

(Monotypic genus, not recorded from the study area: *Psammotorrenticola gracilis* E. ANGELIER, 1951, hyporheobiont, Corsica)

\- Anterior part of Cx-I not projecting and fused (*8-37 d*), gnathosoma not at the tip of a membranous tube. . . . . *Stygomamersopsis* COOK, 1974

(Not recorded from the study area: Two species known from the interstitial habitat of streams in Spain – COOK 1974)

4 (2) Genital field with more than ten pairs of Ac; Cx-IV with a glandular pore; no paramedian suture line separating a quadrangular medial area from lateral part of Cx-III (*8-37 f*); dorsal shield with ornamentation (*8-37 g*). *Sigthoria* (page 89)

\- Genital field with three pairs of Ac; Cx-IV not perforated by a glandular pore; a paramedian suture line separating a quadrangular medial area from lateral part of Cx-III (*8-37 e, 8-38 a, c*); dorsal shield without ornamentation (*8-38 d*). 5

5 (4) Posterior margin of Cx-IV not extending beyond posterior margin of genital field, insertions of IV-L on the level of genital field (*8-38 a, c*). . . . . *Bandakia* (page 81)

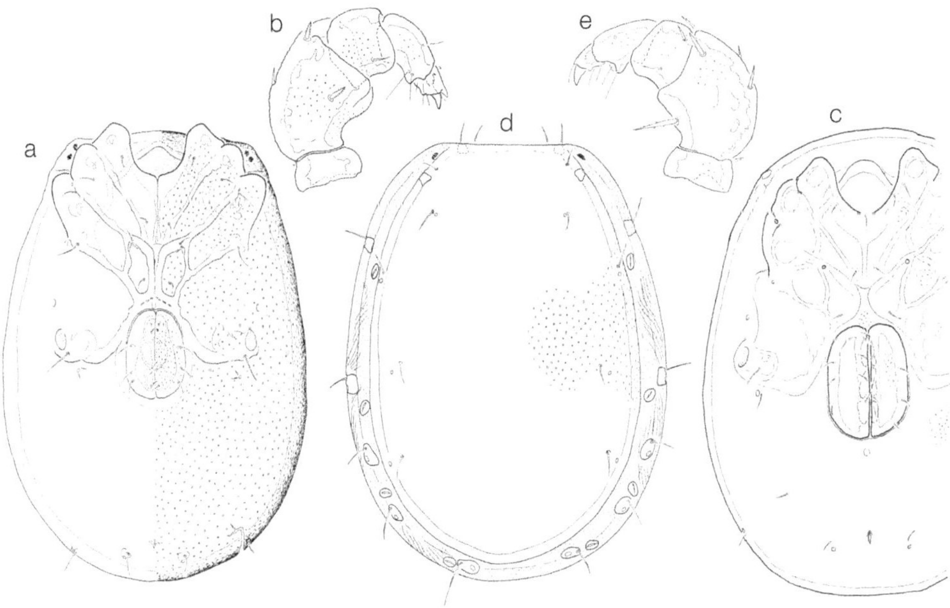

8-38: a-b, *Bandakia speciosa* male; a, venter; b, palp; c-e, *Bandakia concreta* male; c, venter partially; d, dorsal shield; e, palp (a-c from GERECKE 1991a; e from GERECKE 1994c).

- Posterior margin of Cx-IV (visible as a fine suture line) extending beyond posterior margin of genital field, insertions of IV-L anterior to anterior margin of genital field (*8-37 e*). . . . . *Utaxatax* HABEEB, 1964

  (Not recorded from the study area. Holarctic, in Europe only *Utaxatax ventriplax* GERECKE, 1991, Sicily, Sardinia and Spain, crenobiont – GERECKE 1994).

Genus *Bandakia* THOR, 1913

1913c *Bandakia* THOR, Zool. Anz. 43: 40.

Type species: *Bandakia concreta* THOR, 1913c, by monotypy.

Diagnosis: Idiosoma flattened, with dorsal and ventral shields, separated by a narrow dorsal furrow bearing glandularia and slit organs. Lateral eye lenses rarely free in the furrow, generally on sclerites which often fuse to each other or (in all European species) to anterior ventral shield margin. Glandularia on Cx-III at anterior margin, laterally flanked by a suture line, parallel to the median line or anteriorly slightly diverging, extending between anterior and posterior margins of Cx-III. Genital field extending beyond posterior margin of Cx-IV, with three pairs of Ac. Palps five-segmented, various in shape, P-2 with a ventrolateral seta. Legs stocky, without swimming setae, I-III-L claws each with ventral claw blade, and ventral and dorsal clawlets. IV-L with hook-like claws, sometimes bearing a medial clawlet, flanked by three to four strong terminal setae, one of them as long as the claw.

Remarks: Species of the genus are known from the Holarctic and Oriental regions. On the basis of palp morphology they were divided by SMITH (1979) into three groups, the "*anisitsipalpis-*", "*longipalpis-*" and "*vietsi-*group". The latter two grade into each other (GERECKE 1991b); all European species have stout palps and lateral eye platelets fused to the ventral shield, and therefore belong to the *vietsi*-group (PANESAR 2004).

Key to species

1 Anterior margin of Cx-I not reaching level of anterior idiosoma margin (*8-38 c*); P-4 with two well-developed ventral projections (*8-38 e*). . . . . *Bandakia concreta* (page 82)

- Anterior margin of Cx-I extending beyond anterior idiosoma margin (*8-38 a*); P-4 without ventral projections (*8-38 b*). . . . . *Bandakia speciosa* (page 82)

1. *Bandakia concreta* THOR, 1913 (*8-38 c-e*)

1913e *Bandakia concreta* THOR, Zool. Anz. 43: 40

1994 *Bandakia bieberi* BADER, Erg. wiss. Unters. Schweiz. Nationalpark 16: 262 – DI SABATINO et al. (2009).

Description: Colour brownish-yellow, often with distinct white ornamentation provided by the excretory organ. Idiosoma various in shape, often rounded, with maximum width in posterior part. Distal margins of Cx-I-III smoothly rounded, anteriorly not reaching the anterior idiosoma margin. Suture lines flanking glandular pore on Cx-III anteriorly slightly diverging. Cx-IV oblique subrectangular, with medial margins touching anterolateral edges of genital field, posteriorly forming an abrupt angle and in the lateral part directed anteriorly to IV-L insertions. Palp stocky. P-2 ventral margin with a mediodistal, flattened extension and a lateral seta, P-4 very stout, with a pair of prominent ventrodistal extension associated with a group of slender setae; P-5 strong, with a prominent principal claw. Males: Length/width idiosoma 530/420, genital field 125/100 µm, lateral flap margins smoothly convex. Females: Length/width idiosoma 580/470, genital field 170/125 µm, lateral flap margins straight.

Preadult stages: Larva: MARTIN (2003). Deutonymph: K. VIETS (1928a).

Habitat and Biology: Crenobiont. Preferably in weakly seeping helocrenes, generally in low numbers; in exceptional cases interstitial of low order streams (GERECKE & MARTIN 2006). Tenacious of life: K. VIETS (1936d) reports survival, without food, over six months in a small, closed glass tube containing only air and 1 cm^3 of water. Larvae parasitic on chironomids (Tanypodinae, MARTIN & STUR 2006; MARTIN et al. 2010).

Distribution: Europe.

2. *Bandakia speciosa* K. VIETS, 1953 (*8-38 a-b*)

1953a *Bandakia speciosa* K. VIETS, Zool. Anz. 150: 67.

1961a *Bandakia corsica* SCHWOERBEL (*nec* E. ANGELIER), Arch. Hydrobiol., Suppl. 25 (4): 244 – GERECKE (1991b).

Description: Colour unknown. Idiosoma elongated oval, length/width 530-565/390-410 µm. Anterior tips of Cx-I extending beyond anterior idiosoma margin, posterolateral margin Cx-III smoothly concave, Cx-III+IV anterior to genital field closely in touch, without leaving an area of porose sclerotization. Cx-IV medial margins straight, anteriorly touching edge of genital field, posteriorly diverging and forming a smooth curve, posterior margins horizontal to median axis. Genital field length/width 102-128/73-79 µm, in both sexes with four pairs of fine setae on flap margins. P-2 in lateral view with S-shaped ventral margin (proximally concave, distally convex), with a distoventral longitudinal furrow; P-4 stout (length/height ratio 1.3-1.6) and relatively short (length ratio P-3/P-4, 1.4), distoventrally slightly thickened, but without extensions. Males: Genital field length/width ca. 1.4, lateral flap margins smoothly convex. Females: Genital field length/width > 1.5, lateral flap margins straight.

Similar species: By some authors considered a junior synonym of *B. corsica* E. ANGELIER, 1951b (see GERECKE 1991b). This Corsican endemic differs from *B. speciosa* in P-2 ventrally lacking a longitudinal furrow and P-4 being more slender (length/height ratio 1.9) and relatively longer (length ratio P-3/P-4, 2.0).

Preadult stages: Larva unknown. Deutonymph: SCHWOERBEL (1961a).

Habitat and Biology: Hyporheobiont. Sand and gravel substrata in low- and middle-order streams, one record from a spring habitat (BOTOŞANEĂNU & NEGREA 1961).

Distribution: Germany (Harz, Black Forest), Romania, Bulgaria.

Genus *Nilotonia* THOR, 1905

1905b *Nilotonia* THOR, Zool. Anz. 28: 807.

Type species: *Teutonia loricata* NORDENSKIÖLD, 1905, by original designation.

Diagnosis: Sclerotization of idiosoma varying from two small postocularia and a small posterodorsal platelet, with posteroventral area completely smooth to a well developed dorsal shield combined with an extended posteroventral plate; posteroventral sclerotization generally separated by a membranous suture from coxal and genital field (but fused to a complete shield in one species of the subgenus *Nilotonia s. str.* from India). Coxae in four groups or fused to various degree, with suture Cx-III/IV complete or incomplete. Glandularia on Cx-II various in position or absent. Genital field extending between Cx-IV or located posterior to coxal field, with three to four Ac. Palp often slender, P-2 with a ventral seta and often ventrally denticulate, P-4 with a group of fine ventral setae generally associated with pointed denticles in the distal part of segment. Legs with or without swimming setae; I-III-L claws with ventral clawlets often numerous and forming a comb; IV-L claws reduced (small peg-seta-like remnants visible), tips of IV-L-6 often with one or several prominent terminal seta(e).

Remarks: Revising species of the genus, PANESAR (2004) discovered swimming setae in the African type species, *N. loricata*. Consequently, the subgenus *sensu stricto* now includes four species known from Africa and India, the remaining species being attributed to the subgenera *Dartiella* and *Dartia* (both recorded from the area treated); *Manotonia* (Mediterranean area, India, Africa), *Mamersonia* K.VIETS, 1954e (Neotropical), *Dartonia* K.VIETS, 1929h (Turkey, India) and *Tadjikodartia* BADER, 1980h (Central Asia, Iran, India).

Key to subgenera and species

1 IV-L-6 without a terminal or subterminal seta located near tip of segment (*8-40 a*); medial margins of Cx-III anterior to genital field parallel to each other (*8-39 a*); males: genital field posterior to posterior margin of Cx-IV; a dorsal shield with undulating margins (*8-39 b*) covering more than 50 % of the dorsal idiosoma. . . . subgenus *Manotonia* K. VIETS, 1935c

(Not recorded from the area treated. In Europe only *Nilotonia (Manotonia) tegulata* K. VIETS, 1951d, known from springs in the Mediterranean area - GERECKE 1994c).

\- IV-L-6 with a terminal or subterminal seta located near tip of segment (*8-40 b-d*); medial margins of Cx-III anterior to genital field with rounded medial edge (*8-36 f, 8-39 c*); males: anterior part of genital field embraced by Cx-IV; European species with three small dorsal platelets, covering less than 20 % of the idiosoma (*8-36 g*). 2

2 (1) Interspace between anterior and posterior Ac larger than maximum diameter of Ac-1; Ac small, but on extended basal sclerites, three or four pairs of Ac; often posterior Ac basal sclerites fused (*8-39 d-e*); males: II-L-5 ventral margin proximally projecting, distally concave, II-L-6 with a proximoventral kink (*8-40 e*). . . . . subgenus *Dartia*, 3

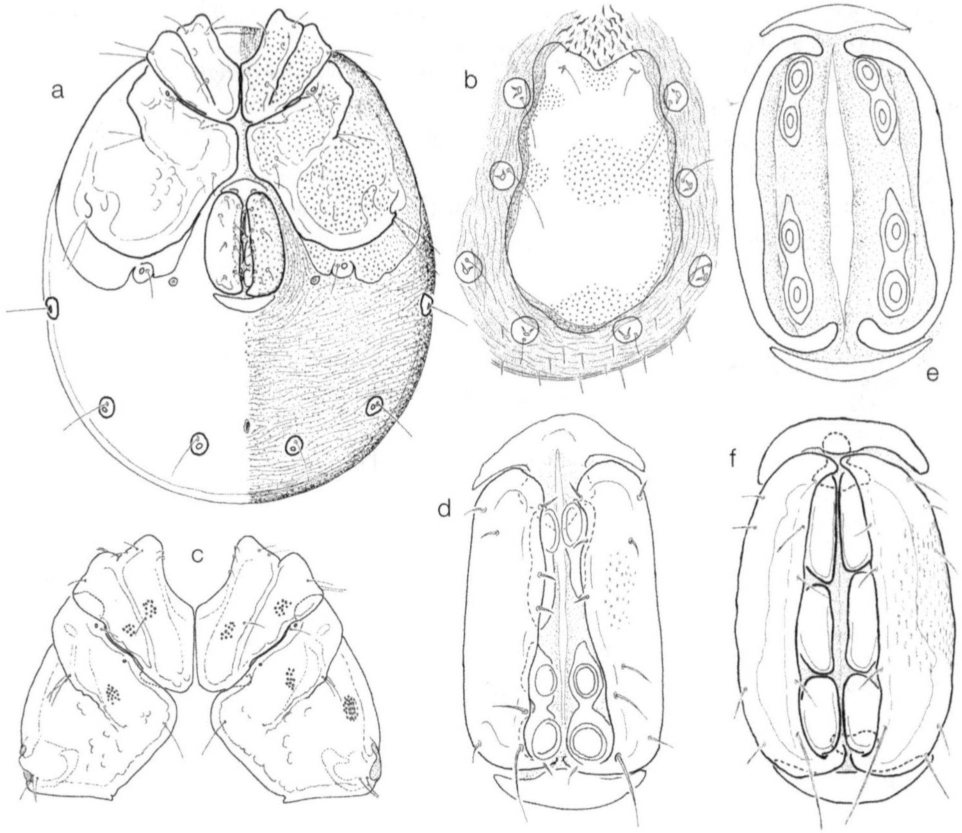

8-39: a-b *Nilotonia (Manotonia) tegulata*; a, female venter, b, male dorsal shield; c, *Nilotonia (Dartiella) pyrenaica*, coxal field; d-f, genital fields; d, *N. (Dartia) borneri* female; e, *N. (Dartia) harrisi* male; f, *N. (Dartiella) pyrenaica* male (from Gerecke 1991a) (a-c, f from Gerecke 1991a, e after Soar 1917a).

-	Interspace between anterior and posterior Ac shorter than maximum diameter of Ac-1; Ac larger, on smaller basal sclerites, three pairs of Ac, Ac-2/3 always separate (*8-39 f*); no sexual dimorphism in II-L.	subgenus *Dartiella*, 4
3 (2)	Genital field with four pairs of Ac (*8-39 e*).	*Nilotonia (Dartia) harrisi* (page 86)
-	Genital field with three pairs of Ac (*8-39 d*).	*Nilotonia (Dartia) borneri* (page 86)
4 (2)	Terminal seta on IV-L long, extending far beyond tip of segment (*8-40 d*).	*Nilotonia (Dartiella) longipora* (page 87)
-	Terminal seta on IV-L short, not, or only slightly, extending beyond tip of segment (*8-40 b-c*).	5
5 (4)	P-2 with distinct ventral denticulation, ventral seta long (> 60 % ventral length); main ventral denticle on P-4 away from distal margin, dividing segment in proximal/distal sectors 1.2-1.8 (*8-40 i*); ventral comb of I-III-L claws formed by > 6 denticles (*8-40 g*).	*Nilotonia (Dartiella) parva* (page 88)
-	P-2 with weakly developed ventral denticulation, ventral seta short (40-50 % ventral length); main ventral denticle on P-4 closer to distal margin, dividing segment in proximal/distal sectors 1.9-2.1 (*8-40 h*); ventral comb of I-III-L claws formed by 3-5 denticles (*8-40 f*).	*Nilotonia (Dartiella) pyrenaica* (page 88)

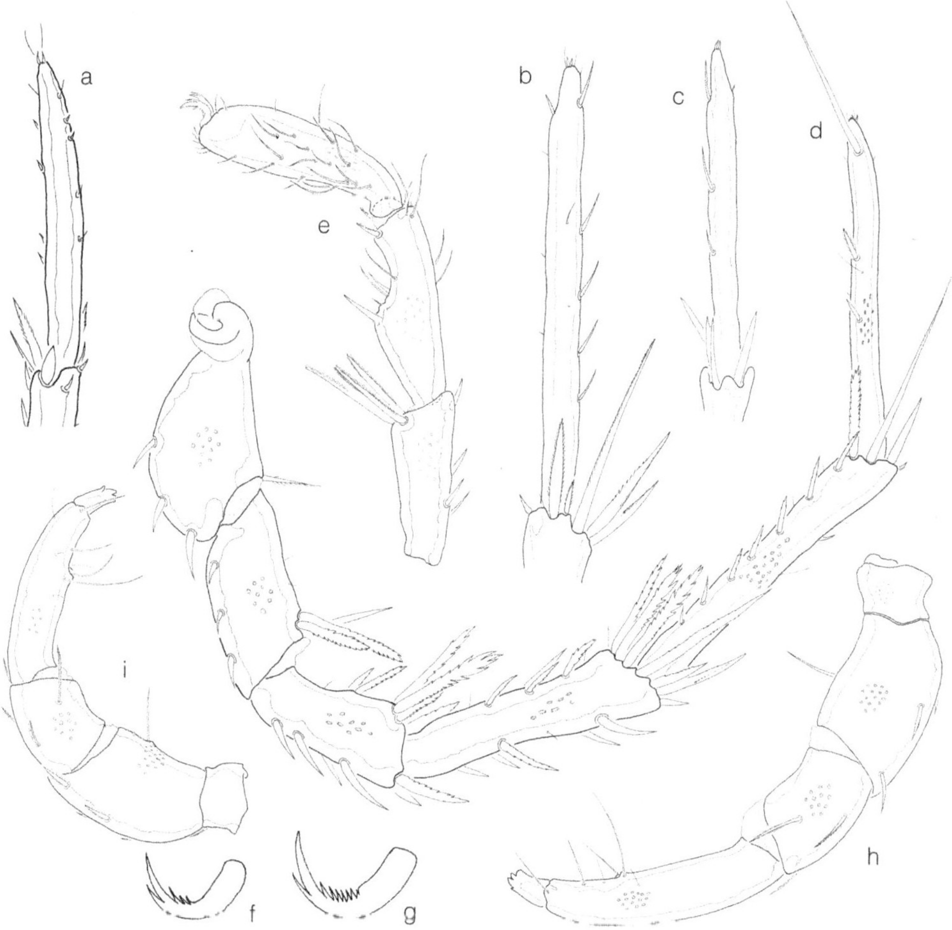

8-40: *Nilotonia*; a-c, IV-L-6, males; a, *N.(Manotonia) tegulata*; b, *N. (Dartiella) pyrenaica*; c, *N. (Dartiella) parva*; d, *N. (Dartiella) longipora*, male IV-L; e, *N. (Dartia) borneri* male II-L-4-6; f-g, leg claws, males; f, *N. (Dartiella) pyrenaica*; g, *N. (Dartiella) parva*; h-i, palps; h, *N. (Dartiella) pyrenaica*; i, *N. (Dartiella) parva* (a-d, h from GERECKE 1991a, f-g, i from GERECKE & MORENO 2008).

Subgenus *Dartia* SOAR, 1917

1917a *Dartia* SOAR, J. Quekett Microsc. Club (s. 2) 13: 277.

Type species: *Dartia harrisi* SOAR, 1917a, by monotypy.

Diagnosis: Idiosoma sclerotization restricted to the pair of postocular platelets and the posterodorsal plate, postgenital integument soft. Cx-III anteriorly not protruding over posterior margin of Cx-II. Anterior to genital field, medial margins of Cx-III rounded, not parallel to each other. Acetabula small, located on extended basal sclerites, often basal sclerites of Ac-1 drop-shaped (posteriorly pointed). Some or all basal sclerites may fuse to form elongate strips; anterior Ac (Ac-1, in four-acetabulate species Ac-1+2) and posterior Ac (Ac-2+3, in four-acetabulate species Ac-3+4) separated by a large interspace (length > length of Ac-1). II-L with sexual dimorphism: in males, II-L-5 ventral margin proximally projecting, distally concave, II-L-6 with a basal kink; in females both segments normal in shape. IV-L with a long subterminal seta extending beyond tip of segment.

Remarks: PANESAR (2004) included in this subgenus *N. turcica* ÖZKAN & SOYSAL, 1989, a species lacking the sexual dimorphism of II-L. However, the latter character is of greater taxonomic weight, while the reduced size of the Ac may have more easily evolved several times independently. Excluding *N. turcica*, the remaining species probably represent a well-defined monophyletic clade.

1. *Nilotonia (Dartia) borneri* (WALTER, 1922) (*8-39 d, 8-40 e*)

1922a *Dartia borneri* WALTER, Rev. Suisse Zool. 29 (7): 279.

Description: Colour brownish-orange to bright red, sclerites bluish. Integument with strong, rugose lineation, mostly in longitudinal direction dorsally, but transverse on flanks and ventrally. Dorsal plate slightly elongated. Some of the distal setae on I-IV-L-4-5 deeply serrated. Leg claws I-III with dorsal and ventral clawlets and a ventral comb formed by 4-5 denticles. IV-L-6 subterminal seta inserting close to tip of segment, length 42-92 µm. Genital field with three pairs of rather small Ac on large, occasionally fused basal sclerites. P-2 with fine distoventral denticulation, ventral seta near centre of segment or in distal half, in length about 50 % ventral segment length; length ratio P-2/P-4, 0.87-0.90. Males: Idiosoma length/width 580-700/430-500 µm; II-L-5 ventral margin proximally inflated, distally concave and flanked by variously-shaped setae; II-L-6 with a basal kink, covered by numerous slender setae; P-2 ventral seta insertion and P-4 ventral denticles further from distal edge than in females. Females: Idiosoma length/width 750-900/600-700 µm; II-L not modified.

Habitat and Biology: Weakly seeping springs and ponds in bogs; in habitats strongly exposed to sunlight. Preadult stages and life cycle unknown.

Distribution: Western, Central and Eastern Europe, Northern Italy.

2. *Nilotonia (Dartia) harrisi* SOAR, 1917 (*8-39 e*)

1917a *Dartia harrisi* SOAR, J. Quekett Microsc. Club (s. 2) 13: 278.

Description: Male (Female unknown): Colour bright red, sclerites slaty blue. Integument finely ridged, lines wide apart. Idiosoma length/width 740/710 µm; posterodorsal plate ovate, with apex directed posteriorly. Coxal field similar to *8-39 c*, but posterior margins of Cx-II more pointed and medial margins of Cx-IV more divergent. Central segments of legs with strong, partly serrated distal setae, claws slender, no information on clawlets and/or denticulation. II-L similar to *8-40 e*; II-L-5 ventral margin proximally bulging, distally concave, II-L-6 with a basal kink, bearing few setae. IV-L-6 with a long subterminal seta located slightly away from tip of segment. Genital field located for about half its length between Cx-IV, with strong, crescent-shaped pre- and postgenital sclerites. Basal sclerites of Ac-1+2 and Ac-3+4 fused to form anterior and posterior acetabular platelets, separated from each other by a large gap. Gnathosoma with a short rostrum, palp in shape and setation similar to *8-40 i*, but presence/absence of ventral denticulation on P-2 unclear.

Similar species: *Nilotonia harrisi* is similar to *N. borneri*, but the presence of four pairs of Ac, instead of three in all other species of the subgenus, is an unmistakable diagnostic character. As it was described from "several specimens" (SOAR 1917a), this character cannot be an individual malformation. The original description leaves questions concerning palp morphology and shape of leg claws. The strongly divergent Cx-III/IV (SOAR 1917a, Fig. 2) suggest that it was figured from a specimen distorted by slide preparation.

Habitat and Biology: Known from a bog habitat. Preadult stages and life cycle unknown.

Distribution: Only known from the type locality in Great Britain (Dartmoor, Broad Down Bog near Postbridge, Devon).

Subgenus *Dartiella* K. VIETS, 1929

1929h *Dartiella* K. VIETS, Zool. Anz. 86: 53.

1980h *Parvadartia* BADER, In: WITTMER, W. & BÜTTIKER, W., Fauna of Saudi Arabia 2: 54 – GERECKE & MORENO (2008).

Type species: *Dartia longipora* WALTER, 1925a, by original designation.

Diagnosis: Idiosoma sclerotization restricted to the pair of postocular platelets and the posterodorsal plate, postgenital integument soft. Cx-III anteriorly not protruding over posterior margin of Cx-II. Anterior to genital field, medial margins of Cx-III rounded or slightly pointed, not parallel to each other. Acetabula generally large and not separated by large interspaces, their basal sclerites not or slightly extending beyond acetabular margin, never fused to each other. II-L without sexual dimorphism. IV-L with one subterminal seta various in length, extending or not extending beyond tip of segment.

Remarks: This subgenus is a paraphyletic dumping ground for all species lacking one of the particular apomorphies that define the remaining subgenera (PANESAR 2004).

3. *Nilotonia (Dartiella) longipora* (WALTER, 1925) (*8-40 d*)

1925a *Dartia longipora* WALTER, Bull. Soc. Hist. natur. Afrique Nord 16: 218.

1950d *Limnesia tuberculata* K. VIETS, J. Quekett Microsc. Club London (s. 4) 3: 151 – GERECKE (1991a).

Description: Colour bright red, in immature specimens pale orange, sclerotized plates blue. Integument smooth. Posterodorsal plate round, slightly elongate or slightly transverse (length/width ratio 0.8-1.2). Legs with strongly serrated distal setae on I-IV-L-3-4, leg claws with a comb of 5-7 ventral denticles. Subterminal seta IV-L-6 long and slender (basal width 4-6 µm, length/width ratio 15.0-25.0), 15-30 µm away from segment tip. Genital field with elongated Ac flanking gonopore all over its length or nearly so. P-2 with a relatively short ventral seta (length < ventral segment length), ventral denticulation indistinct, restricted to the area anterior to ventral seta insertion, P-4 relatively slender (length/height ratio 3.9-4.9), distinctly longer than P-2, location of ventral group of denticles and setae rather variable, dividing segment in ratios 1.3-2.0. Males: Idiosoma length/width 750-900; posterodorsal plate length/width 85-100/75-110; IV-L-6 subterminal seta length 75-105 µm; genital field occasionally with a small gap between Ac-1 and Ac-2, lateral genital flap margins smoothly convex. Females: Idiosoma length/width 570-1150; posterodorsal plate length/width 100-120/75-120; IV-L-6 subterminal seta length 95-135 µm; genital field with Ac-1 and Ac-2 closely adpressed, lateral genital flap margins subparallel.

Similar species: In shape of coxae, gnathosoma and chelicerae, and in setation of legs (with strong serrated distal setae on I-IV-L-3-5, subterminal seta on IV-L-6 long) similar to *N. robusta*, recorded from the Eastern Mediterranean and the Arabian peninsula. This species differs in the stouter subterminal seta on IV-L-6 (basal width 6-8 µm, length/width ratio < 15.0), generally located closer to the segment tip; P-2 with long ventral seta (length > ventral segment length) and strong denticulation extending to the area proximal to the ventral seta insertion, P-4 more robust (length/height ratio 3.2-3.7), as long as P-2.

Preadult stages: Larva: GERECKE & SMITH (1993). Deutonymph: K. VIETS (1950d).

Habitat and Biology: Crenobiont. In weakly seeping helocrenes much exposed to sunlight. Also in habitats deteriorated by pasturing (GERECKE 1991a). Larvae reared in the laboratory selectively attracted by, and attaching to, ceratopogonid pupae (ignoring chironomids and dixids), no further development observed (GERECKE & SMITH 1996).

Distribution: Widespread in the Mediterranean. Presence in the treated area uncertain. Presence in the Alps (Limnofauna area 4) recorded by K. O. VIETS (1978), but no corresponding locality record published. The specimen recorded under this name from Poland (BIESIADKA 1973) depicted with a genital field typical for the subgenus *Dartia*. If sex determination is correct, this would be a male with unmodified II-L, thus representing a species of uncertain subgeneric placement, as in *N. turcica* (see discussion of *Dartia*).

4. *Nilotonia (Dartiella) parva* (WALTER, 1931) (*8-40 c, g, i*)

1931c *Dartia parva* WALTER, Bull. Soc. Hist. natur. Afrique Nord 22: 340.

1967 *Nilotonia gallica* MOTAŞ, & C. ANGELIER Annls. Limnol. 3 (1): 27 – GERECKE & MORENO (2008).

Description: Colour pale orange to red, sclerotized plates greenish to blue. Integument with fine lineation, mostly in a longitudinal direction dorsally, but transverse on the flanks and ventrally. Posterodorsal plate round, slightly elongated or larger than long, occasionally "lemon-shaped" (with blunt anterior and posterior extensions), length/width ratio 0.8-1.2. Legs with serrated or pennate distal setae at I-IV-L-3-4, leg claws with a comb of 7-10 ventral denticles. Subterminal seta IV-L-6 short (length 25-45 µm), slightly extending, or not at all, over tip of segment (seta insertion 25-35 µm away from segment tip). Genital field with elongated Ac flanking gonopore all over its length. P-2 with relatively long ventral seta (ratio seta length/ventral segment length 0.6-1.1), ventral denticulation strong, partly extending to the segment surface proximal to ventral seta insertion, P-4 relatively slender (length/height ratio 4.1-4.3), distinctly longer than P-2, ventral group of denticles and setae away from distal segment margin, dividing segment in ratios 1.5-1.9. Males: Idiosoma length/width 700-1000; posterodorsal plate length/width 95-130/100-120 µm; lateral genital flap margins smoothly convex. Females: Idiosoma length/width 1200-1300/900-1000; posterodorsal plate length/width 145-180/130-155 µm; lateral genital flap margins subparallel.

Habitat and Biology: Springs and low order streams; in the latter, probably bound to riparian seepage affluents. Halotolerant (in a Spanish stream with a mean annual conductivity of 12.8 mS/cm, GERECKE & MORENO 2008). Preadult stages and life cycle unknown.

Distribution: Western Mediterranean, Central Sahara (GERECKE & MORENO 2008).

5. *Nilotonia (Dartiella) pyrenaica* LUNDBLAD, 1956 (*8-39 c, f, 8-40 b, f, h*)

1956a *Nilotonia pyrenaica* LUNDBLAD, Ark. Zool. 10 (1): 82.

1991a *Nilotonia cf. pyrenaica* LUNDBLAD, 1956 – GERECKE, Lauterbornia 7: 97.

Description: Colour bright red, sclerotized plates blue. Integument with strong lineation, lines on dorsum directed longitudinally, on posterior venter transversally. Posterodorsal plate round, slightly elongate or slightly transverse (length/width ratio 0.7-1.1). Legs with strongly serrated distal setae at I-IV-L-3-4. Subterminal seta IV-L-6 short (length 20-40 µm), not or slightly extending beyond segment tip. Genital field with elongated Ac flanking gonopore all over its length. P-2 with short ventral seta (length about 1/2 ventral segment length), ventral denticulation indistinct, restricted to the area anterior to ventral seta insertion, P-4 very slender (length/height ratio 4.1-5.5), distinctly longer than P-2, ventral group of denticles and setae located close to distal segment margin, dividing segment in ratios 1.9-2.6. Males: Idiosoma length/width 870-1050/760-890; posterodorsal plate length/width 120/90 µm; lateral genital flap margins smoothly convex. Females: Idiosoma length/width 700-1250; posterodorsal plate length/width 70-120/80-130 µm; lateral genital flap margins subparallel.

Similar species: Variability described only for two populations collected in Sicily. In the holotype and only specimen known from the Pyrenees, characters such as the leg claw comb and ventral denticles on P-2 are hardly visible due to the deteriorated mounting medium, and the position of the ventral denticle on P-4 is at the variability limit of the species (proximal/distal sector 1.9 – GERECKE 1991a). Thus, *N. pyrenaica* cannot be excluded as a junior synonym of *N. parva*, while the Sicilian specimens could represent an undescribed species absent from the area treated.

Habitat and Biology: Crenobiont. In weakly seeping helocrenes with strong insolation. Preadult stages and life cycle unknown.

Distribution: Pyrenees, Sicily.

Genus *Sigthoria* KOENIKE, 1907

1907a *Sigthoria* KOENIKE, Abh. Naturw. Ver. Bremen 19: 127.

1969 *Pseudatractides* NAYAR, Bull. Ent. 10: 49 – K. O.VIETS (1987).

Type species: *Amasis niloticus* NORDENSKIÖLD, 1905, by original designation.

Diagnosis: Dorsal shield complete and with characteristic sculpturing, with postocularia and five pairs of glandularia. Lateral eyes incorporated into ventral shield. Coxal plates fused with postgenital sclerotization to a ventral shield; Cx-IV subrectangular, closely flanking genital field, each with a large glandularium near the middle, posterior margins anterior to posterior margin of genital field. I-III-L with fine claws bearing ventral denticles, IV-L with a few swimming setae, without claws; IV-L insertions at posterior end of a well-developed ridge, on the level of genital field, covered by a small rounded projection ("condyle"); other legs with small claws. Genital field with 10-17 Ac; hair like setae of ventral margin of P-2 on a small tubercle; basal segment of chelicera without distinct dorsal elevation, proximal end bent ventrally.

Remarks: Monotypic genus of uncertain relationships, from several points of view agreeing with *Rheolimnesia* and *Siboneyacarus*, two neotropical genera of unclear taxonomic placement (GOLDSCHMIDT 2001, PANESAR 2004).

Sigthoria nilotica (NORDENSKIÖLD, 1905) (*8-37 f-h*)

1905 *Amasis niloticus* NORDENSKIÖLD, Results Swedish zool. Exped. Egypt and White Nile 1901 2: 9.

1906 *Amasis minimus* PIERSIG, Zool. Jbch., Syst. 23: 351 – LUNDBLAD (1969).

1907a *Sigthoria nilotica* KOENIKE, Abh. naturw. Ver. Bremen 19 (1): 127.

1969 *Pseudatractides pyriformis* NAYAR, Bull. Ent. 10: 50 – K.O.VIETS (1987).

Description: Idiosoma broadly oval in shape (length/width ratio 1.2-1.3). Swimming setae numbers: IV-L-3, 1-2; IV-L-4, 2-3; IV-L-5, 1-2. Males: Length dorsal shield 510-540; ventral shield 535-555 µm; Females: Length dorsal shield 545-590; ventral shield 595-640 µm.

Similar species: Harvey (1990) noted some minor differences in the setation of IV-L between specimens from Africa, Asia and the Northern Territory on the one hand, and specimens from Queensland and Victoria (Australia) on the other. However, at present, only a single species is recognized.

Preadult stages: Larva unknown. Deutonymph: Harvey (1990).

Habitat and Biology: All kinds of standing and slowly flowing waters, e.g. lakes, ponds, ditches, large rivers. Also reported from a hot spring (Harvey 1990). Life cycle unknown.

Distribution: Widespread in tropical regions: West Africa (Liberia, Ghana), East Africa (Sudan, Uganda, Kenya, Madagascar), Asia (Burma, Indonesia), New Guinea and Australia (Harvey 1990). From the area covered only known from The Netherlands (Smit & Van der Hammen 1992). In the European fauna possibly a neozoan.

Family Lebertiidae Thor, 1900

1900b Lebertiidae Thor, Nyt Mag. Naturv. 38: 2.

Diagnosis: Idiosoma oval or spherical in shape; dorsal and posteroventral integument soft, muscle-attachments generally unsclerotized (rare exceptions: dorsum with small muscle attachments or a complete dorsal shield, coxal sclerites fused into a shield extending to the posteroventral, lateral or even dorsal surface); lateral eyes in capsules; coxal shield covering the anteroventral surface of idiosoma, posteriorly approaching or completely surrounding genital field, with a characteristic Y-shaped suture extending from anterior edge of genital bay to gnathosomal bay; suture line between Cx-II and -III interrupted centrally, opening of glandularium C-4 located anteriorly, near tips of Cx-I and flanking the gnathosomal bay; leg insertions not set far forward, insertions of IV-L not associated with extensions of the coxal surface, easily seen in ventral view; all legs bearing well-developed claws; swimming setae present on II-L-5, III-L-4/5, and IV-L-4/5, or (in many species) completely absent; genital field with movable flaps and three (in one exceptional case six) pairs of elongate Ac flanking the gonopore; gnathosoma bearing slender chelicerae with small claws, palps with P-1 bearing one short (exception: long in *Scutolebertia*) dorsodistal seta; P-2 bearing one (or two in *Paralebertia*) strong ventrodistal seta(e) and several generally short (but long in *Scutolebertia*) dorsal setae, P-3 with four to nine long setae, in most cases restricted to the medial and dorsal surface, but occasionally located laterally or in a dorsal row; P-4 with two fine ventral setae and a short, peg-like mediodistal seta, additional setae in various numbers mostly on the dorsal surface, rarely (e.g. in *Scutolebertia*) on other parts of the segment.

Discussion: Most members of the family Lebertiidae are rather uniform in appearance and easily distinguished from other taxa of Hydrachnidia by the characteristic shape of the coxal shield. Oxidae K.Viets, 1926l, distributed on all continents except Antarctica, the monotypic family Acucapitidae Wiles, 1996, known only from Brunei (Borneo) and Vietnam (Tuzovskij 2009), and Lebertiidae probably represent a monophyletic group of families. Wiles (1996) suggested these three taxa could be included as subfamilies in a single family Lebertiidae. However, as only little information is available on some important aspects of these groups, e.g. concerning larval morphology and life histories, the three families were maintained by Gerecke (2009).

The relationships of these three families to other lebertioid families are still unclear. The probably polyphyletic family Anisitsiellidae includes numerous taxa showing different grades of anterior displacement of C-4. Members of the family Torrenticolidae agree with Lebertiidae, Oxidae and Acucapitidae in having this glandular opening on the tip of Cx-I. The presence of a Y-shaped suture on the coxal field is a character state shared by species of Lebertiidae and Torrenticolidae and obviously secondarily reduced in Oxidae. This may indicate a sister group relation between Torrenticolidae and the Lebertiidae + Oxidae + Acucapitidae lineage (GERECKE 2009).

In the Western Palaearctic, the family is represented exclusively by the species-rich genus *Lebertia*. Two other genera, *Scutolebertia* SMITH, 1991 from North America, and *Paralebertia* TUZOVSKIJ, 2003 from Eastern Russia, are monotypic. They are both characterized by a plesiomorphic setation of palp segments, with setae in higher numbers and irregular arrangement, obviously also subject to considerable individual variability. On the other hand, they display also apomorphic character states such as the development of a dorsal shield and fusion of lateral eye capsules to the coxal shield in *Scutolebertia* (GERECKE 2009).

Genus *Lebertia* NEUMAN, 1880

1880b *Lebertia* NEUMAN, Svensk. Vetensk.-Akad. Handlingar, Stockholm (N. F.) (1879) 17: 68.

1897c *Pseudolebertia* THOR, Arch. Math. Naturv. 20 (3): 19 – GERECKE (2009).

1963c *Estelloxus* HABEEB, Leaflets Acadian Biol. 32: 1 – GERECKE (2009).

Type species: *Pachygaster tauinsignitus* LEBERT, 1879, by original designation.

Diagnosis: Idiosoma generally not or only slightly laterally compressed, with a clear-cut subdivision into an extended membranous area on the dorsal and posteroventral surface, and a well-sclerotized coxal shield more or less embracing the genital field, only in exceptional cases extending on to the dorsal surface; muscle attachments unsclerotized, very rarely (not in European species) appearing as small platelets or forming a dorsal shield; suture lines Cx-II/III forming an acute angle at anterior margin of genital bay and interrupted at level of III-L insertions, suture lines Cx-III/IV originating laterally and directed medially at an obtuse angle, obliterated before reaching genital bay; swimming setae present or absent; if present, arranged as single setae at distal margins or in rows in distal half of II-L-4, III-L-4/5 and IV-L-4/5. Gnathosoma with well developed proximodorsal apodemes, chelicerae long and slender, with thin, weakly curved claws; palp without ventral projections on P-2-4; P-1 bearing one short dorsodistal seta, P-2 one strong, pennate ventrodistal seta and several dorsal setae, P-3, 5-8 long, hair-like setae on medial and dorsal surfaces – one (rarely 2) of them near ventrodistal edge, 2 at distal edge in medial or dorsal position, and another 2 or 3 (rarely up to 5) on dorsal or mediodorsal surface; P-4 with 2 short and slender ventral setae and a short, peg-like mediodistal seta.

Discussion: Compared with other water mite genera, most *Lebertia* species are highly uniform in the shape of coxae, legs and mouth parts. Corresponding to the low degree of morphological diversity in the shape of the coxal shield, the way of insertion of legs is the same in all species, and all leg claws have a large claw blade and a well developed ventral clawlet. Characters useful for discrimination of species and subgenera are mainly integument structure, as well as details in setation and shape of legs and palps. Glandularia are always simple in structure, uniform with both the rather short setae and glandular opening inserting on a round sclerotized platelet. A particular extension, or even fusion of neighbouring glandular platelets is never found, but elements close to the coxal margins can fuse to them when the coxal sclerotization extends. Only a few examples of remarkable sexual dimorphism are known, mostly involving extension of coxae and shape of palps.

General characters of larval *Lebertiae* are described and discussed e.g. by LUNDBLAD (1930a), K.VIETS (1936d), SPARING (1959), PRASAD & COOK (1972), MULLEN (1974), WAINSTEIN (1980) and SMITH (1982). At this stage, there are obviously characters suitable

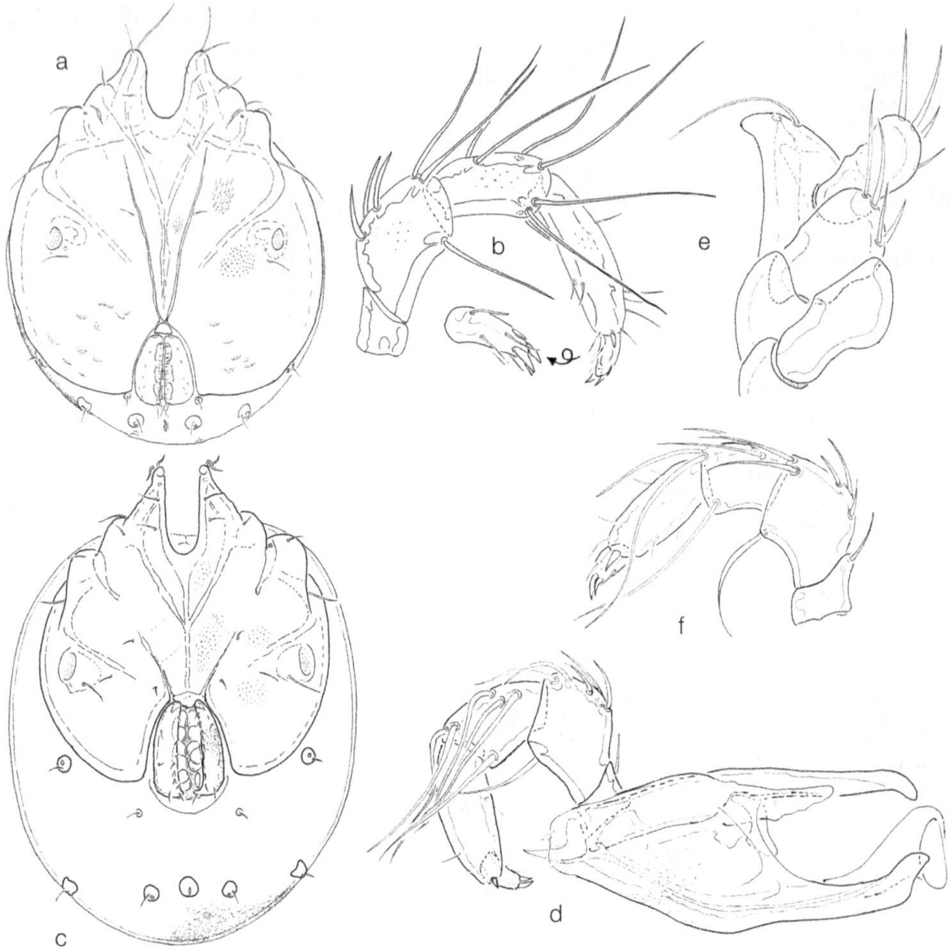

8-41: *Lebertia*, subgenera *Eolebertia* and *Brentalebertia* (from GERECKE 2009, modified); a-b, *L. elsteri*; a, male venter; b, female palp; c-d *L. hygropetrica*; c, female venter; d, gnathosoma and palp; e-f *L. minutipalpis*; e, female tips of anterior coxae, lateral view; f, male palp, medial view.

for separation at species level, but older descriptions are generally incomplete, or the attribution to species remains uncertain.

Mites of the genus reach their highest diversity in the Holarctic, but isolated species have been recorded in several parts of South Asia (India, China), Africa (including South Africa) and South America. This recent distribution pattern suggests that the family originated in Pangaea and diversified under a cold and temperate climate in the Northern Hemisphere. Tundra ponds and helocrenic springs are possible plesiotypic habitats from where radiation and adaptation to other kinds of freshwaters took place.

In a revision of the European lebertiid fauna GERECKE (2009) accepted five of the formerly established subgenera, several of them probably not monophyletic, and introduced one subgenus new to science. *Septlebertia* IMAMURA, 1954b, the only subgenus absent from the area treated, is known only from two Asian species, which, apart from the presence of the name-giving additional seventh seta on P-3, from many points of view agree with typical *Mixolebertiae.*

Deutonymphal morphology of *Lebertia* species can often be deduced from adult character states, and therefore little attention has been paid to describing this stage. In the following treatment, descriptions of deutonymphs are mentioned only if they give important information on species characters at this stage.

Key to subgenera

1 P-3 bearing two ventrodistal setae, located on the medial and lateral surface (*8-41 b*); total setae number on P-3 six to eight; ventral suture lines Cx-II/III in their posterior part forming an acute angle (< 30°, *8-41 a*). . . . . *Eolebertia* (page 96)

\- P-3 bearing generally one ventrodistal seta, if an additional seta is present, both are located on the medial surface of the segment (*8-42 e*), total setae number on P-3 five to six (exceptionally, in a highly variable population of a species of *L. (s. str.)*, elevated numbers of P-3 setae were found; this species differs among others in having deep red sclerotized parts); ventral suture lines Cx-II/III posteriorly more divergent (angle ≥ 30°, e.g. *8-41 c, 8-42 a*). 2

2 (1) Tips of Cx-I pointed and ventrally bent (*8-41 e*). Acetabula may be divided, resulting in six pairs (*8-41 c*). Palp stout (length/height P-2 < 1.3, P-3 < 1.9, P-4 < 2.7); gnathosoma slender and elongate (*8-41 d*). . . . . *Brentalebertia* (page 95)

\- Tips of Cx-I directed anteriorly, not bent ventrally, in general more truncate (*8-42 a, 8-43 e*). Acetabula never subdivided into six pairs. Palp more slender (generally length/height P-2 > 1.2, P-3 > 1.4, always length/height P-4 > 2.5); gnathosoma shorter and stouter (*8-42 c*). 3

3 P-3 bearing six setae (as in *8-41 d*, see also *8-47 e, h*). Excretory pore with or without a sclerotized ring. In males Cx-IV may extend beyond, or embrace, posterior margin of genital field (*8-50 c, 8-51 a*). . . . . *Mixolebertia* (page 117)

\- P-3 bearing five setae (*8-42 c* – exceptionally, in a highly variable population of *L. rufipes*, elevated numbers of P-3 setae were found; this species differs from most *Mixolebertia* species in having deep red sclerotized parts). In most species, excretory pore smooth and Cx-IV not extending beyond posterior margin of genital field in both sexes (*8-42 a, 8-43 e*). . . 4

4 P-4 distally narrowed, with dorsal setae concentrated in distal quarter of segment (*8-42 e*). Swimming setae numerous (minimum numbers: II-L-5, 2; III-L-4, 2; III-L-5, 3; IV-L-4, 2; IV-L-5, 5). . . . . *Pilolebertia* (page 140)

\- P-4 various in shape, but always some of the dorsal setae further from tip, in the centre or proximal half of segment (*8-42 c*). II-III-L only rarely with numerous swimming setae (apart from the exceptional cases of *Lebertia natans*, *L. oblonga* and *L. brigantina*, maximum numbers are: II-L-5, 1; III-L-4, 2; III-L-5, 6; IV-L-4, 2; IV-L-5, 5). . . *Lebertia s. str.* (page 97)

Subgenus *Brentalebertia* Gerecke, 2008

2008 *Brentalebertia* Gerecke, Fund. Appl. Limnol., Arch. Hydrobiol. 170 (4): 325.

Type species: *Lebertia hygropetrica* Gerecke, 2008, by monotypy.

Diagnosis: Coxal field anteriorly narrowed, and Cx-I with distinct downwardly curved tips (*8-41 e*). Legs robust, swimming setae absent or weakly developed; distal segments of I-L with numerous fine ventral setae, II/III-L-3/4 distally with numerous strong and stout sword setae. Leg claws with ventral clawlet and claw blade. Acetabula simple, or doubled, but resulting 3 pairs on each side of gonopore on a common basal sclerite. Excretory pore sclerotized or smooth. Gnathosoma flat and slender in lateral view, palp stout (length/height P-2, 1.2-1.3; P-3, 1.3-1.5; P-4, 1.9- 2.5; total length < 270), P-3 bearing 5-6 medial

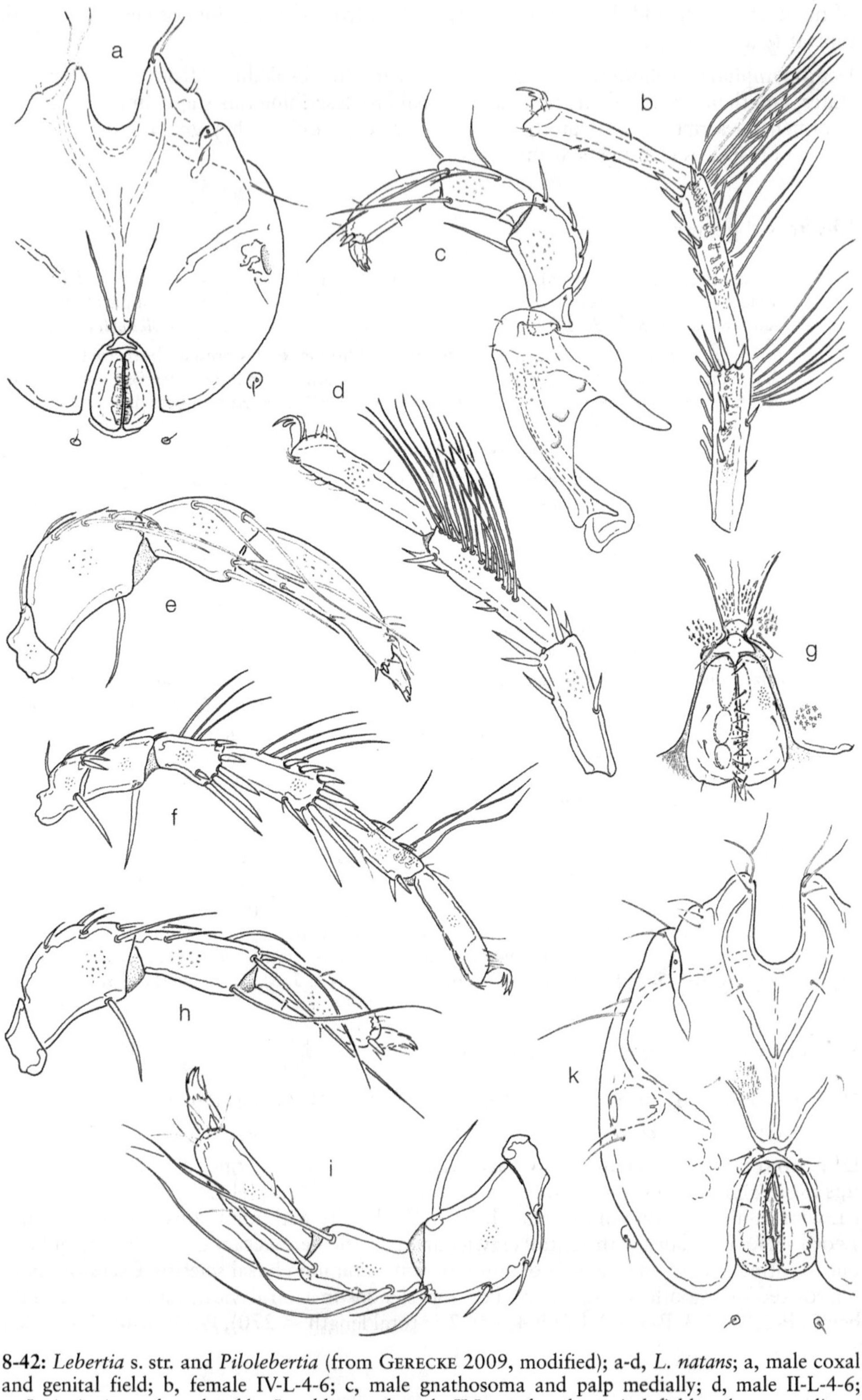

8-42: *Lebertia* s. str. and *Pilolebertia* (from GERECKE 2009, modified); a-d, *L. natans*; a, male coxal and genital field; b, female IV-L-4-6; c, male gnathosoma and palp medially; d, male II-L-4-6; e, *L. insignis*, male palp; f-h, *L. oblonga*; f, male II-L; g, female genital field and surroundings; h, male palp; i-k, *L. brigantina*; i, female palp; k, male coxal and genital field.

setae, P-4 shortened, with convex dorsal and ventral margins, maximum height in centre of segment near insertion of proximoventral seta.

Similar species: Members of the subgenus are distinguished from representatives of all other subgenera of *Lebertia* in the tips of Cx-I, distinct and downwardly curved, and the elongated gnathosoma. A palp with similar small dimensions (total length < 270) is only found in representatives of the *oblonga*-group of *Lebertia* s. str., but differs in P-2 and P-3 being stouter. A similarly stout P-4 is found in representatives of the *cuneifera*-group of *Mixolebertia* which however differ from *Brentalebertia* in the distally narrowed, curved shape of this segment.

Key to species

1 Excretory pore sclerotized. Acetabula subdivided, resulting in six pairs (*8-41 c*). P-3 with six setae; P-3 (length/height > 1.5) and P-4 (length/height > 2.5) relatively slender, length ratio P-2/P-4, 0.7 (*8-41 d*). . . . . *Lebertia hygropetrica* (page #)

\- Excretory pore unsclerotized. Acetabula not subdivided, but often slightly indented laterally. P-3 with five setae. P-3 (length/height < 1.5) and P-4 (length/height < 2.5) stout, length ratio P-2/P-4, 0.8-1.1 (*8-41 f*). . . . . *Lebertia minutipalpis* (page #)

1. *Lebertia (Brentalebertia) hygropetrica* GERECKE, 2008 (*8-41 c-d*)

2008 *Lebertia (Brentalebertia) hygropetrica* GERECKE, Fund. Appl. Limnol., Arch. Hydrobiol. 170 (4): 325.

Description: Colour yellowish to brown. Dorsal integument lines short, locally disintegrated to longish papillosity. Lateral eye capsules fused with neighbouring glandularia. Medial sutures Cx-I/II equal in length or Cx-I > Cx-II. Acetabula doubled (occasionally trebled) by transversal constriction, but resulting 6 pairs on each side of gonopore maintaining a common basal sclerite. Excretory pore sclerotized. P-2 with a robust seta arranged on the bulging ventral surface; P-3 dorsal setae insertions in a triangle in proximal half of segment, mediodistal seta close to dorsodistal seta; P-4 with convex dorsal and ventral margins, maximum height in centre of segment, ventral sectors 3:2:2. P-2/4 length ratio 0.7-0.8. Males: Idiosoma length/width 570-600/350-480 µm; Cx-IV with pointed medioposterior tips including V-1, embracing genital field; genital flap length 120-125; gnathosoma length 173 µm. Females: Idiosoma length/width 670-930/410-605 µm, Cx-IV with rounded medioposterior margins, V-1 on small, round platelets; genital flap length 159-180, gnathosoma length 200 µm.

Similar species: *Lebertia hygropetrica* differs strikingly from all other species of the genus in the doubled Ac.

Preadult stages: Larva unknown. Deutonymph: At least Ac-1 doubled (GERECKE 2008).

Habitat and Biology: Mossy cascade spring with strong discharge and extended hygropetric margins. Life cycle unknown.

Distribution: Italian Alps, Adamello-Brenta Natural Park; only known from the type locality.

2. *Lebertia (Brentalebertia) minutipalpis* K.Viets, 1920 (*8-41 e-f*)

1920b *Lebertia (Neolebertia) minutipalpis* K. Viets, Arch. Hydrobiol. 12 (4): 811.

Description: Integument smooth, with a very fine striation; Cx-II very short (medial length ratio Cx-I/II > 2.0) and posteriorly wide (width 50-60 µm); III-L-5 and IV-L-5 each with one short "swimming" seta (about 2/3 length terminal segment); genital flaps with medial setae low in number (about 20 on each side), Ac-1-3 rather short and placed on elevated and strong basal sclerites, at least some of them slightly constricted in the centre; excretory pore smooth; gnathosoma elongated; palp stout and compact (total length 230-250 µm), but P-1 relatively slender; P-2 humped (length/height < 1.3), P-3 length/height 1.3-1.4; P-4 ventral sectors 3:2:2 or 2:1:1, P-5 stout and short. Males: Idiosoma length/width 800/600 µm, genital flap setae scattered in anterior part, more dense, but only slightly elongated, in posterior part. Females: Idiosoma length/width 1000/700 µm; genital flap setae not more dense posteriorly than anteriorly.

Similar species: *Lebertia (Brentalebertia) hupalupu* Gerecke, 2009 from the Canary island of Gomera, similar in the smooth excretory pore and P-3 bearing only 5 setae, differs in Cx-I and Cx-II medially nearly equal in length, Cx-II with narrower posterior margin, and P-2 more slender (length/height 1.3), as long as P-4.

Habitat and Biology: Crenobiont, in weakly seeping helocrenes. Preadult stages and life cycle unknown.

Distribution: Central and Western European lowlands, Carpathians, rare.

Subgenus *Eolebertia* Gerecke, 2009

2009 *Eolebertia* Gerecke, Abh. Senck. Naturf. Ges. 566: 132.

Type species: *Lebertia elsteri* Schwoerbel, 1957a, here designated.

Diagnosis: Coxal field elongated and slender; Cx-I/II posteriorly forming an acute angle; IV-L-4 bearing one, IV-L-5 two short hair-like setae. Posteromedial margins of Cx-IV not including V-1; excretory pore not sclerotized; gnathosoma and chelicera robust and stout; P-2 with short and stout ventral seta, 3 robust dorsocentral setae and two finer and longer dorsodistal setae; P-3 with 6-8 long, basally very robust setae (2 to 3 dorsocentrally – at least one of them on lateral surface, 2 dorsodistally – one laterally, one medially, and 2 to 3 ventrodistally – at least one laterally, the other medially); P-4 with 5-6 irregularly scattered, hair-like dorsal setae, both ventral setae in distal segment half, arranged close to each other, or apart and dividing the segment 2:1:1; peg-like seta well developed, long and pointed. P-5 long and slender.

Similar species: Obviously an early derivative clade with members differing from all other *Lebertia* species in the plesiomorphic presence of two ventrodistal setae on P-3, as in the sister genus *Paralebertia*. In that genus, as well as in the subgenera *Eolebertia* and *Brentalebertia*, a certain variability in palp setation has been preserved, while all other subgenera evolved towards a highly stable scheme from which only single taxa or populations may differ under particular conditions. *Eolebertia* species are furthermore distinguished by a rather long terminal palp segment, elongated coxae and a relatively short genital field. The Eastern Asiatic species of the subgenus *Septlebertia* Imamura, 1954b agree in the elevated number of setae on P-3, but they are similar to typical *Lebertia s. str.* from all other points of view. Only one species in the area treated.

3. *Lebertia (Eolebertia) elsteri* SCHWOERBEL, 1957 (*8-41 a-b*)

1957a *Lebertia (Septlebertia?) elsteri* SCHWOERBEL, Mitt. Bad. Landesver. Naturkde. Naturschutz (N. F.) 7 (1): 43.

Description: Colour pale orange to brownish. Integument with fine short lines alternating with porose interspaces or smooth; Cx-I/II elongated and slender, suture lines Cx-II/III forming a very acute posterior angle; median length ratio Cx-I/II 0.8-1.1; Cx-IV enlarged (*8-41 a*); I-L and II-L terminal segments stout, IV-L-6 with 3-6 ventral setae; genital field in both sexes relatively small, embraced by strongly concave posteromedial margins of Cx-IV; posteromedial edges of Cx-IV right-angled; gnathosoma and chelicera robust and stout; P-2 ventrally concave, length/height ratio 1.6-1.7, P-2 and P-3 nearly equal in length; P-3 robust (length/height 1.9-2.1), P-4 maximum height in distal part, ventral margin straight; P-2/P-4 length ratio 0.7-0.8. Males: Idiosoma length/width 800/650-660 µm; Cx-II posteriorly pointed, pregenital sclerite triangular; gnathosoma ventral length 140-150, length/height P-2, 63-65/38-40; P-3, 63-65/30-33; P-4, 80-85/25 µm. Females: Idiosoma length/width 1000/800 µm; Cx-II sometimes pointed; pregenital sclerite rhombic; gnathosoma ventral length 150-155, length/height P-2, 68-73/35-38; P-3, 68-75/35-38; P-4, 88-98/28 µm.

Similar species: *Lebertia danielei* GERECKE, 2009 from a spring in Northern Italy, the only other known species of the subgenus, differs in shape of coxal field (Cx-IV not extended, sutures Cx-II/III forming a less acute angle, Cx-II medially shorter, posteromedial angles of Cx-IV not pointed) and palp (P-2 inflated and relatively longer – length ratio P-2/4, 1.0-1.1, setae of P-3 extending far beyond tip of P-5).

Habitat and Biology: Crenophilous, with a preference for interstitial habitats in the ecotone between spring and first order stream (GERECKE & MARTIN 2006). Preadult stages and life cycle unknown.

Distribution: Central-southern Europe (GERECKE 2009).

Subgenus *Lebertia s. str.* NEUMAN, 1880

Diagnosis: Medial surface of P-3 with five setae. P-4 bearing fine dorsal setae in proximal and distal part of segment. Legs with or without swimming setae, but only in exceptional cases with high numbers of such setae. Species generally little sclerotized, Cx-IV in males rarely extending posteriorly beyond genital field, excretory pore in most species unsclerotized.

Discussion: The subgenus *Lebertia s. str.* includes species combining a palp with setae of P-3 reduced to the lowest number possible in *Lebertia*, five (apomorphic), and dorsal setae on P-4 not restricted to a distal pit near insertion of P-5 (plesiomorphic). As demonstrated both by the variable P-3 setation in *L. s. str. rufipes*, as well as by the divergent evolution of this character in representatives of *Brentalebertia* and among two sister species from Madeira, *L.* (*Mixolebertia*) *madericola* LUNDBLAD, 1942c and *L.* (*M.*) *maderigena* LUNDBLAD, 1942c, the number of five setae may be reached independently in distant clades and is not a reliable basis for interpreting phylogeny within the genus. Probably, *Lebertia s. str.* is a paraphyletic or polyphyletic clade, a dumping ground for species from all parts of the distribution area of the genus, not to be attributed to one of the other subgenera.

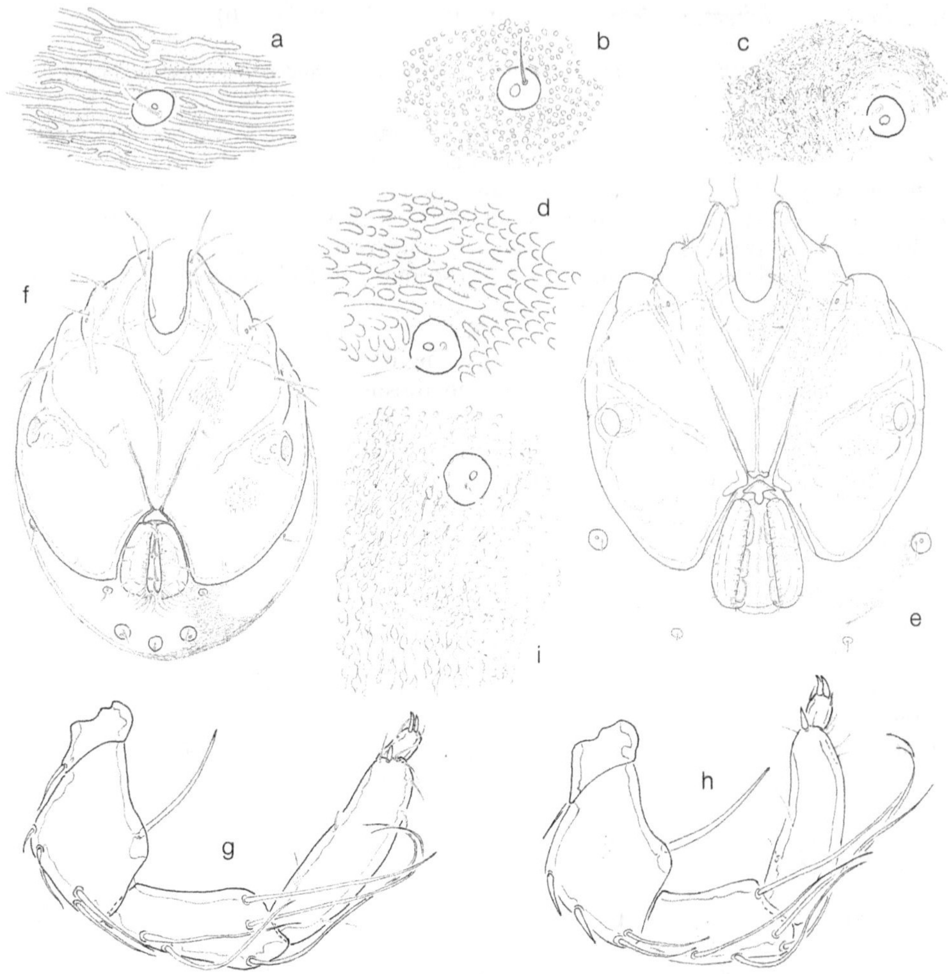

8-43: *Lebertia* s. str. (from Gerecke 2009, modified); a-e, integument structure; a, *L. glabra* male; b, *L. schechteli* male; c, *L. salebrosa* male; d, *L. maculosa* male; e, *L. fimbriata* female, coxal and genital field; f-g, *L. lusitanica* male; f, venter; g, palp; h-i, *L. guttata* female; h, palp; i, integument structure.

Key to species

The following (sub)species are excluded as *species incertae* (Gledhill & K. O.Viets 1976; Gerecke 2009): *Lebertia areolata* Halbert, 1944; *L. brunnea* Halbert, 1944; *L. compacta* Halbert, 1944; *L. dalmatica fragilis* Halbert, 1944; *L. dresdensis* K.Viets, 1928f; *L. elliptica* Husiatinschi, 1937d; *L. flumenia* Halbert, 1944; *L. gladiator* Thor, 1913a; *L. hirtipalpis* Halbert, 1944; *L. komareki* Thor, 1914c; *L. lacustris* Koenike, 1914f; *L. laticoxalis* K.Viets 1949b; *L. minuta* Halbert, 1944; *L. mollis* Halbert, 1944; *L. monardi* Migot, 1926a; *L. multicincta* Thor, 1914b; *L. pulchella* K.Viets, 1925g; *L. schechteli globifera* Thor, 1914c; *L. tauinsignita* (Lebert, 1879); *L. tenuipalpis* Halbert, *1944; L. walteri* Thor, 1906e.

1 Numerous swimming setae on II-III-IV-L. II-L-5 bearing at least two, often three to ten, swimming setae on posterior distal surface (*8-42 d*); these setae in general as long as, but occasion-

ally only 1/3 length of subsequent segment II-L-6. III-L-4 generally with five or more setae in a line in distal part of posterior surface, as long as subsequent segment, III-L-5. 2

- Swimming setae reduced in number or absent. II-L-5 at most with one short distal hair-like seta on posterior surface. III-L-4 bearing at most five fine distal setae on posterior surface. 4

2 (1) I-L-3/4 and II-L-3/4 without swimming setae on dorsal surface, distal setae on II-L-3/4 (II-L-4: *8-42 d*); and III-L-3/4 poorly developed. Numbers of posterior swimming setae: II-L-4, > 8; III-L-4, > 5; III-L-5, > 10; IV-L-4, > 5; IV-L-5, > 10. P-2 shorter than P-4 (*8-42 c*, length ratio P-2/P-4 < 0.9). . . . . . *Lebertia natans* (page 107)

- I-L-3/4 and II-L-3/4 each provided with three to four long swimming setae on dorsal surface; II-L-3/4 and III-L-3/4 with a collar of strong distal sword setae (*8-42 f*). Numbers of posterior swimming setae: II-L-4, < 8; III-L-4, < 5; III-L-5, < 10; IV-L-4, < 5; IV-L-5, < 10. P-2 and P-4 equal in length (length ratio P-2/P-4 > 0.9, *8-42 h-i*). 3

3 (2) Cx-II with shorter medial suture (length < 120 µm, Cx-I/II median length ratio 1.0-1.4), posteriorly truncated (*8-42 k*). P-3 relatively stout (length/height < 2.1, *8-42 i*). In males, posterior margin of Cx-IV extending beyond posterior margin of genital field. . . . . . *Lebertia brigantina* (page 103)

- Cx-II with longer medial suture (length > 130 µm, Cx-I/II median length ratio 0.8-1.0), posteriorly pointed (*8-42 g*). P-3 more slender (length/height > 2.1, *8-42 h*). In both sexes, posterior margin of Cx-IV anterior to posterior margin of genital field. . . . . . *Lebertia oblonga* (page 111)

4 (1) Integument surface sculptured (lines: *8-43 a*, papillae: *8-43 b, d*, reticulation: *8-43 c*); interspace between surface structures finely porose. 5

- Integument surface smooth, generally finely porose, occasionally finely striated (*8-43 e*). 11

5 (4) Excretory pore surrounded by a sclerotized ring (*8-43 f*). 6

- Excretory pore unsclerotized. 7

6 (5) Cx-I/II nearly equal in medial length (ratio 1.0-1.1, *8-43 f*). Mediodistal seta on P-3 away from dorsodistal seta (*8-43 g*). . . . . . *Lebertia lusitanica* (page 108)

- Cx-I medially longer than Cx-II (Cx-I/II median length ratio > 1.2). Mediodistal seta on P-3 close to dorsodistal seta (*8-43 h*). . . . . . *Lebertia guttata* (page 107)

7 (5) P-3 basally slender, equally enlarged towards distal margin, as long as, or even longer than, P-4 (length ratio P-3/P-4, 0.96-1.10, *8-44 a*). Ac prominent (*8-44 b*). Dorsal integument generally covered by cupuliform papillae which, in the centre are well separated, laterally and ventrally often forming ridges (*8-43 b*); occasionally a pattern of curved hooks and short lines may be present. . . . . . *Lebertia schechteli* (page 115)

- P-3 basally thicker, shorter than P-4 (length ratio P-3/P-4 < 0.95, *8-44 c*). Ac narrower. Dorsal integument covered by elongate tubercles (*8-43 d*), long lines well separated from each other (*8-43 a*), or fine lines forming a net-like structure (*8-43 c*). 8

8 (7) Cx-II medially equal in length as, or only slightly shorter than, Cx-I (median length ratio Cx-I/II, 0.7-1.2). Integument with fine lines dorsally forming an irregular often net-like structure (*8-43 c*). . . . . . *Lebertia salebrosa* (page 114)

- Cx-II medially distinctly shorter than Cx-I (median length ratio Cx-I/II > 1.2). Integument various. 9

9 (8) Integument covered by fine lines dorsally forming a net-like structure (*8-44 d*). . . . . . *Lebertia reticulata* (page 112)

- Integument covered by drop-shaped or elongate tubercles (*8-43 d*) or long lines well separated from each other (*8-43 a*). 10

10 (9) IV-L stout, IV-L-6 distally notably enlarged (*8-44 e*, length/height IV-L-5 ♂ < 5.6, ♀ < 5.8; IV-L-6 ♂ < 5.4, ♀ < 5.0); IV-L-5 bearing three to four ventral setae. Generally, Cx-I medially by far longer than Cx-II (median length ratio Cx-I/II > 1.6), posterior margin Cx-II broad (♂ > 40, ♀ > 60 µm). . . . . . *Lebertia maculosa* (page 109)

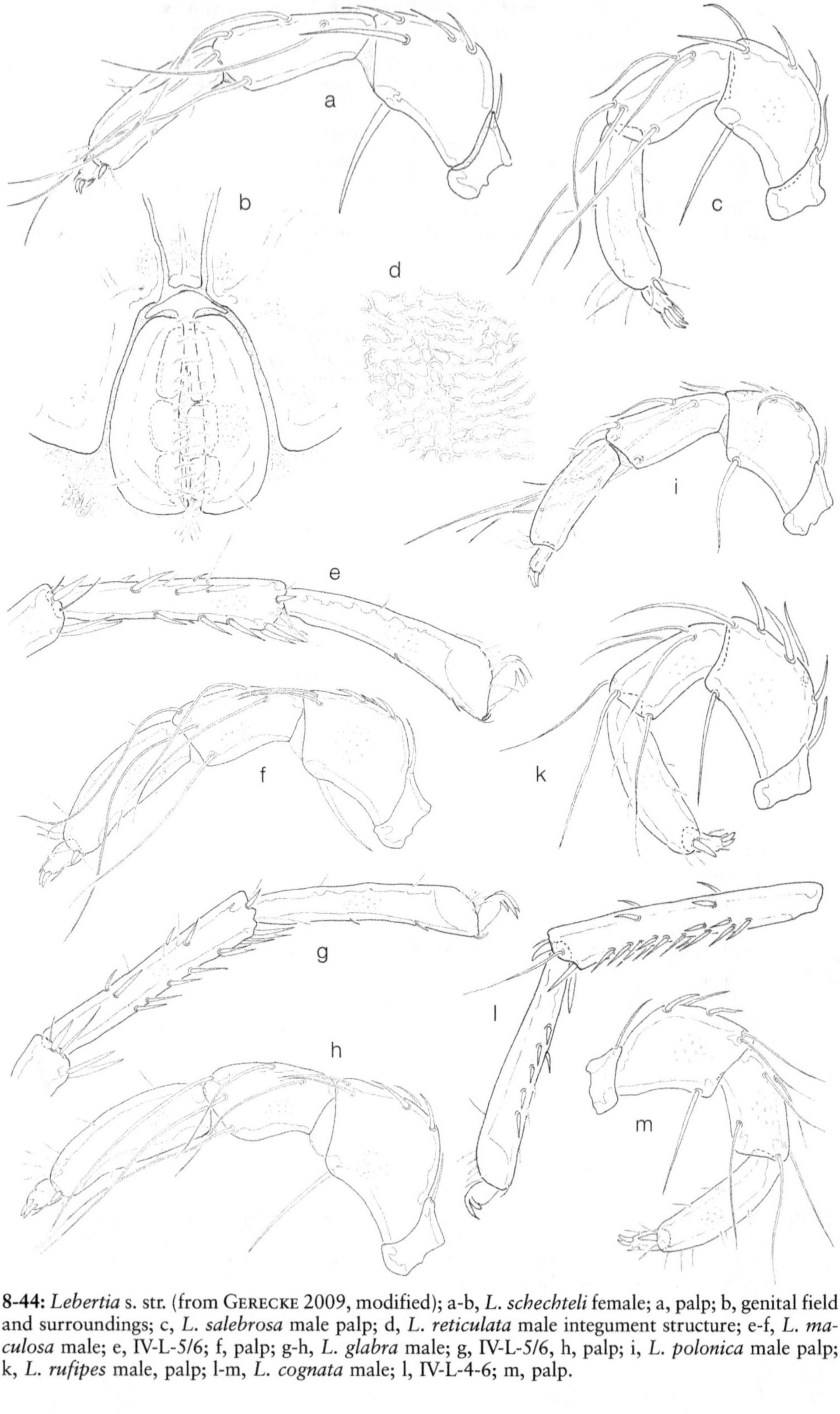

8-44: *Lebertia* s. str. (from GERECKE 2009, modified); a-b, *L. schechteli* female; a, palp; b, genital field and surroundings; c, *L. salebrosa* male palp; d, *L. reticulata* male integument structure; e-f, *L. maculosa* male; e, IV-L-5/6; f, palp; g-h, *L. glabra* male; g, IV-L-5/6, h, palp; i, *L. polonica* male palp; k, *L. rufipes* male, palp; l-m, *L. cognata* male; l, IV-L-4-6; m, palp.

- IV-L more slender, IV-L-6 not notably enlarged distally (*8-44 g*, length/height IV-L-5 ♂ > 5.6, ♀ > 5.8; IV-L-6 ♂ > 5.4, ♀ > 5.5); IV-L-5 bearing five to nine ventral setae. Generally, Cx-I medially shorter (median length ratio Cx-I/II < 1.6 – but in exceptional cases up to 2.0), posterior margin Cx-II narrower (♂ < 45, ♀ < 50 µm). . . *Lebertia glabra* (page 107)

11 (4) P-3 basally slender, enlarging towards distal margin (length/height 2.5-2.6), as long as, or even longer than, P-4 (length ratio P-3/P-4, 1.0-1.1) (*8-44 i*). Ac prominent (as in *8-44 b*). . . . . *Lebertia polonica* (page 111)

- P-3 basally less slender, and stouter (length/height < 2.5), shorter than P-4 (length ratio P-3/P-4 < 1.0, e.g. *8-44-m*). Ac slender. 12

12 (11) Ventral surface of IV-L-6 with 5-10 well developed setae (*8-44 l*). 13

- Ventral surface of IV-L-6 with one to two, rarely up to four setae, these setae generally small. 14

13 (12) Colour red, with intense colouration of sclerites, generally also visible in slide-mounted specimens; IV-L-5 with two long and slender distal setae; ventrodistal seta on P-3 close to distal edge of segment (*8-44 k*). Usually, palp slender (length/height P-3, 2.0-2.4; P-4, 3.4-4.4) and Cx-I medially longer (Cx-I/II median length ratio 1.5-2.0). . . . . *Lebertia rufipes* (page 114)

- Colour brown, sclerites pale green, IV-L-5 bearing only one long and slender distal seta, ventrodistal seta on P-3 away from distal edge of segment (*8-44 m*). Usually, palp stouter (length/height P-3, 1.9-2.1; P-4, 3.2-3.8) and Cx-I medially shorter (Cx-I/II median length ratio 1.2-1.6). . . . . *Lebertia cognata* (page 105)

14 (12) Distal margins of III-L-5 and IV-L-5 each bearing three or more long swimming setae, often reaching, or extending beyond, tip of terminal segment (*8-45 b*). 15

- Distal margins of III-L-5 and IV-L-5 bearing at most two swimming setae, generally distinctly shorter than terminal segment (e.g. *8-46 g*). 16

15 (14) III-L-5 and IV-L-5 each bearing three to four swimming setae (*8-45 b*). Palp shorter (P-4 length < 100 µm); P-3 with dorsal setae, as well as medio- and dorsodistal setae close to each other; P-4 maximum height distally, ventral setae mostly equidistant (sectors 1:1:1) (*8-45 a*). . . . . *Lebertia pusilla* (page 112)

- III-L-5 and IV-L-5 each bearing five swimming setae (*8-45 d*). Palp longer (P-4 length > 100 µm); P-3 with dorsal setae, as well as medio- and dorsodistal setae further apart; P-4 maximum height proximally, proximoventral seta shifted proximally (sectors 1:3:3) (*8-45 c*). . . . . *Lebertia rivulorum* (page 113)

16 (15) Male genital field with high numbers of medial setae (about 50 on each flap) in the posterior part long and arranged in two to three parallel lines (setae longer than genital flap width, *8-45 e*). Dorsal setae on P-3 apart (distance about 20 µm, *8-45 f*). Median length ratio Cx-I/II > 1.3; legs without swimming setae (questionable species). . . . . *Lebertia subtilis* (page 117)

- Male genital field with fewer medial setae, arranged in one single line over the whole length, setae not longer than genital flap width (e.g. *8-46 b*). Arrangement of dorsal setae on P-3, proportions of Cx-I/II, and swimming setae various. 17

17 (16) P-4 maximum height in proximal part, rarely in the centre, segment distally narrowed, ventral margin with a kink near insertion of distoventral seta, proximoventral seta insertion shifted proximally; P-3 with paired dorsal and medio-/dorsodistal setae apart (*8-46 a*, *c-d*). III/IV-L-5 with one swimming seta (always shorter than 50 % of the following segment) or lacking swimming setae. 18

- P-4 maximum height in distal part, rarely in the centre, segment distally not narrowed, ventral margin straight; P-3 with paired dorsal and medio-/dorsodistal setae close to each other (*8-46 e-f*). III/IV-L-5 with one or two swimming setae (*8-46 g* – in length sometimes more than 50 % of the following segment) or lacking swimming setae. 20

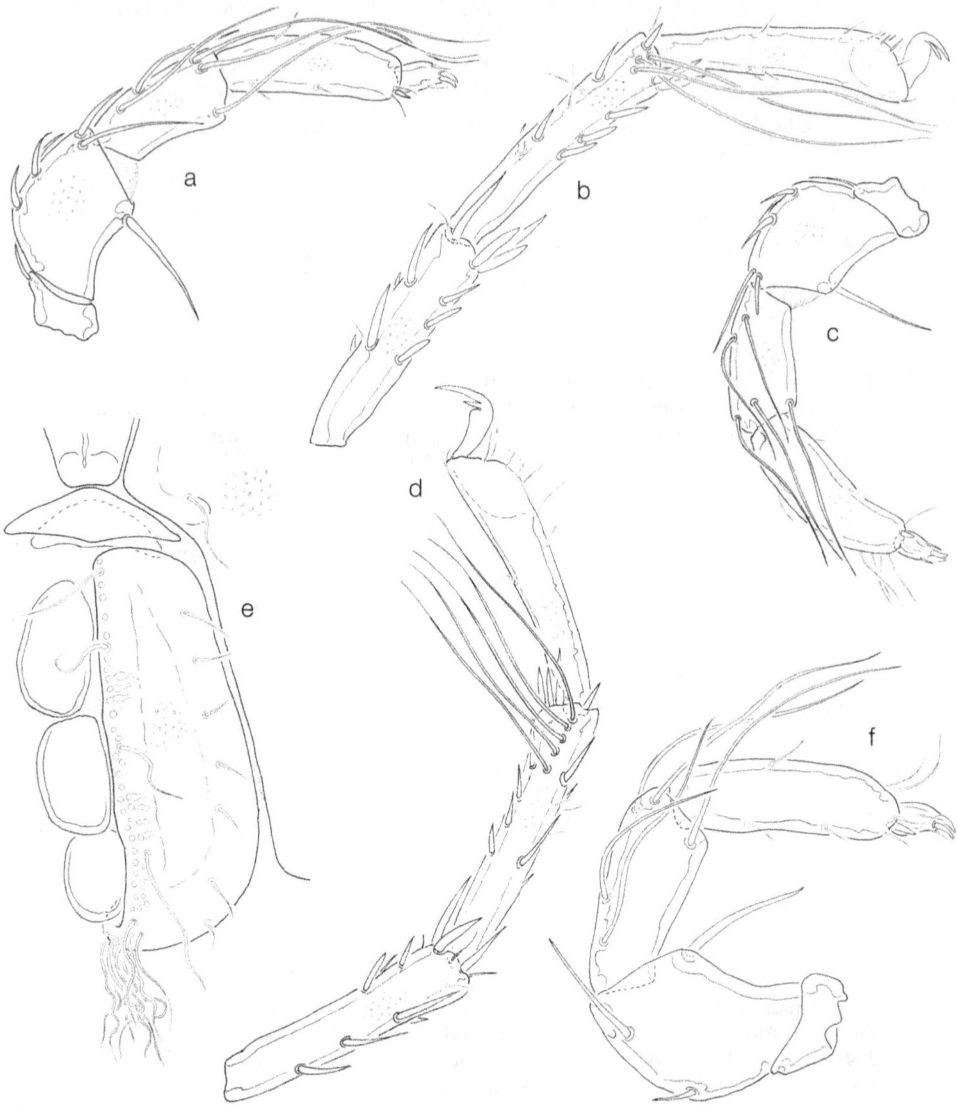

8-45: *Lebertia* s. str. (from Gerecke 2009, modified); a-b, *L. pusilla* male; a, palp; b, IV-L-4-6; c-d, *L. rivulorum* male; c, palp; d, IV-L-4-6; e-f, *L. subtilis* male; e, genital field partially (most setae given as insertion points only); f, palp.

18 (17) Cx-I/II medially equal in length, or Cx-I slightly longer than Cx-II (median length ratio 0.9-1.1). Distance between dorsal setae on P-3 < 20 µm (*8-46 a*). . . . . *Lebertia maglioi* (page 109)

\- Cx-I medially distinctly longer than Cx-II (median length ratio 1.2-2.1). Separation of dorsal setae on P-3 various. 19

19 (18) Cx-II medially longer (length 110-160 µm, Cx-I/II median length ratio 1.2-1.7). P-3 length > 110 µm, distance between dorsal setae < 20 µm, these far from distal setae (*8-46 d*). Large in dimensions (e. g., genital flap length ♂ > 170, ♀ > 220 µm). . . . . *Lebertia groenlandica* (page 107)

\- Cx-II medially shorter (length 100-105 µm, Cx-I/II median length ratio 1.7-2.1, *8-46 b*). P-3 length < 110 µm), distance between dorsal setae > 20 µm, these closer to distal setae (*8-46 c*); smaller in dimensions (e. g., genital flap length ♂ < 170, ♀ < 220 µm). . . . . *Lebertia harenarum* (page 108)

20 (17) Large in dimensions (Cx-I median length ♂ > 150, ♀ > 175; IV-L-6 length ♂ > 245, ♀ > 270, genital flap length ♂ > 165, ♀ > 190, P-4 length ♂ > 100, ♀ > 105 µm – in case of doubt, consider all these measurements). Idiosoma slender, sclerites deep purple in colour. . . . . *Lebertia sparsicapillata* (page 116)

\- Smaller in dimensions (Cx-I median length ♂ < 150, ♀ < 175; IV-L-6 length ♂ < 245, ♀ < 270, genital flap length ♂ < 165, ♀ < 190, P-4 length ♂ < 100, ♀ < 105 µm). Idiosoma slender or more rounded, sclerite colour from deep purple to pale green. 21

21 (20) III/IV-L-5 without swimming setae (at most two short setae, the longer of which not exceeding one third length of terminal segment, *8-46 h*). . . . . *Lebertia castalia* (page 105)

\- III/IV-L with two, more rarely only one swimming seta(e) exceeding one third length of terminal segment (*8-46 g*) . . . . *Lebertia fimbriata* (page 106)

4. *Lebertia (Lebertia) brigantina* K.Viets, 1933 (*8-42 i-k*)

1933e *Lebertia (Lebertia) brigantina* K.Viets, Zool. Anz. 104: 261.

Description: Integument finely striated, with underlying porosity; coxal field rather slender, anterior tips of Cx-I obtuse, equally rounded; median suture Cx-II relatively short (Cx-I/II median length ratio 1.2-1.4); legs with characteristic, strong blade setae arranged at distal segment borders of I-L-2/3, II-L-2-4, III-L-2-4; I-L-6 with maximum height at base of segment, distally equally narrowed; I- and II-L-4 both with 2-3 long (swimming seta-like) dorsal setae; numbers of swimming setae at distal segment margins: II-L-4, 1; II-L-5, 3; III-L-4, 2-3; III-L-5, 6-8; IV-L-4, 2-3; IV-L-5, 4; IV-L-6 with 2 ventral setae near the centre; P-3 rather stout (length/height ratio 1.8-2.0) and shorter than P-4 (length ratio P-3/P-4 ca. 0.8), with long setae (but not extending beyond tip of P-5), dorsal setae apart (proximal closer to proximal segment edge than to distal), mediodistal seta close to dorsodistal seta; P-4 maximum height in distal part of segment, ventral sectors 1:1:2, in exceptional cases 2:1:1, these may be very faint and often observed with difficulty; P-2/P-4 length ratio 0.93-0.96. Males: Idiosoma length/width 700-730/550 µm; posterior margins of Cx-IV generally extending far beyond posterior margin of genital field, but occasionally not even reaching this level; genital flap length 130-140 µm. Females: Idiosoma length/width 700-750/600; posterior margins of Cx-IV not reaching level of posterior genital field margin; genital flap length 150 µm.

Preadult stages: Larva: Wainstein (1980) – *sub nom. L. gladiator* (?): Gerecke 2009.

Similar species: *Lebertia oblonga*, a species very similar in shape and setation of palp and legs (with swimming seta-like dorsal setae on I-L-4 and II-L-3/4), differs in having P-3 as long as P-4 and more slender (length/height > 2.2). This species is without sexual dimorphism in shape of coxal field.

Habitat and Biology: High order lowland streams and clear lakes, often on sandy bottoms; life cycle unknown.

Distribution: Europe, scattered records, more frequent in northern and northeastern parts.

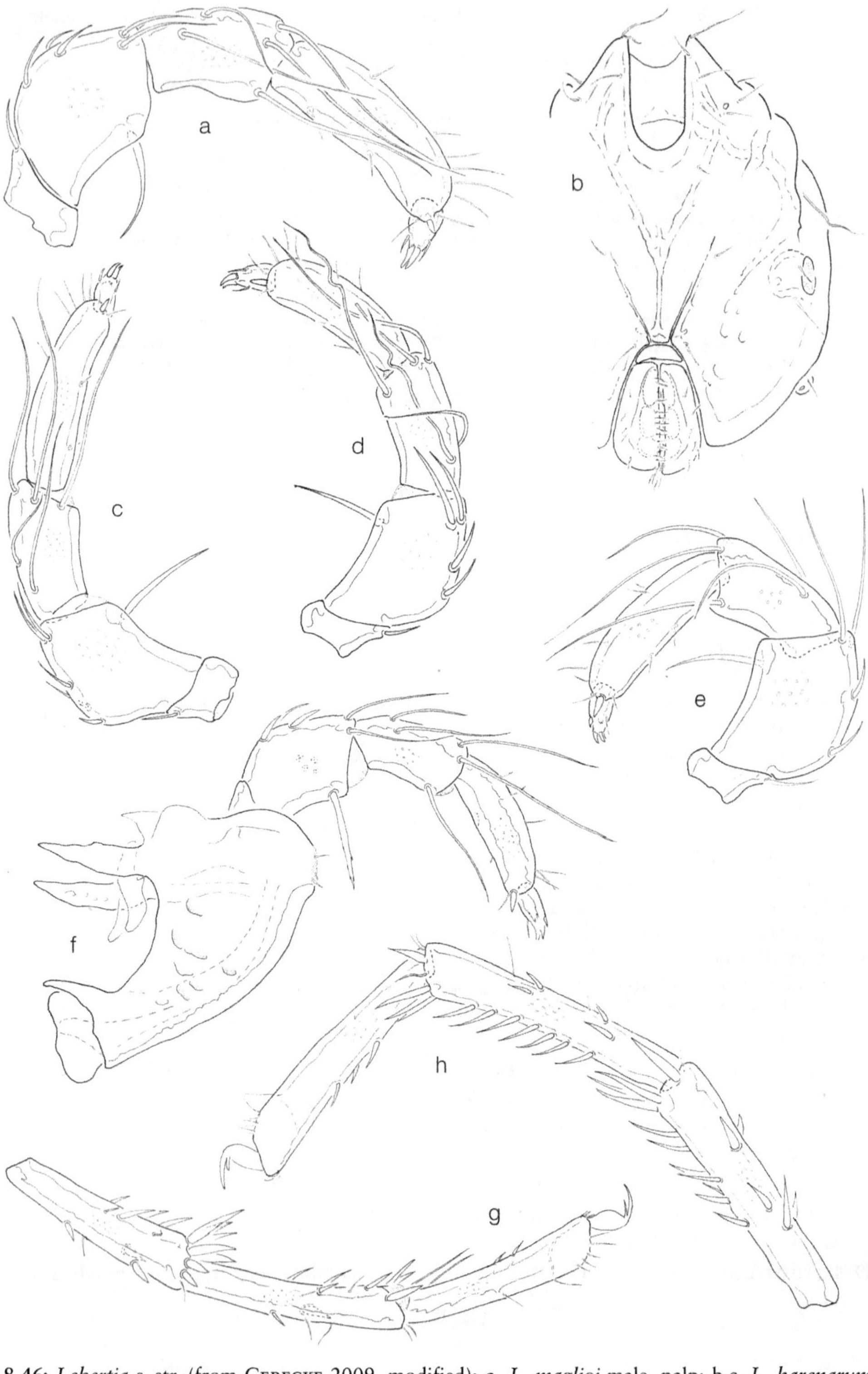

8-46: *Lebertia* s. str. (from Gerecke 2009, modified); a, *L. maglioi* male, palp; b-c, *L. harenarum* male; b, venter, partial view; c, palp; d, *L. groenlandica* male palp; e, *L. sparsicapillata* male, palp; f-g, *L. fimbriata* male; f, gnathosoma and palp; g, IV-L-5/6; h, *L. castalia* male IV-L-5/6.

5. *Lebertia (Lebertia) castalia* K.Viets, 1925 (*8-36 b, 8-46 h*)

1925g *Lebertia (Lebertia) castalia* K.Viets, Zool. Jbch. Syst. 50 (4-6): 521.

Description: Colour of sclerotized parts in mounted specimens red; integument finely striated; Cx-I medially hardly longer than Cx-II (Cx-I/II median length ratio 1.1-1.2), posterior width of Cx-II 30-35 µm; legs without swimming setae (one slender distal seta on IV-L-5 reaching 1/3 length IV-L-6; IV-L-6 with 2-3 stout ventral setae; excretory pore smooth; P-3 dorsal setae generally close to each other (but with considerable variability), mediodistal seta close to dorsodistal seta, P-4 with maximum height in distal part, ventral sectors 2:3:2. Males: Idiosoma length/width 530/510; genital flap length 125 µm. Females: Idiosoma length/width 780/620, genital flap length 170 µm.

Similar species: In measurements, shape and setation of palps and legs, *L. castalia* agrees with *L. fimbriata*, only differing in the absence of paired swimming setae on IV-L-5 and the presence of an additional dorsal seta on IV-L-1. Swimming setae are easily lost during preparation, or overlooked if they are in an unsuitable position. Consequently, records of *L. castalia* could refer to *L. fimbriata*, of which *L. castalia* could be a junior synonym, or to the generally larger, but from all other points of view similar *L. sparsicapillata*.

Habitat and Biology: Crenobiont. Preadult stages and life cycle unknown.

Distribution: Central, Eastern and Northern Europe, Turkey; scattered records.

6. *Lebertia (Lebertia) cognata* Koenike, 1902 (*8-44 l-m*)

1902b *Lebertia (Lebertia) cognata* Koenike, Zool. Anz. 25: 612.

1919 *Lebertia (Lebertia) inversa* Koenike, Arch. Hydrobiol. 12: 562 – Gerecke (2009).

1925g *Lebertia (Lebertia) obesa* K.Viets, Zool. Jbch. Syst. 50: 528 – Gerecke (2009).

Description: Colour yellow-brown, sclerites generally pale green; integument surface smooth or finely striated; Cx-I/II median length ratio 1.0-1.9 (in most cases 1.1-1.6); posterior width Cx-II 40-90 µm; IV-L-6 with 5-6 (rarely 3-7) ventral setae; number of swimming setae: III-L-5, 2; IV-L-5, 2; P-2 distally not narrowed; P-3 equally enlarged from base to distal margin, dorsal setae close to each other, mediodistal seta close to dorsodistal seta, often ventrodistal seta and, to a lesser extent dorsodistal seta, away from segment edge, P-4 maximum height in distal part, ventral setae insertions dividing segment into sectors 2:1:1, occasionally 1:1:1; length/height P-3, 1.9-2.3; P-4, 3.2-3.3; length ratio P-2/P-4, 0.8-0.9; P-3/P-4, 0.8. Males: Idiosoma length/width 900-1400/700-1000; genital flap length 200-260 µm. Females: Idiosoma length/width 1000-1500/800-1000; genital flap length 200-270 µm.

Similar species: In addition to the details given in the key, *Lebertia cognata* differs from *L. rufipes* in a relatively shorter median suture of Cx-I (Cx-I/II median length ratio in *L. rufipes* 1.5-2.0) and often stouter palp segments.

Habitat and Biology: Crenophilous. Rheo- and limnocrenes and pools of spring streams. At lower altitude replacing the closely related, but more montane *L. rufipes*, occasionally coexisting with the related species, *L. reticulata* (Gerecke 2006). Preadult stages and life cycle unknown.

Distribution: Central and Eastern Europe including the Alps.

7. *Lebertia (Lebertia) fimbriata* THOR, 1899 (*8-43 e, 8-46 f-g*)

1899a *Lebertia (Lebertia) fimbriata* THOR, Kristiania (O. Norli): 5.

1926c *Lebertia (Lebertia) africana* WALTER, Bull. Soc. Sci. natur. Maroc 6: 142 – SMIT (1995).

1927c *Lebertia (Lebertia) affinis* SOKOLOW, Trav. Stat. biolog. Caucase Nord, Gorsky Inst. Agronom., Vladicaucase 5: 61 – GERECKE (2009).

1937c *Lebertia (Lebertia) tenuistriata* K.VIETS, Arch. Hydrobiol. 31: 554 – GERECKE (2009).

1957a *Lebertia (Lebertia) elongata* BESSELING, Ent. Berichten, Amsterdam 17: 46 – SMIT (1992).

Description: Colour often pale yellowish to brown, sclerotized parts greyish or greenish, in older specimens, dark brown, but also reddish with deep purple sclerotization; idiosoma elongate or rounded, integument finely striated; Cx-I/II medially nearly equal in length in most cases, but with considerable variability. Posterior margin of Cx-II rather narrow; Cx-IV with obtuse posteromedial angles flanking the genital field; one or two swimming setae on III-L-5, IV-L-4, and IV-L-5; ventral setae numbers: IV-L-5, 3-6 (generally 4-5, robust); IV-L-6, 2 (small ones); genital field flanked by concave posteromedial margins of Cx-II-IV, posterior margin of Cx-IV equally rounded or laterally slightly concave, forming two edges near posterior muscle apodeme and at transition to genital bay; excretory pore smooth; P-2 ventral seta about same as ventral length of segment or slightly longer; P-3 dorsal setae close to each other, inserted proximally from centre of segment, distal setae not very elongated (not overreaching the tip of P-5), medio- and dorsodistal setae close to each other, ventrodistal seta inserted slightly away from ventrodistal segment edge; P-4 in most cases rather stout (but occasionally more slender and elongated), proximal and distal edges of the segment nearly equal in height, maximum height distally from, in rare cases in, centre of segment, ventral margin straight, generally divided by seta insertions into equal sectors, dorsal margin convex, mediodistal peg-like seta stout, in dimensions similar to one of the big claws of P-5, but not curved. Males: Idiosoma length 600-950; genital flap length 95-176 µm, posteromedial setae not particularly elongated. Females: Idiosoma length 650-1000, genital flap length 135-185 µm.

Preadult stages: Larva: MARTIN (2000).

Similar species: Occasionally, specimens are found with an additional 3rd swimming seta on IV-L-5 (clearly distanced from the other two, inserting more proximal), a longer swimming seta on IV-L-4 and an additional one on III-L-5. These specimens often have the dorsal setae on P-3 more proximal than "typical" *fimbriata* specimens. The separation from *L. sparsicapillata* is problematic. In southern Germany, typical *sparsicapillata*-like populations were found distinctly larger than *fimbriata*-specimens and the following diagnostic standards were established for *L. fimbriata* (*L. sparsicapillata* in parentheses): Cx-I median length males 120-160 (150-200), females 135-180 (170-200), genital flaps length males 100-175 (160-190), females 135-190 (290-210); Ac-1 length 30-60 (60-70), IV-L-5 length males 250 (250-310), females 260 (300-320); P-4 length males 71-100 (98-118), females 76-106 (113-123). However, in several parts of Southern Europe, populations intermediate in measurements were found. Two further similar species from the Mediterranean, *L. intronata* GERECKE, 2009 and *L. vegacabrerae* GERECKE, 2009, differ in sculptured integument and angled posteromedial edges of Cx-IV.

Habitat and Biology: Low to high-order streams, on mosses and under stones, in northern Europe also in oligotrophic lakes (SOKOLOW 1930a, LUNDBLAD 1968, MARTIN 1997). In southern Germany overwintering at the adult stage in pool habitats, during other seasons preferring lotic conditions (GERECKE 2002). Larvae parasitic on chironomid Diptera (Orthocladiinae, Chironominae, Tanytarsini; attachment sites: thorax, rarely cervix – MARTIN 2000).

Distribution: Western Palaearctic.

8. *Lebertia (Lebertia) glabra* Thor, 1897 (*8-43 a, 8-44 g-h*)

1897c *Lebertia (Pseudolebertia) glabra* Thor, Arch. Math. Naturv. 20: 19.

1906e *Lebertia lineata* Thor, Zool. Anz., 31: 69 – Gerecke (2009).

Description: Colour yellow to brown, sclerites often greenish; idiosoma length/width 600-1000/500-1000 µm; dorsal integument with prominent lines; Cx-I medially longer than Cx-II; distal margins of III/IV L 2-5 with groups of fine setae, and in most specimens with IV-L-4/5 having one seta reaching $^1/_4$-$^1/_3$ length of the following segment; IV-L-5 with 5-8 ventral setae, IV-L-6 ventrally with 2 small and short setae, distally not strongly thickened; genital flaps with rather few medial setae; excretory pore smooth; length P-2 ≤ P-3; P-2 ventral seta broad and short; P-3 with dorsal setae, as well as mediodistal and dorsodistal setae separated from each other; P-4 distally narrowed and with a slight kink, maximum height in basal third, proximoventral seta close to segment base (seta insertions dividing segment into sectors 1:2:2). Males: Genital flap length 160; Females: Genital flap length 180 µm.

Preadult stages: Larva: Lundblad (1930c), Martin (2000).

Habitat and Biology: In the area treated a rhithrobiont, in the Mediterranean crenophilous. Spring-dwelling populations in Sweden (Lundblad 1968) and Luxembourg (Gerecke et al. 2005a) with a preference for mineral substrata and elevated conductivity. Larva parasitic on chironomid Diptera (Orthocladiinae, Prodiamesinae, Tanytarsini; attachment sites: thorax, rarely abdomen – Efford 1963, Martin 2000).

Distribution: Western Palaearctic.

9. *Lebertia (Lebertia) groenlandica* Lettevall, 1962 (*8-46 d*)

1962 *Lebertia (Pseudolebertia) groenlandica* Lettevall, Meddel. Groenland 170 (1): 20.

Description: Integument dorsally completely smooth or with indistinct and irregular sculpturing; medial suture Cx-I longer than Cx-II (ratio 1.2-1.7); legs without swimming setae; IV-L-6 bearing 3-4 fine ventral setae; excretory pore smooth; P-2/P-4 length ratio ca. 0.9, dorsal setae on P-3 far from distal setae; P-4 ventral sectors 1:2:2. Males: Idiosoma length/width 1100-1300/900-1000; genital flap length 190-200 µm, with about 30 densely arranged medial setae. Females: Idiosoma length/width 1400-1500/900-1100; genital flap length 235 µm, with 17-20 medial setae.

Preadult stages: Larva: Lettevall (1962). Deutonymph: Lettevall (1962).

Similar species: Differing from other species of the *L. maglioi*-group (with ventral setae of P-4 inserting proximally and swimming setae absent or weakly developed) in larger dimensions.

Habitat and Biology: Rhithrobiont? Life cycle unknown.

Distribution: Greenland.

10. *Lebertia (Lebertia) guttata* K.Viets, 1952 (*8-43 h-i*)

1952c *Lebertia (Pseudolebertia) guttata* K.Viets, Zool. Anz. 149 (11/12): 245.

Description: Female (male unknown): Sclerotized parts dark red; integument ventrally lineate, with short lines at regular distances forming drop-shaped extensions; these extensions more prominent laterally, dorsally appearing as globular papillae arranged along remnants of more or less obliterated lines; Cx-I medially longer than Cx-II (ratio 1.36), posterior width of Cx-II 30 µm; legs without swimming setae; IV-L rather stout (length/height IV-L-4, 200/50; IV-L-5, 215/45; IV-L-6, 210/60; IV-L-5 with 3 strong, IV-L-6 with 2 small ventral setae; genital flap length 170 µm; excretory pore embraced by an anterior and a posterior sclerotized piece; P-3 with dorsal setae far apart (insertions dividing margin into sectors 1:1:1), mediodistal and dorsodistal setae close; P-4 maximum height proximally, ventral sectors 2:2:3, peg-like seta slender; length ratio P-2/P-4 and P-3/P-4 both 0.85.

Similar species: *Lebertia schechteli* is similar in having dorsal lines almost completely disintegrated into pearl-like papillae, but papillae not connected by line remnants; furthermore different in palp morphology (P-3 as long as, or longer than, P-4). *Lebertia glabra* and *L. maculosa* differ in having more slender leg segments, the latter also in Cx-II, having a relatively shorter medial suture and broader posterior margin.

Habitat and Biology: Crenobiont? Preadult stages and life cycle unknown.

Distribution: French Alps, only known from the type locality (Villar d'Arène).

11. *Lebertia (Lebertia) harenarum* LÁSKA, 1953 (*8-46 b-c*)

1953b *Lebertia (Lebertia) harenarum* LÁSKA, Acta Soc. zool. Bohemoslovenicae 17 (4): 289.

Description: Integument locally with fine striation, everywhere with a rather strong porosity (in tangential view, lines of pores may suggest a "lineated" aspect); Cx-I medially much longer than Cx-II (ratio 1.7-2.1), posterior margin Cx-II broad; III-L-5 and IV-L-5 each bearing one very short swimming seta (1/4 length terminal segment); IV-L-6 with 2-3 short ventral setae; excretory pore smooth, P-3 dorsal setae apart (distance: 23 µm), mediodistal seta away from dorsodistal seta (distance: 10-13 µm); P-4 maximum height in basal part, ventral sectors 1:2:2. Males: Idiosoma length/width 800/600, Cx-I/II median length 180/105, posterior width Cx-II, 35; genital flap length 155 µm, each bearing about 15 medial setae, also the posterior medial setae very short. Females: Idiosoma length/width 900/800, Cx-I/II median length 210/100, posterior width Cx-II 40; genital flap length 200 µm, each bearing only about 12 medial setae.

Similar species: *Lebertia maglioi*, the most closely related species, differs in having Cx-I/II medially nearly equal in length, and dorsal setae on P-3 closer (distance: 13-15 µm). Representatives of the *L. glabra*-species group with reduced integumental sculpturing may assume an aspect similar to *L. maglioi* and *L. harenarum*. Such specimens differ from *L. harenarum* in P-3 dorsal setae and dorsodistal/mediodistal setae being closer together and P-4 less tapering distally.

Habitat and Biology: Rhithrobiont. Preadult stages and life cycle unknown.

Distribution: Only recorded from Silesia, Jeseniky in the Czech Republic.

12. *Lebertia (Lebertia) lusitanica* LUNDBLAD, 1956 (*8-43 f-g*)

1956a *Lebertia (Pseudolebertia) lusitanica* LUNDBLAD, Ark. Zool. 10 (1): 110.

Description: Colour unknown; integument lineate, lines ventrally long (up to 50 µm), laterally shorter, dorsally formed into minute papillae; Cx-I/II medially nearly identical in length, Cx-II posteriorly narrow (width 20-25 µm); III-L-5 with one short seta, IV-L-5 with one long swimming seta; IV-L-6 with 2 small ventral setae; excretory pore surrounded by a sclerotized ring; P-3 dorsal setae apart, mediodistal seta close to ventrodistal seta; P-4 ventral margin with a slight kink near insertion of distoventral seta, ventral sectors 1:2:2. Males: Idiosoma length/width 750/600, genital flap length 135 µm, each bearing about 20 medial setae. Females: Idiosoma length/width 1000/750, genital flap length 175 µm.

Similar species: Within the subgenus, *L. lusitanica* is unique in the ventral position of the mediodistal seta; a sclerotized excretory pore is found only in one other species, *L. guttata*, apart from P-3 setation, differing in a medially relatively longer Cx-I.

Habitat and Biology: Rhithrobiont? Preadult stages and life cycle unknown.

Distribution: Western Europe, only known from the type locality in Portugal and a site in the French Pyrenees (ANGELIER et al. 1985).

13. *Lebertia (Lebertia) maculosa* KOENIKE, 1902 (*8-43 d, 8-44 e-f*)

1902b *Lebertia maculosa* KOENIKE, Zool. Anz. 25: 613.

1902b *Lebertia zschokkei* KOENIKE, Zool. Anz. 25: 614 – GERECKE (2009).

1922a *Lebertia (Pseudolebertia) zermattensis* WALTER, Rev. Suisse Zool. 29: 322 – GERECKE (2009).

1926g *Lebertia (Pseudolebertia) arndti* K. VIETS, Zool. Anz. 67: 12 – GERECKE (2009).

1926g *Lebertia (Pseudolebertia) schischkoffi* K. VIETS, Zool. Anz. 67: 16 – GERECKE (2009).

1975 *Lebertia (Pseudolebrtia) pseudotuberosa* BADER, Ergebn. wiss. Unters. schweiz. Nationalpark 14: 123 – GERECKE (2009).

Description: Colour red, with deep purple sclerites; integument dorsally with variously sized papillae and lines (length at least 10 µm, sometimes, especially on ventral surface, considerably longer); Cx-II medially short (Cx-I/II median length ratio 1.5-2.1), posterior margin Cx-II broad; IV-L-5/6 stout (length/height ratio IV-L-5, 4.7-5.6; IV-L-6, 4.3-5.2); IV-L-5 bearing 3-4 ventral setae; IV-L-6 distally enlarged; P-3 with dorsal setae rather variable in position (close to each other or the proximal seta halfway between segment base and dorsodistal seta), mediodistal and dorsodistal setae apart; P-4 ventral sectors 1:1:1 or 1:2:2; length ratio P-2/P-4, 0.77-0.83, P-3/P-4, 0.80-0.89; P-5 stout. Males: Idiosoma length/width 650-900/600-800; Cx-II posterior width 30-50; genital flap length 145-220 µm. Females: Idiosoma length/width 700-1200/ 600-900; Cx-II posterior width 40-70; genital flap length 210-230 µm.

Preadult stages: Larva: MARTIN (*sub nom. L. zschokkei*: 2003).

Similar species: Often confused with *L. schechteli*, a species clearly different in the shape of integument papillae and proportions of palp (P-3 as long as, or longer than P-4, length ratio 0.95-1.30).

Habitat and Biology: Crenophilous. Usually in strongly flowing rheocrenes and their spring streams (GERECKE & MARTIN 2006). Larvae parasitic on Chironomidae (Orthocladiinae: MARTIN et al. 2010); prefered attachment site: thorax.

Distribution: Central, Western and Southeastern Europe; widely distributed in higher mountains from the Pyrenees to the Balkans and Appennines, one isolated record from southern Italy.

14. *Lebertia (Lebertia) maglioi* THOR, 1907 (*8-46 a*)

1907d *Lebertia (Neolebertia) maglioi* THOR, Zool. Anz. 31 (4): 902.

1940a *Lebertia (Lebertia) balcarica* SOKOLOW, in SERNOW & STACKELBERG, Faune de l'URSS, Archnides 5 (2): 217 – GERECKE (2009).

1954c *Lebertia (Lebertia) apposita* LÁSKA, Zool. Anz. 153 (7-8): 159 – GERECKE (2009).

1954d *Lebertia (Lebertia) acuta minor* LÁSKA, Acta Soc. Zool. Bohemoslovenicae 18 (4): 275 – GERECKE (2009).

1955a *Lebertia (Lebertia) husmanni* K. VIETS, Arch. Hydrobiol. 50 (1): 36 – GERECKE (2009).

1955 *Lebertia (Lebertia) scutellata* K. O. VIETS, Zool. Anz. 154 (3-4): 72 – GERECKE (2009).

Description: Colour red, integument smooth, locally with very fine striation and fine porosity; Cx-I medially as long as Cx-II, posterior margin Cx-II relatively large (width 40-50 µm); Cx-IV with distinct, nearly right-angled posteromedial edges; leg setae rather short and "horn"-shaped, in particular on II/III-L-3/4 , swimming setae not exceeding 50 % length of next segment; number: III-L-5 one, short, IV-L-5 one longer and one shorter; IV-L-5 with 3-4 ventral setae, IV-L-6 with 2 minute ventral "nail" setae; excretory pore smooth; palp stout, P-2 with short dorsodistal setae, P-3 club-shaped, generally stout (length/height about 1.9), but occasionally more slender (length/height 1.9-2.3), with separated dorsal (distance 10-15 µm) and medio-/dorsodistal (distance 10-15 µm) setae; P-4 maximum height in basal part of segment, ventral margin with a slight kink near insertion of ventrodistal setae, ventral sectors 1:2:2 or 1:1:2, peg like seta small, dorsodistal setae particularly long; length ratio P-2/P-4, 0.73-0.79; P-3/P-4, 0.75-0.83; P-5 slender. Males: Idiosoma length/width 700-1000/550-700, genital flap length 130-160 µm, genital setae dense and long. Females: Idiosoma length/width 950-1200/700-1000, genital flap length 170-190 µm, genital setae scattered, few in number.

Similar species: *Lebertia subtilis* is similar in having the dorsal setae of P-3 separated, but differs in Cx-I medially being distinctly longer than Cx-II and the presence of double and triple lines of genital setae in males. *Lebertia variolata* GERECKE, 2009 from intermittent streams in the Central Mediterranean area, is very similar in palp morphology, but differs in having a papillate integument.

Habitat and Biology: Middle order streams with year-round surface flow. Preadult stages and life cycle unknown.

Distribution: Western Palaearctic, also including the extreme borders (Faeroes, Maghreb, Iran, Ukraine), so far not recorded from Eastern and Northeastern Europe, but the most frequent *Lebertia* species in Southern Europe. The presence of a sister species on Greenland, *L. groenlandica* (see above) suggests that this island has been colonized by populations of Nearctic origin.

15. *Lebertia (Lebertia) natans* K.VIETS, 1926 (*8-42 a-d*)

1926c *Lebertia (Neolebertia) natans* K.VIETS, Abh. naturw. Ver. Bremen 26 (1): 65.

1926b *Lebertia (Neolebertia) multipilosa* THOR, Arb. Biol. Wolgastation Saratow 9 (1-2): 19 – GERECKE (2009).

Description: Integument smooth, locally with remnants of fine striation and fine porosity in an inner layer; Cx-I medially longer than Cx-II, Cx-II posteriorly pointed; swimming setae very long, extending far beyond end of following segments and located in long rows all over the distal part of segments – numbers: II-L-5, 11-14; III-L-4, 7-8; III-L-5, 14-15; IV-L-4, 10-12; IV-L-5, 12-14; IV-L-6 with 3 tiny ventral setae; palp slender, P-3 dorsal setae (distance 5-8 µm) and dorsodistal and mediodistal seta close to each other (distance 5-13 µm); P-4 ventral sectors 1:2:2, peg-like mediodistal seta small; length ratio P-2/P-4, 0.76-0.80; P-3/P-4, 0.81-0.91. Males: Idiosoma length/width 1000/800, genital flap length 155 µm. Females: Idiosoma length/width 1000-1100/800-900, genital flap length 160-180 µm.

Similar species: Swimming setae in similar high numbers are found otherwise only in *Pilolebertia* species, from which *L. natans* differs in a palp typical for species of the subgenus *Lebertia s. str.* (P-4 not narrowed distally, dorsal setae not restricted to the distal segment sector).

Habitat and Biology: High order streams. Preadult stages and life cycle unknown.

Distribution: Central and Northeastern European lowlands, scattered records, rare.

16. *Lebertia (Lebertia) oblonga* KOENIKE, 1911 (*8-42 f-h*)

1911c *Lebertia oblonga* KOENIKE, Abh. naturw. Ver. Bremen 20 (2): 239.

1922a *Lebertia (Neolebertia) sublitoralis* WALTER, Rev. Suisse Zool. 29 (7): 294 – GERECKE (2009).

1940a *Lebertia (Lebertia) shadini* SOKOLOW (= "*L. subtilis var.*" SOKOLOW, 1925b), in SERNOW & STACKELBERG, Faune de l'URSS, Arachnides 5 (2): 217 – GERECKE (2009).

Description: Colour unknown (sclerotized parts pale red in some slide mounted specimens); integument finely striated, porosity visible both in membranous and sclerotized integument areas; anterior tips of Cx-I rounded, stout; glandularium C-4 distinctly set off; Cx-II with a long medial suture, posterior margin narrow, slightly away from pregenital sclerite and with a fine medial indentation; a distinct convex extension associated dorsally with insertions of IV-L; legs with characteristic, strong blade setae arranged at distal segment borders of I-L-2/3, II-L-2-4, III-L-2-4; I-L-4 with particularly elongated ventrodistal setae, I-L-6 with parallel dorsal and ventral margins; swimming setae at distal segment margins: absent from I-L; II-L-5, 3; III-L-4, 3, III-L-5, 4; IV-L-4, 3; IV-L-5, 5; additional long, swimming hair-like dorsal setae on I-L-4 (2-3), II-L-3 (2-3) and II-L-4 (2-3); genital field subtriangular, with lateral margins converging anteriorly, posterior margins of Cx-IV equally rounded at posterior edge of genital bay; excretory pore smooth; P-2 and P-3 basally narrowed, P-3 slender (length/height ratio 2.3-2.6), as long as P-4, with dorsal setae apart (distance 23-38 µm), mediodistal and dorsodistal setae close (distance 5-8 µm); P-4 maximum height in distal part, medial peg-like seta strong, ventral sectors 1:1:2 or 1:1:1. Males: Idiosoma length/width 800/600 µm; genital flap length 130-160 µm, with medial setae fine and long, often spiralling, especially in posterior part. Females: Idiosoma length/width 800/600 µm; interspace between Cx-II and pregenital sclerite filled by converging medial tips of Cx-III, genital flap length 160, lateral margins straight and distinctly converging anteriorly, with a low number of medial setae which are equally short all over flap length.

Similar species: See discussion of *L. brigantina*.

Habitat and Biology: High order lowland streams and clear lakes, often on sandy bottom. Preadult stages and life cycle unknown.

Distribution: Europe, scattered records, rare.

17. *Lebertia (Lebertia) polonica* BIESIADKA, 1972 (*8-44 i*)

1972 *Lebertia (Lebertia) polonica* BIESIADKA, Bull. Acad. Pol. Sci., Ser. Sci. Biol., Zool., Cl. II 20 (4): 251.

Description: Colour dark brown; integument smooth, finely porose; medial suture Cx-I longer than Cx-II (ratio 1.1-1.4); posterior margin of Cx-II narrow; legs without swimming setae (one single fine distal seta, relatively short, on III/IV-L-4/5), IV-L-5 bearing 5-7, IV-L-6, 2 small ventral setae; Ac enlarged, Ac-2 subrectangular in shape; excretory pore smooth; P-3 slender (length/height 2.4-2.6), basally narrow, the two dorsal setal groups apart from each other (distance: 43-53 µm), mediodistal and dorsodistal setae close together; P-4 nearly equal in height from base to tip (length/height 3.0-3.3), maximum height in distal half; ventral sectors 1:2:1, mediodistal peg-like seta slender and pointed, the 2 dorsal setae close to each other in basal part of segment; length ratio P-2/P-4 ca. 0.9, P-3/P-4, 1.0-1.1; P-5 short. Males: Idiosoma length/width 745-900/600-700; genital flap length 122-145 µm, with about 40 densely arranged medial and 3-4 lateral setae. Females: Idiosoma length/width 830-1050/610-700; genital flap length 165-180 µm, with about 12 medial and 2 lateral setae.

Similar species: Closely related to *L. schechteli* (genital field with very large Ac, male genital flaps with dense medial setation; P-3 elongated and basally distinctly narrowed; P-4 robust with maximum height in the distal half; P-5 short), differing only in the smooth, not papillate integument.

Preadult stages: Larva: WAINSTEIN (1980).

Habitat and Biology: Crenobiont. Life cycle unknown.

Distribution: Carpathians, only known from the type localities in Poland.

18. *Lebertia (Lebertia) pusilla* KOENIKE, 1911 (*8-45 a-b*)

1911c *Lebertia (Lebertia) pusilla* KOENIKE, Abh. naturw. Ver. Bremen 20 (2): 236.

1930a *Lebertia (Neolebertia) olonensis* SOKOLOW, Zool. Jbch. Syst. 59 (2-3): 174 – GERECKE (2009).

1954c *Lebertia (Lebertia) pallida* LÁSKA, Zool. Anz. 153 (7-8): 161 – GERECKE (2009).

1955c *Lebertia (Lebertia) silesiaca* LÁSKA, Přirodověd. Sborn. Ostrv. kraje 16 (1): 46 – GERECKE (2009).

Description: Colour brownish or greenish (?); relatively small in dimensions (maximum length 800 µm); integument smooth, with fine striation; sclerites of body and appendages porose; Cx-I/II medially equal in length or Cx-II longer than Cx-I (median length ratio 0.8-1.0), posterior margin Cx-II narrow; III-L-5 and IV-L-5 each with 3-4 long swimming setae arranged in a longitudinal line near distal segment margin, in length similar to, or exceeding, terminal segment; IV-L-5 with 3-4 strong ventral setae, IV-L-6 distally thickened, with 2 small ventral setae; excretory pore smooth; P-3 length/height 2.0-2.3, with dorsal setae, as well as dorso- and mediodistal setae, close (distance 3-8 µm), distal setae long, occasionally extending beyond tip of P-5; P-4 robust (length/height 3.2-3.6), maximum height in, or distally from, centre of segment, mediodistal peg-like seta slender, sharply pointed, ventral sectors 1:2:2, rarely 1:1:1; length ratio P-2/P-4 and P-3/P-4 both ca. 0.8. Males: Idiosoma length/width 650/500; posterior margin Cx-II, 25; genital flap length 120; length/height P-2, 60/43; P-3, 60/30; P-4, 73/23 µm. Females: Idiosoma length/width 700-800/450-600; posterior margin Cx-II, 30-40; genital flap length 140-155; length/height P-2, 65-70/43-48; P-3, 65-75/33; P-4, 83-90/25 µm.

Similar species: Differing from most species of the subgenus in the presence of three long swimming setae on III/IV-L-5. *Lebertia ignatowi* SOKOLOW, 1930b (Russia) has ventral sectors P-4 equal in length and a broader posterior margin of Cx-II (width ♂ > 30, ♀ > 40 µm).

Habitat and Biology: Rhithrobiont; mountain streams with extended pool areas. Preadult stages and life cycle unknown.

Distribution: Europe except for the northern parts.

19. *Lebertia (Lebertia) reticulata* KOENIKE, 1919 (*8-44 d*)

1919 *Lebertia (Neolebertia) reticulata* KOENIKE, Arch. Hydrobiol. 12 (3): 579.

1919 *Lebertia (Neolebertia) zachariasi* KOENIKE, Arch. Hydrobiol. 12 (3): 595 – GERECKE (2009).

1928f *Lebertia (Lebertia) rufipes reticulata* K. VIETS, in: BROHMER, EHRMANN & ULMER (eds), Die Tierwelt Mitteleuropas. Leipzig (Quelle & Meyer): 26.

Description: Membranous integument with an irregular fine striation, in centre of dorsum the lines combining to form a reticulate framework; in lower integument layer, groups of fine pores corresponding to the reticulation. Cx-I medially distinctly longer than Cx-II

(Cx-I/II median length ratio 1.4-2.5); III-IV-L-5 with one swimming seta each; IV-L-1 with 2 dorsal and 2 dorsodistal setae; IV-L-6 with 6-8 ventral setae; excretory pore smooth; P-3 dorsal setae close to each other, mediodistal seta close to dorsodistal seta, ventrodistal seta normally at ventrodistal segment edge, but sometimes shifted dorsally or proximally; P-4 slender, generally ventral sectors 2:1:1, peg seta stout; P-2/P-4 length ratio 0.80-0.95. Males: Idiosoma length/width 1000-1300/800-1200, Cx-II posterior width 35-60, genital flap length 190-220 µm. Females: Idiosoma length/width 1100-1400/800-1100, Cx-II posterior width 40-60, genital flap length 210-250 µm.

Similar species: In the deep wine-red coloration, with sclerotized parts violet to blue, the relatively long medial suture of Cx-I and the presence of only one swimming seta on III/IV-L-5, this species is most similar to *L. rufipes*. Apart from the integument sculpturing, *L. reticulata* differs only slightly in having 2 (instead of one) dorsal setae on IV-L 1 and slightly smaller measurements of palp and leg segments. As representatives of both taxa have never been found together, but colonize habitats similar in typology, there is some reason to consider them as subspecies ("varieties" sensu LUNDBLAD 1956a). *Lebertia maderigena*, another species with sculptured integument, but lacking swimming setae, differs in having P-4 with maximum height in the basal quarter, ventral sectors 4:2:1, and all dorsal setae concentrated near the tip of the segment.

Habitat and Biology: Limnocrenes and cold mountain lakes, generally in similar habitats, but at lower altitude as *L. rufipes*. Preadult stages and life cycle unknown.

Distribution: Alps and central European mountains.

20. *Lebertia (Lebertia) rivulorum* K.VIETS, 1933 (*8-45 c-d*)

1933c *Lebertia (Lebertia) rivulorum* K.VIETS, Arch. Hydrobiol. 25 (4): 663.

Description: Colour unknown; integument smooth, with fine lines, in deeper layers finely porose; Cx-I and Cx-II nearly equal in median length (ratio 1.0-1.1), posterior margin of Cx-II narrow; swimming setae numbers: II-L-5, 1-2; III-L-5, 3-5; IV-L-5, 4-5, but without such setae on segments 4, IV-L-1 with one dorsal and 2 dorsodistal setae; IV-L-6 bearing 2-3 minute ventral setae; P-3 with dorsal setae separated (distance 15-20) and mediodistal seta nearly in centre of distal margin (distance ratio 1.3-1.5); P-4 maximum height in basal part of segment, distally distinctly narrowed, dorsal setae long, ventral margin with a kink at the insertion of distoventral seta, ventral sectors 1:3:3. Males: Idiosoma length/width 700-850/600-720; genital flap length 135; length/height P-2, 83/58; P-3, 88/43; P-4, 108/35 µm. Females: Idiosoma length/width 1000-1050/700-800; genital flap length 180; length/height P-2, 105/66; P-3, 108/50; P-4, 130/40 µm.

Preadult stages: Larva: MARTIN (2000).

Similar species: *Lebertia pusilla*, most similar in swimming setation (absent from III/IV-L-4; 3, rarely 4 on III/IV-L-5) differs in having a generally lower swimming seta number, shorter palp and leg measurements (e. g. length P-2, P-3 < 80, P-4 < 100 µm), P-3 with dorsal setae, as well as medio- and dorsodistal setae close to each other, and P-4 maximum height in distal part, ventral sectors 1:1:1. *Lebertia maglioi*, similar in shape and setation of the palp, is distinguished by the larger posterior margin of Cx-II and the presence of only one, shortened, swimming seta on III/IV-L-5.

Habitat and Biology: Rhithrobiont; preferably in mosses (DITTMAR 1955) and *Alnus* roots in sandy lowland streams (MARTIN & SPETH 1996). Larval host range unknown, but obviously parasitic (detached in an emergence trap: MARTIN 2000).

Distribution: Central, Eastern and Southern Europe, most frequent in central mountains and lowlands, rare in the south and west.

21. *Lebertia (Lebertia) rufipes* KOENIKE, 1902 (*8-44 k*)

1902b *Lebertia rufipes* KOENIKE, Zool. Anz. 25: 611

1954c: *Lebertia (Lebertia) slovenica* LÁSKA, Zool. Anz. 153: 163 – GERECKE (2009).

Description: Colour red, with deep red sclerotized parts; integument smooth or with fine striation; Cx-I medially long (Cx-I/II median length ratio 1.5-2.1); Cx-II posterior width 40-70 µm; swimming setae numbers: III-L-5, 0; IV-L-5, 1; IV-L-1 with one dorsal and 2 dorsodistal setae; IV-L-6 with 6-9 ventral setae; excretory pore smooth; P-2 in distal half with subparallel dorsal and ventral margins (distally not narrowed); P-3 equally enlarged from base to distal margin, dorsal setae close to each other, mediodistal seta close to dorsodistal seta, ventrodistal seta away from distal edge, P-4 club-shaped, with straight ventral margin, divided by setae insertions into sectors 2:1:1, rarely 1:1:1; length/height P-3, 2.0-2.4; P-4, 3.4-4.4; length ratio P-2/P-4, 0.7-0.9; P-3/P-4, 0.8. Males: Idiosoma length/width 950-1300/800-1000; genital flap length 200-225 µm. Females: Idiosoma length/width 1000-1200/800-900; genital flap length 230-260 µm.

Preadult stages: Larva: WALTER (1922c).

Similar species: On many occasions, *L. rufipes* has been confused with *L. sparsicapillata*, differing in a lower ventral setae number on IV-L-6. In this character, *L. rufipes* is most similar to *L. cognata*. The two species form a group defined also ecologically, with a preference for cold standing or slowly moving water.

Habitat and Biology: Large rheocrenes, limnocrenes and cold lakes in high mountains. In Lake Tovel (Trentino, Italy) with high water level fluctuations; restricted to deeper layers not exposed to seasonal drought (BERTUZZI et al. 2004). Larvae parasitic on chironomid Diptera (Tanytarsini – BERTUZZI et al. 2004, unpubl.)

Distribution: Central and Eastern Europe, possibly restricted to the Alps, Pyrenees, Tatra and Carpathians (older records from central European mountains and lowlands questionable due to confusion with *L. sparsicapillata* and *L. reticulata*).

22. *Lebertia (Lebertia) salebrosa* KOENIKE, 1908 (*8-43 c, 8-44 c*)

1908d *Lebertia salebrosa* KOENIKE, Abh. naturw. Ver. Bremen 19 (2): 343.

1911a *Lebertia (Pseudolebertia) salebrosa* THOR, Zool. Anz. 37 (19-20): 386.

Description: Colour yellow to brown; integument lineate, lines finer and denser than in *L. glabra*, dorsally forming interlocked "amoeba-like" structures; between these, numerous irregularly arranged pores (exceptionally, sculpturing restricted to lateral and ventral surface, with the dorsum simply smooth and porose, or some of the lines more pronounced, forming a reticulation with a larger mesh size); in general Cx-I medially as long as Cx-II (length ratio 0.8-1.3); legs without swimming setae proper, IV-L-5 with 1-2 fine terminal setae, the longer one in length about 1/3 , rarely 1/2, IV-L-6); IV-L-5 with 3-5 rather strong ventral setae, IV-L-6 ventrally with 2-4 very tiny nail-setae; excretory pore smooth or faintly sclerotized near anterior edge of slit; P-3 with the 2 dorsal setae not far apart (distance generally 8-25 µm) and in basal part of segment, mediodistal and dorsodistal setae not far apart (distance 8-10 µm); P-4 ventral sectors generally 1:1:2, rarely 1:1:1; peg seta small, dorsal setae of P-4 rather long, in particular one or two within the group arranged near distal end of segment; two dorsal setae near centre of segment. Males: Idiosoma length/width 750-800/600-700, Cx-II posterior width 20-35; genital flap length 133-155, length/height P-2, 84-90/58-60; P-3, 90-100/40-45; P-4, 108-113/35-38 µm. Females: Idiosoma length/width 800-1100/650-900, Cx-II posterior width 30-40; genital flap length 160-205, length/height P-2, 88-103/58-68; P-3, 95-118/43-45; P-4, 113-120/33-40 µm.

Preadult stages: Larva: ULLRICH (1978).

Similar species: *Lebertia glabra* and *L. maculosa*, two other species with integument sculpturing, differ in the linear arrangement of cuticular structures, the dorsal and distal setae of P-3 being further from each other, and P-4 bearing a stronger peg-like seta, shorter dorsal setae and ventrally divided in sectors 1:1:1. *Lebertia maderigena*, a species similar in integument structure, differs in P-4 (with maximum height in the basal quarter, ventral sectors 4:2:1, and all dorsal setae concentrated near tip of segment).

Habitat and Biology: Stream-dwelling in Western and Central Europe, exclusively in springs at the southern margin of its area. Larvae parasitic on chironomid Diptera (preferably Orthocladiinae and Tanytarsini, also Metriocnemini and Chironominae; parasitic larvae from May to November, with maximum intensity between August and October; generally 1-2 larvae per host, rarely up to 9, attachment sites: thorax, mostly near coxae of forelegs or on the cervix (data from Breitenbach, Germany: ULLRICH 1978).

Distribution: Frequent in Western and Central European mountains and lowlands, patchy records from the Western Alps (Dauphiné: MOTAŞ 1928e), Pyrenees (MOTAŞ & ANGELIER 1927d) and North Africa.

23. *Lebertia (Lebertia) schechteli* THOR, 1913 (*8-43 b, 8-44 a-b*)

1913b *Lebertia (Pseudolebertia) schechtelii* THOR, Zool. Anz. 42 (4): 181.

1914b *Lebertia (Pseudolebertia) tuberosa* THOR, Zool. Anz. 45 (1): 29 – GERECKE (2009).

Description: Colour orange to red; integument papillate, papillae ventrally elongated and flattened, arranged along regular lines (distance about 10 µm), dorsally more raised and circular, generally as large in diameter as, or less than, distance between papillae; Cx-I medially longer than Cx-II (Cx-I/II median length ratio 1.0-1.6, in most cases 1.2-1.3); posterior margin of Cx-II narrow, slightly separated from pregenital sclerite; legs without swimming setae (one slender, short distal seta on III/IV-L-4/5); ventral margin of IV-L-6 with 2 small setae; Ac rather large, Ac-2 subrectangular; excretory pore smooth; P-3 slender, basally narrow, the 2 dorsal setae separate from each other (distance 11-28 µm, proximal seta often nearly halfway between segment base and distal seta, mediodistal and dorsodistal setae close to each other); P-4 nearly equal in height from base to tip, maximum height in distal half, ventral sectors 1:2:2, mediodistal peg-like seta slender and pointed, the 2 dorsal setae close to each other in basal part of segment; P-5 rather short. Males: Idiosoma length/width 750-900/600-800; posterior width Cx-II 20-25 µm; posteromedial and medial edges of coxae anterior to pregenital sclerite sometimes fused by secondary sclerotization, forming a continuous anterior margin of genital bay; genital flaps length 135-190 µm, with a particularly dense medial seta cover (35-40 on each side). Females: Idiosoma length/width 800-1200/600-900; posterior width Cx-II 25-35; genital flaps length 190-210 µm, bearing about 20 medial setae regularly arranged over flap margin.

Preadult stages: Larva: WALTER (*sub nom. L. tuberosa*: 1922a).

Similar species: From Southeastern Europe, specimens are recorded having integumental papillae flattened, elongate and joining with neighbouring papillae to form lines and net-like structures (GERECKE 2009), but agreeing from all other points of view, and co-existing, with typical *L. schechteli*. *Lebertia rubella* K. VIETS, 1926 (Bulgaria) is a sister species only differing in the relatively longer Cx-I (median length ratio Cx-I/II 1.7). Occasionally, specimens of *L. maculosa* occur with dorsal lines locally subdivided into isodiametric papillae. However, at least in some integument areas papillae remain longish, and if they are rounded, they are close to each other, nearly in touch. Furthermore, the species differs in the broader posterior margin of Cx-II, more slender Ac, a shorter P-3 with dorsodistal and mediodistal setae placed apart and P-4 with its maximum height proximally.

Habitat and Biology: Crenobiont or crenophilous: In Romania found in higher order streams far distant from springs, but in other parts of the area the species is restricted to the eucrenal (Cîmpean & Gerecke 2006). In montane and alpine springs, mostly at higher elevations, often above the tree line, without a clear preference for a selected spring typology (Gerecke & Martin 2006). Larvae parasitic on chironomid Diptera (Tanytarsini), preferred attachment sites: cervix and thorax (Martin et al. 2010); possibility of facultative reduction of the parasitic larval stage discussed by Gerecke & Martin (2006).

Distribution: Western, Central and Southeastern Europe, restricted to higher mountain ranges; isolated populations recorded from Calabria and Corsica.

24. *Lebertia (Lebertia) sparsicapillata* Thor, 1905 (*8-46 e*)

1905c *Lebertia (Neolebertia) sparsicapillata* Thor, Zool. Anz. 28 (26): 823.

1919 *Lebertia (Neolebertia) westfalica* Koenike, Arch. Hydrobiol. 12 (3): 598 – Gerecke (2009).

1952a *Lebertia (Lebertia) rufipes cylindrica* Láska, Publ. Fac. Sci. Univers. Masaryk 5: 112 – Gerecke (2009).

1953b *Lebertia (Lebertia) longiepimerata* Láska, Acat Soc. zool. Bohemoslovenicae 17 (4): 290 – Gerecke (2009).

1954c *Lebertia (Pseudolebertia) corsica* E. Angelier, Vie et Milieu 5 (1): 98 – Gerecke (2009).

Description: Colour red, with deep purple sclerotized parts; idiosoma slender; integument striated, at least in the postgenital area, inner layer finely porose; Cx-I medially as long as, or longer than, Cx-II (ratio 0.9-1.3), posterior margin of Cx-II variable in width (30-60 µm); Cx-IV irregular, often appearing truncated in the area between posteromedial edge and a knob-shaped apodeme in centre of posterior margin; numbers of swimming setae: III-L-5, 1-2; IV-L-4, 0; IV-L-5, 2 (one longer, one shorter); ventral setae: IV-L-5, 4-5 (rarely 3-6), rather robust; IV-L-6, 2-3 (rarely up to 5), small; excretory pore smooth; P-2 ventral seta as long as ventral length of segment or slightly longer; P-3 dorsal setae close together, inserted proximally from centre of segment, distal setae not very elongated (not overreaching tip of P-5), medio- and dorsodistal setae close to each other, ventrodistal seta at a slight distance from ventrodistal segment edge; P-4 proximal and distal edges nearly equal in height, maximum height distally from, or rarely in, centre of segment, ventral margin straight, sectors 2:1:1, occasionally 1:1:1, dorsal margin convex, mediodistal peg-like seta stout, in size similar to one of the big claws of P-5. Males: Idiosoma length/width 850-1300/650-1000; genital flap length 160-190 µm, with posteromedial setae not particularly elongated (long as maximum width of a genital flap). Females: Idiosoma length/width 850-1200/600-900; genital flap length 190-210 µm.

Similar species: The *fimbriata/sparsicapillata* species complex is not well understood and needs further revision: *Lebertia fimbriata*, in most characters and measurements very similar, apart from the often greenish body coloration, differs in generally smaller measurements (e.g. Cx-I medial length ♂ 120-160, ♀ 135-180; genital flap length ♂ 100-175, ♀ 135-190 µm). In Southern and Southeastern Europe populations may be found intermediate in dimensions between the two species. Three further species representing this group, *L. jadrensis* K. Viets, 1936b from Croatia, the Mediterranean *L. intronata* Gerecke, 2009 and *L. vegacabrerae* Gerecke, 2009, endemic on Tenerife (Canary Islands), differ in various forms of integument sculpturing.

Habitat and Biology: Rhithrobiont; more frequent in waters draining crystalline bedrock. On Sicily in higher order streams, at middle and low elevation. Preadult stages and life cycle unknown.

Distribution: Europe except for most northern and eastern parts.

25. *Lebertia (Lebertia) subtilis* KOENIKE, 1902 (*8-45 e-f*)

1902b *Lebertia subtilis* KOENIKE, Zool. Anz. 25: 614.

Description: No information on colour; integument finely striated, Cx-I medially distinctly longer than Cx-II (ratio 1.3-1.5); legs without swimming setae; IV-L-6 bearing only 2 minute ventral setae; excretory pore smooth; P-3 dorsal setae apart (distance about 20 µm), mediodistal and dorsodistal setae close to each other; P-4 parallel-sided, ventral sectors 2:3:2, P-5 strong, with curved claws. Males: Cx-I/II median length 200/130, Cx-II posterior width 38; genital flap length 160 µm, with 7 lateral setae and numerous medial setae (about 50 on each side), in the posterior part densely arranged in double rows; length/height P-2, 93/60; P-3, 103/43; P-4, 108/33. Females: Idiosoma length/width 1000/850, Cx-I/II median length 170/130 µm, Cx-II posterior width 30, genital flap length 200 µm, with 15 setae on each side, not particularly dense in posterior part; length/height P-2, 98/63; P-3, 110/43; P-4, 113/33.

Similar species: Male distinct in the dense medial setation of the genital flaps, particularly in the posterior part. *Lebertia fimbriata* is very similar from all other points of view, differing only in Cx-I medially being slightly shorter, the presence of 1-2 elongated swimming setae on III/IV-L-5, and dorsal setae on P-3 less separated (distance 5-10 µm). In view of the scanty documentation and, apart from the male genital setation, hardly significant diagnostic characters, not allowing for recognizing the female, it cannot be excluded that this species is taxonomically insignificant, only based on a misshaped individual.

Habitat and Biology: Standing water. Preadult stages and life cycle unknown.

Distribution: Alps, with certainty recorded only from the type locality, a lake in Rhätikon, Switzerland.

Subgenus *Mixolebertia* THOR, 1906

1906a *Mixolebertia* THOR, Zool. Anz. 29 (25-26): 769.

1906e *Hexalebertia* THOR, Zool. Anz. 31 (2-3): 69 – NEWELL (1959).

Type species: *Lebertia brevipora* THOR, 1899a, by original designation.

Diagnosis: Medial surface of P-3 with six setae. Legs with, or (more frequently) without swimming setae. Species having a general tendency towards reduction in size and stronger sclerotization (extension of Cx-IV, occasionally in males resulting in the genital field surrounded by the coxal field and sclerotization of the excretory pore).

Discussion: *Mixolebertia* species are rather inhomogeneous in their morphology, but the subgenus includes numerous species agreeing in having a total absence of swimming setae. These share a rather uniform habitat preference, being restricted to spring habitats or lowest order streams, while representatives of a small group of swimming setae-bearing species are found in higher order streams and cold lakes. An origin from a common, possibly swimming setae- bearing ancestor is possible, but little supported by available data. More probably, *Mixolebertia* is a paraphyletic or polyphyletic clade.

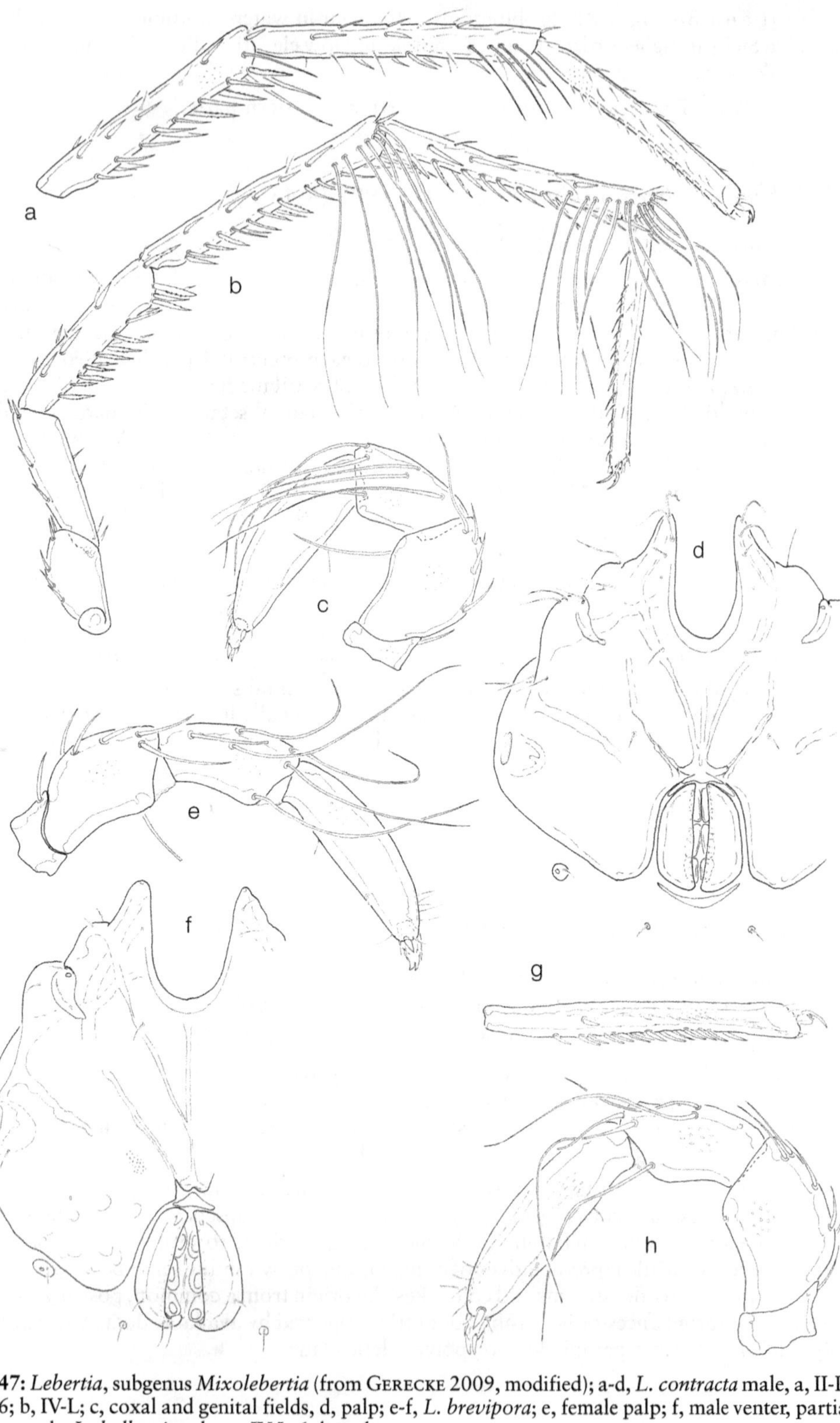

8-47: *Lebertia*, subgenus *Mixolebertia* (from GERECKE 2009, modified); a-d, *L. contracta* male, a, II-L-4-6; b, IV-L; c, coxal and genital fields, d, palp; e-f, *L. brevipora*; e, female palp; f, male venter, partial view; g-h, *L. halberti* male; g, IV-L-6; h, palp.

Key to species

The following (sub)species are excluded as *species incertae* (GLEDHILL & K. O.VIETS 1976; GERECKE 2009): *Lebertia aspera* WALTER, 1922a; *L. costata* KOENIKE, 1908d; *L. cylindrata* BADER, 1955; *L. duricoria* KOENIKE, 1911a; *L. incognita* K. VIETS, 1922d; *L. inflexa* THOR, 1913b; *L. longipes* MONTI, 1904a; *L. novipalpis* HALBERT, 1944; *L. pyrenaica* MIGOT, 1926c; *L. sigthori* MAGLIO, 1908b.

1 Numerous swimming setae on II-III-IV-L (*8-47 a-b*). II-L-5 bearing at least 2, often 3 to 10, swimming setae on the posterior distal surface. III-L-4 generally with 5 or more setae in a line in distal part of posterior surface, nearly as long as the subsequent segment III-L-5. . 2

- Swimming setae reduced in number or absent. II-L-5 at most with one short distal hair-like seta on posterior surface. III-L-4 bearing at most 4 distal setae on posterior surface. . 4

2 (1) Ac shortened, oval in shape: maximum diameter less than, or equal to, distance between Ac (*8-47 f* – note that Ac are situated on larger basal sclerites). . . . . *Lebertia brevipora* (page 127)

- Ac elongated, maximum diameter more than interspaces between Ac. 3

3 (2) Excretory pore smooth. Median suture Cx-II very short or absent (median length ratio Cx-I/II > 3.0). Swimming setae on II-L-5 shortened (*8-47 a*). Some ventral setae on IV-L-6 fine, hair-like (*8-47 b*). . . . . *Lebertia contracta* (page 128)

- Excretory pore with a sclerotized ring. Median suture Cx-II relatively long (median length ratio Cx-I/II < 2.5). Swimming setae on II-L-5 as long as the following segment. All ventral setae on IV-L-6 spine-like (*8-47 g*). . . . . *Lebertia halberti* (page 132)

4 (1) Cx-II with a relatively short medial suture line (median length ratio Cx-I/II > 1.8, in most cases > 2.0) and a broad posterior margin (normally > 50 µm, *8-48 a, c-d*) and P-4 distally equally narrowed, with strong, pointed mediodistal peg seta (*8-48 b, e*). 5

- Cx-II with a relatively longer suture line (median length ratio Cx-I/II < 1.7) and in most cases a more narrow posterior margin (e. g. *8-49 f*). Exceptions (median length ratio Cx-I/II 1.7-2.0) may be found in the species *L. sefvei* and *L. crenophila*, both characterized by a P-4 distally not equally narrowed, but with subparallel dorsal and ventral margins and a very small mediodistal peg seta (*8-52 c, e*). 8

5 (4) Excretory pore surrounded by a sclerotized ring. Integument lineated (as in *8-43 a*). . 6

- Excretory pore unsclerotized or (rarely) posteriorly embraced by a fine, horseshoe-shaped sclerite. Integument finely striated or smooth. 7

6 (5) Palp stout (length/height ratio P-3, 1.9-2.0; P-4, 2.9-3.3), ventral setae of P-4 further apart than the length of either the proximal or distal ventral sectors (*8-48 b*). 7-9 ventral setae on IV-L-6. . . . . *Lebertia dubia* (page 130)

- Palp more slender (length/height P-3 > 2.1, P-4 > 3.5), ventral setae of P-4 dividing segment in sectors equal in length, or inserting closer to each other (*8-48 e*). 9-11 ventral setae on IV-L-6. . . . . *Lebertia sinuata* (page 139)

7 (5) Median suture of Cx-II shortened or nearly absent (length < 60 µm, Cx-I/II median length ratio > 3.9 – *8-48 c*). Posterior end of medial apodemes of Cx-III shifted posteriorly (in ♂ posterior to centre of Ac-1, in ♀ in anterior third of Ac-1). . . . . *Lebertia oudemansi* (page 136)

- Median suture of Cx-II longer (length > 60 µm, Cx-I/II median length ratio < 3.5 – *8-48 d*). Posterior end of medial apodemes of Cx-III posteriorly not extending beyond level of anterior edge of Ac-1. . . . . *Lebertia helvetica* (page 133)

8 (4) Excretory pore without a sclerotized ring (*8-49 f*). 9

- Excretory pore with a sclerotized ring (*8-49 k*). 12

9 (8) Dorsal integument reticulated (*8-49 a*). P-4 relatively long (length ratio P-2/P-4, 0.50-0.55) and slender (length/height ratio 3.5-4.3) (*8-49 b*). Male posterior margin of Cx-IV partly embracing genital field (*8-36 a*). . . . . *Lebertia holsatica* (page 134)

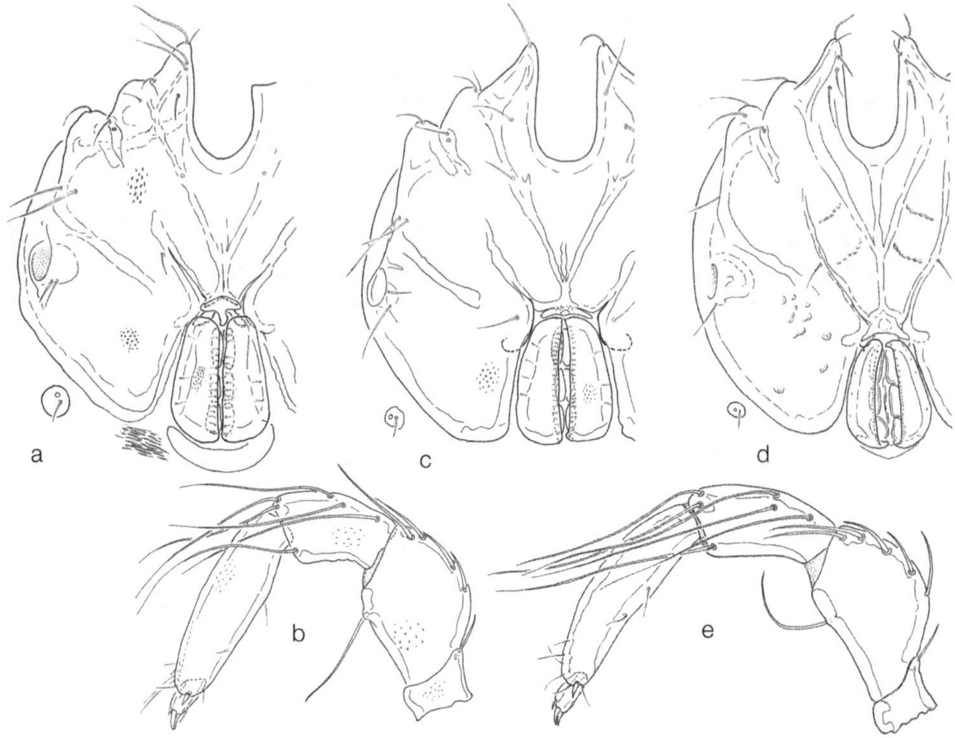

8-48: *Lebertia*, subgenus *Mixolebertia* (from Gerecke 2009, modified); a-b, *L. dubia* female; a, venter, partial view; b, palp; c, *L. oudemansi* male venter, partial view; d, *L. helvetica* male venter, partial view; e, *L. sinuata* male palp.

-	Dorsal integument lineated or with longish papillae arranged in rows, originating from disintegration of lines (*8-49 c*). Palp, and male coxal field, various in shape.	10
10 (9)	Medial suture of Cx-II short (length ♂, 105; ♀, 90 µm; Cx-I/II median length ratio ♂, 1.3; ♀, 1.8). P-3 proximally not strongly narrowed and in males relatively shorter (*8-49 e*). Males with posterior margin Cx-IV extending far over genital field and including V-1.	*Lebertia excellens* (page 131)
-	Medial suture of Cx-II longer (length ♂, 110-180; ♀, 110-160 µm; Cx-I/II median length ratio ♂, 1.0-1.2; ♀, 1.1-1.6). P-3 relatively long and proximally strongly narrowed (*8-49 d, g*). Males (known in only one species) with posterior margin Cx-IV extending only partly over genital field, or ending at the same level, not including V-1 (*8-49 f*).	11
11 (10)	P-2 more slender (length/height > 1.4); P-4 more stout (length/height ratio < 3.7), length ratio P-2/P-4, 0.65-0.70 (*8-49 g*).	*Lebertia lativentris* (page 135)
-	P-2 more stout (length/height < 1.4), P-4 more slender (length/height ratio > 3.7), length ratio P-2/P-4, 0.61 (*8-49 d*).	*Lebertia depressostriata* (page 130)
12 (8)	Dorsal integument with irregular rugose sculpturing (*8-49 h*) or reticulated (*8-49 i*). .	13
-	Dorsal integument lineated (as in *8-43 a*) or smooth.	14
13 (12)	Dorsal integument with irregular bulges formed by the fusion of longitudinal lines; a fine porosity on these bulges (*8-49 h*).	*Lebertia macilenta* (page 136)
-	Dorsal integument reticulated, a fine porosity in the honeycomb-like meshes (*8-49 i*).	*Lebertia semireticulata* (page 137)

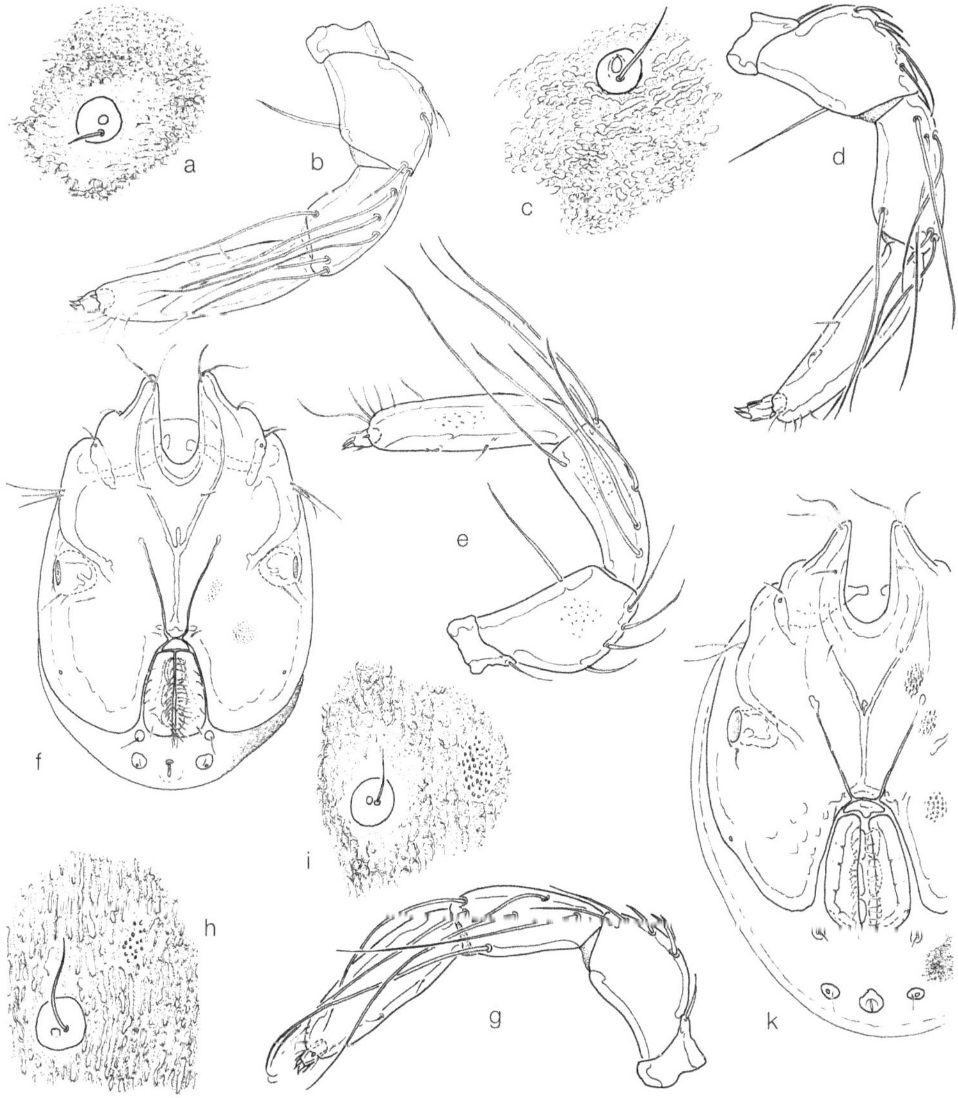

8-49: *Lebertia*, subgenus *Mixolebertia* (from GERECKE 2009, modified); a-b, *L. holsatica* male; a, integument structure; b, palp; c-d, *L. depressostriata* female; c, integument structure; d, palp; e, *L. excellens* female palp; f-g, *L. lativentris* male; f, venter; g, palp; h, *L. macilenta* female integument structure; i-k, *L. semireticulata* male; i, integument structure; k, female venter, partial view.

14 (12) P-4 ventral setae separated (distoventral seta may be further away from proximoventral seta than from distal segment edge – *8-50 d-e, 8-51 b, f-g*); segment relatively short and in most species bent (length ratio P-2/P-4, 0.7-0.9; length/height ratio P-4, 2.6-3.4). 15

\- P-4 ventral setae closer to each other (distoventral seta never further from proximoventral seta than from distal segment edge – *8-52 c, 8-53 b-c*); segment longer and straighter (length ratio P-2/P-4, 0.5-0.7; length/height ratio P-4, 3.2-4.2). 24

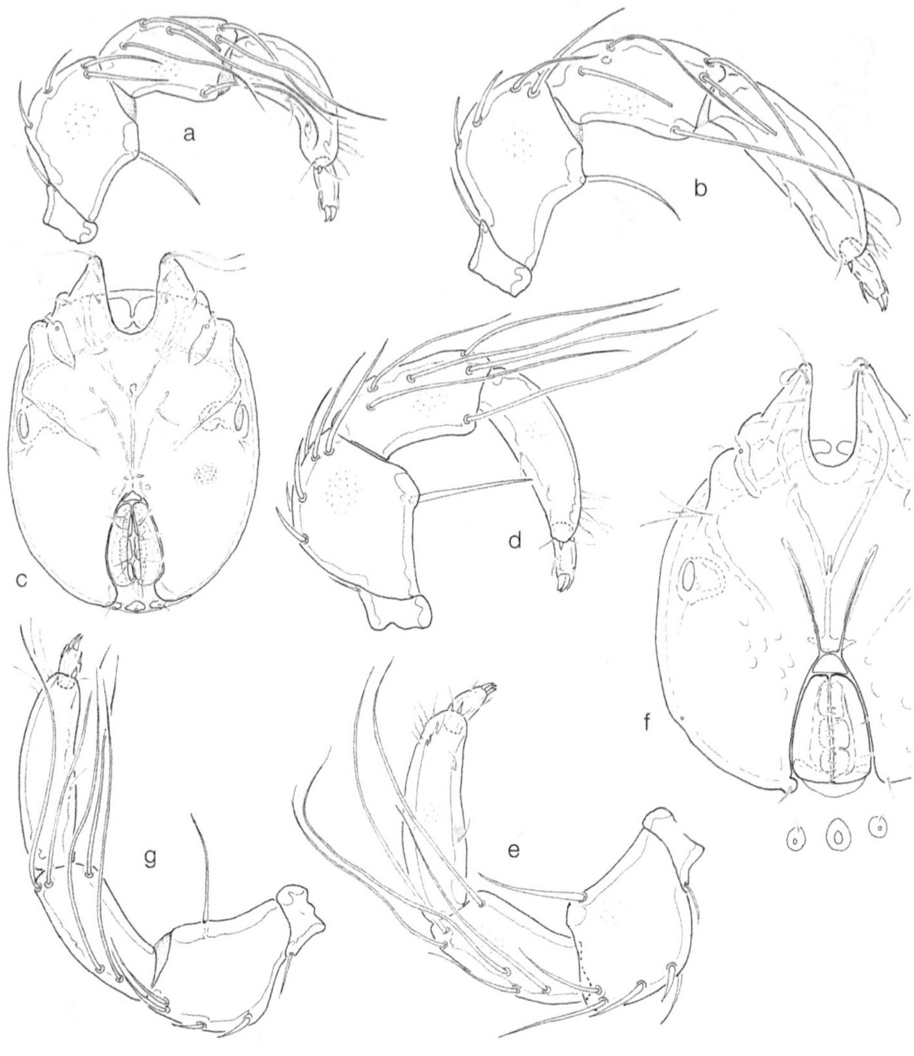

8-50: *Lebertia*, subgenus *Mixolebertia* (from GERECKE 2009, modified); a-b, *L. aberrata*; a, male palp; b, female palp; c-e, *L. cuneifera*; c, male venter; d, male palp; e, female palp; f-g, *L. bisbisetosa* male; f, venter, partial view; g, palp.

15 (14) Posterior margin of pregenital sclerite straight, without posteromedial extension (*8-50 c, f*), posterior margin of Cx-IV extended at least to level of posterior margin of genital field and including V-1, often partly (*8-50 c*) or completely (*8-51 a*) embracing the genital field. Ejaculatory complex present: Males. 16

\- Posterior margin of pregenital sclerite with a medial extension (*8-51 c*); posterior margin of Cx-IV not reaching the level of posterior margin of genital field, not including V-1. Ejaculatory complex absent: Females. 20

16 (15) P-4 shortened and bent (length/height ratio P-4 < 3.0; length ratio P-2/P-4, 0.9 – *8-50 a, d*). IV-L relatively stout (length/height IV-L-4 < 3.8). 17

\- P-4 longer and straight (length/height ratio P-4, > 3.0; length ratio P-2/P4, < 0.9 – *8-50 g, 8-51 b, f*). IV-L more slender (length/height IV-L-4 > 3.8). 18

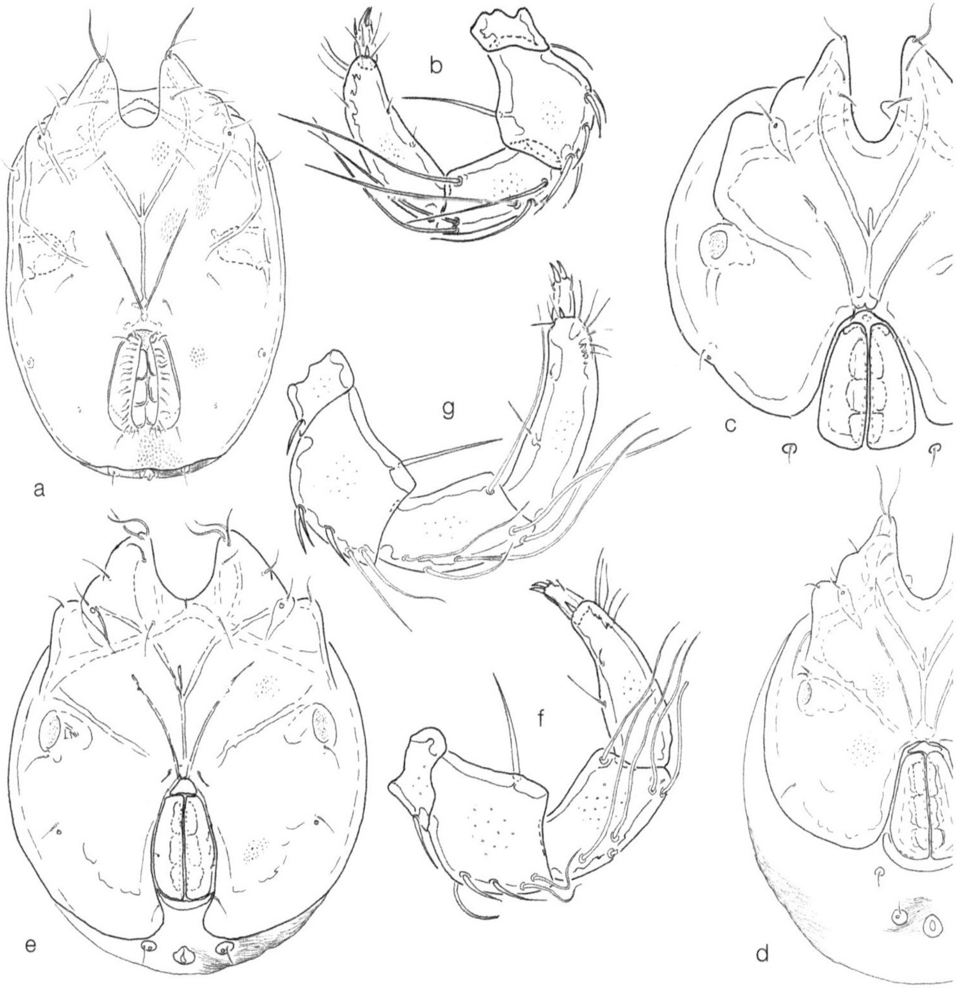

8-51: *Lebertia*, subgenus *Mixolebertia* (from GERECKE 2009, modified); a-c, *L. giardinai*; a, male venter; b, male palp; c, female venter, partial view; d-g, *L. fontana*; d, female venter, partial view; e, male venter; f, male palp; g, female palp.

17 (16) Palp very stout (length/height ratio P-3 1.8), P-4 strongly bent ventrally (*8-50 a*). IV-L-6 with 9-15 ventral setae. . . . . *Lebertia aberrata* (page 126)

- Palp more slender (length/height ratio P-3 > 1.8), P-4 less strongly bent ventrally (*8-50 d*). IV-L-6 with 4-7 ventral setae. . . . . *Lebertia cuneifera* (page 129)

18 (16) Posterior margin of Cx-IV not or only partly extending beyond posterior margin of genital field; its posteromedial edges not embracing the genital flaps posteriorly; posterior margin Cx-II relatively large (width 30 µm – *8-50 f*). Palp slender (length/height ratio P-2 1.6; P-3, 2.4 – *8-50 g*). . . . . *Lebertia bisbisetosa* (page 126)

- Posterior margin of Cx-IV distinctly extending beyond posterior margin of genital field and embracing it posteriorly, posterior margin Cx-II narrow or not recognizable due to fusion of all sclerotized elements of pregenital coxal field (*8-51 a, e*). Palp less slender (length/height P-2 < 1.5; P-3 < 2.3 – *8-51 b, f*). 19

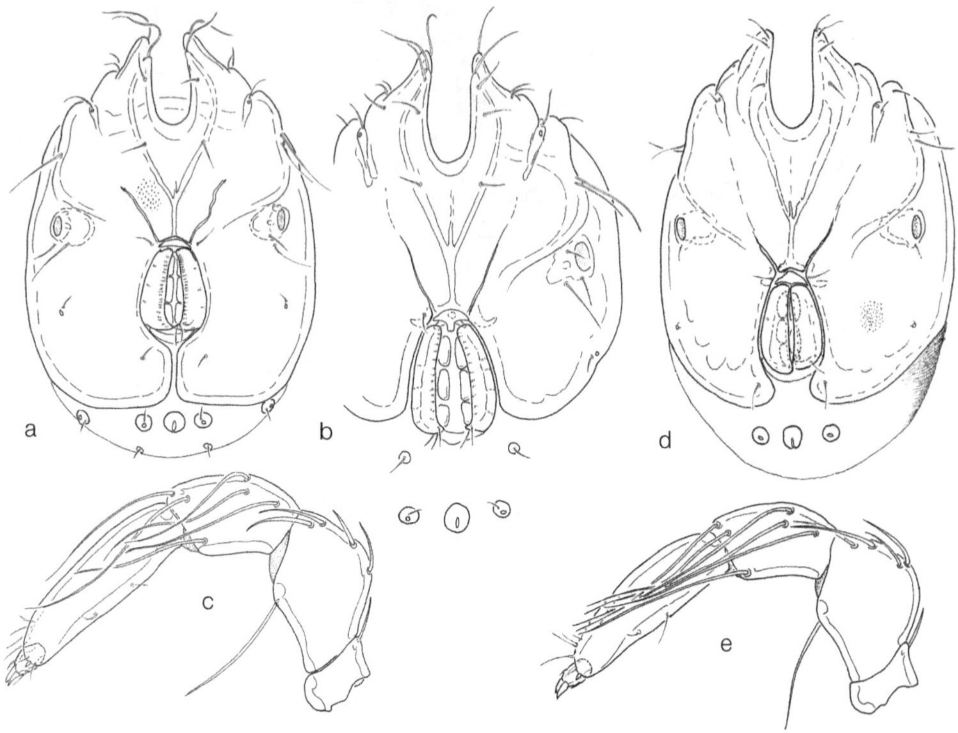

8-52: *Lebertia*, subgenus *Mixolebertia* (from GERECKE 2009, modified); a-c, *L. sefvei*; a, male venter; b, female venter partial view; c, female palp; d-e, *L. crenophila* male; d, venter; e, palp.

19 (18) Genital field completely surrounded by a closed ventral shield (*8-51 a*). . . . . *Lebertia giardinai* (page 132)

\- Posteromedial extensions of Cx-IV leaving a broad strip of membranous integument connecting the genital field with the area around the excretory pore (*8-51 e*). . . . . *Lebertia fontana* (page 131)

20 (15) Ventral seta of P-2 located on a more or less pronounced nose-shaped projection, segment margin convex distally from the seta insertion (*8-50 b, e*). 21

\- Ventral margin of P-2 straight, not projecting near the seta insertion, forming an abrupt ventrodistal angle (*8-51 g*). 22

21 (20) Palp stout (length/height ratio P-3, 2.0 – *8-50 b*). IV-L-6 with 9-15 ventral setae. . . . . *Lebertia aberrata* (page 126)

\- Palp more slender (length/height ratio P-3 $>$ 2.3 – *8-50 e*). IV-L-6 with 4-7 ventral setae. . . . . *Lebertia cuneifera* (page 129)

22 (20) P-2 slender (length/height ratio $>$ 1.5 – as in *8-50 g*). Smaller in dimensions (median length Cx-II $<$ 90, length P-4 $<$ 120 µm). . . . . *Lebertia bisbisetosa* (page 126)

\- P-2 stouter (length/height ratio $<$ 1.5 – *8-51 g*). Larger in dimensions (median length Cx-I, $\geq$ 90, length P-4 $\geq$ 120 µm). 23

23 (22) Distance between ventral setae of P-4 greater than distance of proximal seta from segment base (*8-51 g*). . . . . *Lebertia fontana* (page 131)

\- Distance between ventral setae of P-4 less than distance of proximal seta from segment base. . . . . *Lebertia giardinai* (page 132)

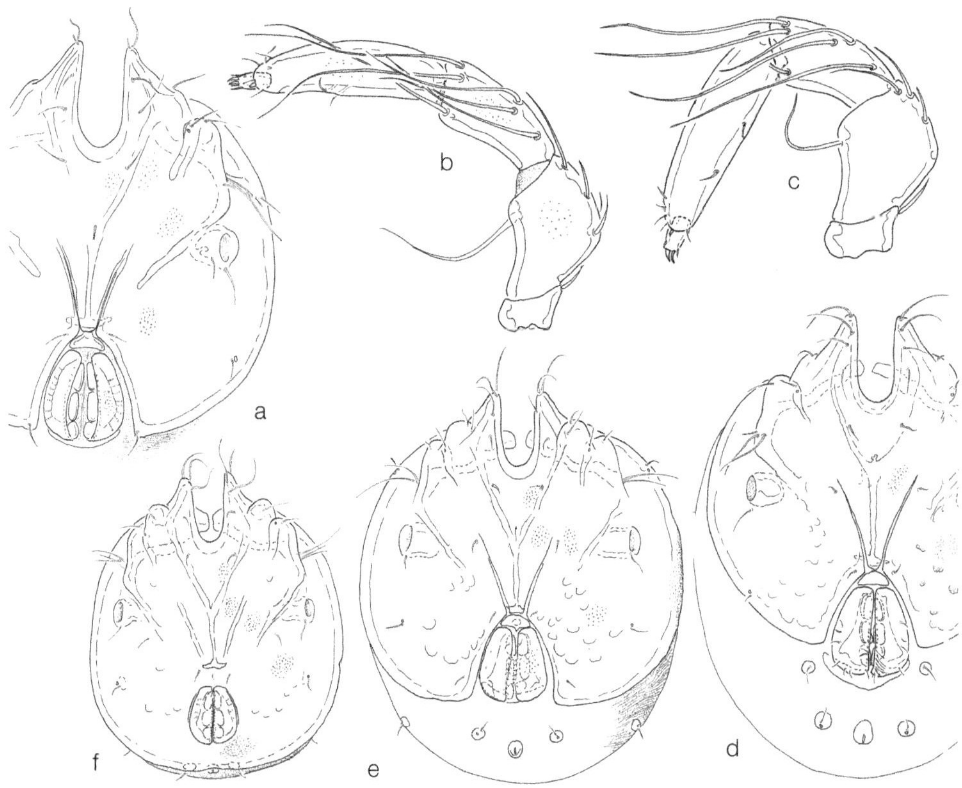

8-53: *Lebertia*, subgenus *Mixolebertia* (from GERECKE 2009, modified); a-b, *L. stigmatifera* male; a, venter, partial view; b, palp; c, *L. bracteata* male palp; d, *L. separata* male venter; e-f, *L. helocrenica* venter; e, female; f, male.

24 (14) Posterior margin of Cx-II broad (width ♂ > 40; ♀ > 50 µm) (*8-52 a-b*). 25

\- Posterior margin of Cx-II narrower (width ♂ < 40; ♀ < 50 µm), often Cx-II posteriorly pointed, occasionally invisible due to fusion of all sclerites to an unique coxal shield (*8-53 a, d, e-f*). 26

25 (24) P-4 slender (length/height ratio > 3.3, *8-52 c*); male Cx-IV posteriorly embracing the genital field almost completely, leaving only a narrow strip of membranous integument (*8-52 a*). . . . . *Lebertia sefvei* (page 137)

\- P-4 stouter (length/height ratio < 3.3, *8-52 e*); male Cx-IV posteriorly less extensive, leaving a large area of membranous integument (distance between posteromedial coxal edges ca. 70 µm – *8-52 d*). . . . . *Lebertia crenophila* (page 128)

26 (24) Coxal field enlarged and extending onto flanks of idiosoma, including at least one lateroglandularium. Male genital field completely enclosed by coxal shield (*8-53 f*). In females posterior margin of Cx-IV (not extending beyond genital field) hardly curved, reaching lateral idiosoma walls at level of genital field centre (*8-53 e*). . . . . . . *Lebertia helocrenica* (page 133)

\- Coxal field not extending onto flanks of idiosoma (*8-53 d*), all lateroglandularia unfused. Male genital field not enclosed by coxal shield (*8-53 a, d*). Female posterior margin of Cx-IV (not extending beyond genital field) strongly curved, reaching lateral idiosoma wall at level of anterior genital field margin. 27

27 (26) P-2 stout (length/height ratio < 1.4), P-4 ventral setae away from each other (distance > 20 µm, > 20 % length ventral margin) (*8-53 c* – check both palps! classification of single specimens may be problematic due to overlapping variability ranges). Male Cx-IV may extend far beyond posterior margin of genital field, posteromedial edges may be pointed or truncated. . . . . *Lebertia bracteata* (page 127)

\- P-2 more slender (length/height ratio > 1.4), P-4 ventral setae closer to each other (distance < 20 µm, < 20 % length ventral margin) (*8-53 b*). Male Cx-IV posteromedially with pointed tips, at most slightly extending beyond level of posterior genital field margin. 28

28 (27) Males (females not distinguishable): Posterior margin of Cx-IV reaching, or slightly extending beyond posterior margin of genital field, V-1 generally fused to the posteromedial edge of Cx-IV (*8-53 a*). . . . . *Lebertia stigmatifera* (page 139)

\- Males: Posterior margin of Cx-IV not reaching level of posterior margin of genital field, V-1 not fused to the posteromedial edge of Cx-IV (*8-53 d*). . . . *Lebertia separata* (page 138)

26. *Lebertia (Mixolebertia) aberrata* K. Viets, 1922 (*8-50 a-b*)

1922d *Lebertia (Hexalebertia) aberrata* K.Viets, Arch. Naturg. 88 (A 9): 62.

1925g *Lebertia (Hexalebertia) cuneifera aberrata* K.Viets, Zool. Jbch. Syst. 50: 500.

Description: Integument lined; medial suture Cx-I long (ratio Cx-I/Cx-II, 1.1-1.5); legs without swimming setae; IV-L-6 with 9-15 ventral setae, these setae stouter in the central part, more hair-like in the distal part of segment; genital bay with sexual dimorphism; excretory pore sclerotized; palp robust, with strong sexual dimorphism. Males: Idiosoma length/width 600/500, Cx-II posterior width 10-20 µm; posterior margin of Cx-IV with extended sclerotization, posteriorly embracing genital field, but leaving a strip of membranous integument (width 70 µm); genital flap length 140 µm, about 15 pairs of medial setae, in the centre arranged more densely than anteriorly and posteriorly; P-2 notably inflated and with a nose-like ventrodistal projection near ventral seta insertion; P-3 stout (length/height 1.8), proximal dorsal seta inserting halfway between segment base and other dorsal setae, distal setae away from distal margin; P-4 basally inflated, very narrowed in distal part, dorsal margin smoothly rounded, ventral margin with a distinct kink, distal part almost forming a right angle with segment base; length ratio P-2/P-4, 0.9, P-3/P-4, 1.0. Females: Idiosoma length 800, Cx-II posterior width 45; genital flap length 180 µm; P-2 with a nose-like projection as in male, but P-3 less stout (length/height 2.0), P-4 less strongly narrowed in distal part, but without a ventral kink; length ratio P-2/P-4, 0.7; P-3/P-4, 0.8.

Similar species: In the shape of palp and coxal field and the excretory pore with a sclerotized ring, *L. aberrata* is most similar to *L. cuneifera* which differs in a more slender palp with length/height P-3 in males > 1.8, in females > 2.3, the less strongly bent male P-4, and a lower number of 4-7 ventral setae on IV-L-6.

Habitat and Biology: Crenobiont. Preadult stages and life cycle unknown.

Distribution: Central European mountains, Alps.

27. *Lebertia (Mixolebertia) bisbisetosa* K.Viets, 1930 (*8-50 f-g*)

1930d *Lebertia (Hexalebertia) bisbisetosa* K. Viets, Arch. Hydrobiol. 21 (2): 368.

Description: Integument with rather fine lines separate from each other; Cx-I/II median length ratio 1.5-1.9, width posterior margin Cx-II ca. 30 µm; legs without swimming setae, IV-L-6 with 5-9 ventral setae; excretory pore with sclerotized ring; P-2 length/height ratio > 1.5, ventral seta slender, dorsodistal setae close to each other, P-3 dorsal setae in proximal half of segment, the two distal ones close to each other, P-2/P-4 length ratio 0.7-0.8, P-4 slender, slightly curved, maximum height in the centre, distally equally narrowed, ventral setae insertions dividing segment into sectors 1:1:1 or 3:2:2, peg seta well developed and pointed. Males: Idiosoma length/width 700-750/500-650 µm, posteromedial margins of Cx-IV reaching level of posterior margin of genital field, including V-1 in a pointed angle; genital flap length 160 µm, with 30 very densely arranged medial setae and ca. 12 lateral setae. Females: Idiosoma length 700 µm, Cx-IV not reaching level of posterior margin of genital field, not including V-1; genital flap length 195 µm, with 13 scattered medial setae and 3 lateral setae.

Similar species: Species of the *giardinai*-group are similar in having a lineate integument, legs without swimming setae, excretory pore sclerotized, and in the shape and setation of P-4, but differ in having a stouter palp (e. g., P-2 length/height ratio < 1.5). Males of this group differ furthermore in having the posteromedial edges of Cx-IV embracing the genital field, and larger posterior margin of Cx-II.

Habitat and Biology: Rhithrobiont? Preadult stages and life cycle unknown.

Distribution: Pyrenees (France, Spain).

28. *Lebertia (Mixolebertia) bracteata* K.Viets, 1925 (*8-53 c*)

1925g *Lebertia (Hexalebertia) bracteata* K.Viets, Zool. Jbch. Syst. 50: 511.

Description: Integument lined; coxal field with sexual dimorphism; posterior margin Cx-II width 30-40 µm; legs without swimming setae, IV-L-6 with 3 ventral setae; excretory pore with a sclerotized ring; P-2 stout (length/height < 1.35), dorsal setae on P-3 may be close to each other or further apart, with distal one closer to distal setae, P-4 slender, distally equally narrowed, peg seta small, ventral sectors 3:2:2 or 2:1:1; P-2/P-4 length ratio 0.62-0.63, P-3/P-4 length ratio 0.67, P-5 small and short. Males: Idiosoma length/width 700/540 µm; Cx-I medially as long as Cx-II; Cx-IV including V-3 and V-1, the contour of the insertion platelets of the latter visible as knob-like extensions; posteromedial extensions of Cx-IV reaching level of, but not embracing posterior margin of genital field; genital flap length/width 165/130 µm, with 25 medial setae mostly in posterior part, and 6 fine lateral setae; P-3 length/height 2.0). Females: Idiosoma length/width 800/600; genital flap length 200 µm, with 12 medial setae in centre of plate margin arranged more densely than in anterior and posterior parts, and 4 lateral setae.

Similar species: *Lebertia stigmatifera* is similar in idiosoma morphology, but differs in having a more slender P-2 (length/height ratio > 1.35), ventral setae of P-4 close to each other (sectors 3:1:3), and male P-3 more slender (length/height ratio > 2.0). *Lebertia mediterranea* Gerecke, 2009 from springs in S Europe can be distinguished only in the male sex (with Cx-I medially longer than Cx-II, Cx-IV posteriorly extending beyond posterior genital field margin, their posteromedial tips truncated, in general including also the glandularia flanking the excretory pore).

Habitat and Biology: Crenobiont? Preadult stages and life cycle unknown.

Distribution: Central, Western and Southeastern Europe, poorly documented.

29. *Lebertia (Mixolebertia) brevipora* Thor, 1899 (*8-47 e-f*)

1899a *Lebertia brevipora* Thor, Kristiania (O. Norli): 3.

Description: Integument ventrally lined, laterally and dorsally with minute short ridges arranged in lines, locally similar to papillae; Cx-I medially long (Cx-I/II median length ratio

1.5-2.6), posterolateral margin Cx-IV concave close to V-3; legs with swimming setae extending beyond end of following segment – numbers: II-L-5, 2-4; III-L-3, 1; III-L-4, 2-5; III-L-5, 4-8; IV-L-4, 4-7; IV-L-5, 9-11; IV-L-6 with ca. 10 ventral setae; genital field with Ac reduced in diameter, but on enlarged basal sclerites; excretory pore sclerotized; P-4 ventral sectors 1:1:1, peg-like mediodistal seta strong, all fine dorsal setae in a longitudinal distal groove. Males: Idiosoma length/width 1000/800-850, Cx-II posterior width 45-50; genital flap length 200-210 µm. Females: Idiosoma length/width 1200/1000, Cx-II posterior width 60; genital flap length 275 µm.

Habitat and Biology: Standing waters, e.g. pools in tundra. Preadult stages and life cycle unknown.

Distribution: Scandinavia, only known from a few sites in northern Norway and Sweden (LUNDBLAD 1968).

30. *Lebertia (Mixolebertia) contracta* THOR, 1900 (*8-47 a-d*)

1900a *Lebertia (Pilolebertia) contracta* THOR, Nyt Mag. Naturv. Oslo 38 (3): 274.

Description: Large in general dimensions, integument finely striated, posterior end of Cx-I pointed; Cx-II medially very short, occasionally Cx-II medially not touching (Cx-I/II median length ratio > 2.5); Cx-II posterior width 70-90; legs with long swimming setae – numbers: II-L-5, 4 (but distinctly shorter and finer than those of III-IV-L); III-L-4, 4; III-L-5, 9-10; IV-L-4, 7; IV-L-5, 11; IV-L-6 very slender, bearing numerous ventral setae and minute claws; genital flaps with nearly straight lateral margins, pre- and postgenital sclerites large and narrow; excretory pore smooth; P-2/4 length ratio ca. 0.65; P-4 distally strongly narrowed, ventral sectors 1:1:1; P-5 small. Males: Idiosoma length/width 1100-1200/900 µm, pregenital sclerite fused to posterior margin of Cx-II, genital flap length 190-207 µm, with about 15 pairs of short medial setae arranged at regular intervals. Females: Idiosoma length 1250; genital flap length 203 µm, pregenital sclerite separate from posterior margin of Cx-II.

Habitat and Biology: Mostly in lakes; in the southern part of the area found only in deep water (MUCKLE 1942). Preadult stages and life cycle unknown.

Distribution: Apart from the single outpost in Lake Constance (MUCKLE 1942), only records from northern Europe: Norway (THOR 1906d), northern Sweden (LUNDBLAD 1968), Karelia (SOKOLOW 1926a); rare.

31. *Lebertia (Mixolebertia) crenophila* K.VIETS, 1920 (*8-52 d-e*)

1920b *Lebertia (Hexalebertia) crenophila* K.VIETS, Arch. Hydrobiol. 12 (4): 807.

Description: Integument lined, lines rather short, often not exceeding glandularium diameter; coxal field with sexual dimorphism; Cx-I medially relatively long (Cx-I/II median length ratio 1.7-1.9); Cx-II posterior margin width 50-70 µm; insertion of IV-L very close to suture Cx-III/IV; legs without swimming setae, IV-L-6 with 4 ventral setae; excretory pore with sclerotized ring; P-2 ventral seta slightly away from distal edge, ventral segment margin here slightly convex; P-3 proximal dorsal seta near segment base, both dorsal and dorsodistal pairs of setae apart, P-2/P-4 length ratio 0.62-0.65, P-4 straight, maximum height in proximal part, length/height ratio 3.1-3.3, distally equally narrowed, mediodistal peg seta tiny, not pointed, P-4 ventral sectors 3:2:2 or 3:1:2; P-5 stout. Males: Idiosoma length/width 800/650 µm, coxal field extending onto lateral idiosoma flanks and to posteroventral surface by secondary sclerotization, Cx-IV including setae V-3 and V-1, their forceps-like posteromedial edges embracing genital field (distance of posteromedial tips

70 µm); genital flap length 160 µm, with 15 pairs of medial setae in anterior half located further apart than in posterior half; genital bay maximum width 125 µm, with slightly concave lateral margins. Females: Idiosoma length/width 1000/750 µm, lateral and posterior margins of Cx-IV not extended by secondary sclerotization, not including V-1; genital flap length 200 µm, with 10 pairs of regularly arranged medial setae.

Preadult stages: Larva: K.VIETS (1923a).

Similar species: *Lebertia sefvei* (similar in the wide posterior margin of Cx-II, a straight P-4 with ventral setae close to each other and posterior margin of male Cx-IV embracing genital field), differs in having a more slender P-4 (length/height > 3.3), IV-L insertions further from suture Cx-III/IV and male Cx-IV extending more posteromedially, nearly completely embracing genital field and leaving only a very narrow membranous strip.

Habitat and Biology: Cold-stenotherm; crenophilous or crenobiont. In Luxembourg in (rheo)helocrenes rich in macrophytes and fine detritus (GERECKE et al. 2005b). Life cycle unknown.

Distribution: Central and Eastern Europe, rare, most records questionable due to uncertain species definition.

32. *Lebertia (Mixolebertia) cuneifera* WALTER, 1922 (*8-50 c-e*)

1922a *Lebertia (Hexalebertia) cuneifera* WALTER, Rev. Suisse Zool. 29 (7): 360.

Description: Integument with long lines regularly arranged on dorsal surface; coxal field with sexual dimorphism; Cx-I medially as long as, or slightly longer than, Cx-II (Cx-I/II median length ratio 0.9-1.4), Cx-II posteriorly narrow; legs without swimming setae, IV-L-6 robust, with 4-6 ventral setae; excretory pore sclerotized; palp stout, with strong sexual dimorphism; P-2 ventral margin forming a distal edge – in males, often visible as a nose-shaped projection, in females smoother, but always recognizable; P-4 curved, ventral setae separated. Males: Idiosoma length/width 650-750/550-600 µm; Cx-IV posterior margin embracing genital field, but leaving a broad strip of membranous integument; genital flap length 135-160 µm, 20-25 pairs of densely arranged medial setae; often postgenital sclerite with a posterior extension and excretory pore sclerite with an anterior extension, or these two sclerites fused by a bridge; P-2 ventral margin forming a distal hump near insertion of ventral seta; P-3 with dorsal setae arranged in a triangle or distal seta away from the other 2; P-4 short and stout, equally narrowed and curved, ventral sectors 2:2:1, distoventral seta away from ventral margin, peg-like mediodistal seta pointed; P-2/P-4 length ratio ca. 0.9; P-5 long and slender. Females: Idiosoma length/width 800-900/750-800 µm; Cx-IV posterior margin not extending beyond genital field; genital flap length 190-200 µm, with 12-15 medial setae more densely arranged in centre; palp far more slender than that of male; P-4 equally curved but less narrowed than in male, ventral sectors 1:1:1; P-2/P-4 length ratio 0.77-0.78.

Similar species: Different from all *Mixolebertia*-species in the shape of P-2, forming a distinct edge distally from the insertion of the stout ventral seta. In the rather short P-4 (P-2/P-4 length ratio > 0.8 in males, > 0.75 in females), most similar to *L. fontana*, a species in the male differing in stouter segments of IV-L (length/height ratio IV-L-4 ≥ 4.0; IV-L-5, ≥ 5.1; IV-L-6 ≥ 4.6) and the longer P-4 (P-2/P-4 length ratio 0.66-0.88. Also female *L. fontana* have a slightly longer P-4 (P-2/P-4 length ratio 0.69-0.76), but overlap in all other proportions listed for males. See also discussion of *L. aberrata*.

Preadult stages: Larva: MARTIN (2003).

Habitat and Biology: Crenobiont; preferably in (rheo)helocrenes with much detritus (GERECKE & MARTIN 2006). Deutonymphs and adults feed on chironomid larvae (MARTIN

2005). Larvae parasitic on chironomid Diptera (Orthocladiinae), preferred attachment site: thorax, near frontleg coxae (MARTIN et al. 2010).

Distribution: Central and Southern Europe, Carpathians; most records from the Alps, also reported from southern Italy and Montenegro.

33. *Lebertia (Mixolebertia) depressostriata* K.VIETS, 1952 (*8-49 c-d*)

1952c *Lebertia (Hexalebertia) depressostriata* K.VIETS, Zool. Anz. 149 (11/12): 245.

Description: Female (male unknown): Integument with prominent, very short lines which in some parts of dorsum disintegrate into longish papillae; idiosoma length/width 750/550 µm, Cx-I/II median length 210/130 µm (ratio 1.2); posterior width Cx-II 30 µm; legs without swimming setae; IV-L-6 with 5 ventral setae; genital flap length 180 µm; excretory pore with a small sclerotized point at anterior edge, but without sclerotized ring; palp slender; length/height P-2, 83/63; P-3, 98/43; P-4, 135/35 µm; P-2/P-4 length ratio 0.61; P-2 with a slender, relatively short ventral seta, dorsal setae on P-3 close to proximal margin, P-4 maximum height in the proximal part of segment, distally weakly narrowed, peg seta small, knob-shaped, ventral margin divided by seta insertions into sectors 3:2:2.

Similar species: *Lebertia lativentris* and *L. longipalpis* K.VIETS, 1936b, the latter only known from Macedonia, agree in the combination of an unsclerotized excretory pore and integumental sculpturing made up by stout, shortened lines. *Lebertia lativentris* differs in a more slender P-2 and P-3, but stouter P-4 and a wider posterior margin of Cx-II; *L. longipalpis* in the relatively longer P-4 and a lower number of ventral setae on IV-L-6 (see also discussion of *L. excellens*).

Habitat and Biology: Hygropetric spring. Preadult stages and life cycle unknown.

Distribution: French Alps; only known from the type locality (Villar d'Arène).

34. *Lebertia (Mixolebertia) dubia* THOR, 1899 (*8-48 a-b*)

1899a *Lebertia dubia* THOR, Kristiania (O. Norli): 4.

1902b *Lebertia densa* KOENIKE, Zool. Anz. 25 (679): 611 – GERECKE (2009).

Description: Integument dorsally with short, undulating parallel lines; coxal area of older specimens surrounded by a border of secondary sclerotization; Cx-II medially short (Cx-I/II median length ratio 2.2-3.3) posterior margin of Cx-II wide, medially indented in an obtuse angle; medial apodemes of Cx-III posteriorly ending on the level of anterior margin of Ac-1; legs without swimming setae (only one slightly elongated seta anteriorly on IV-L-5), IV-L-5 with 9, IV-L-6 with 6-8 ventral setae; excretory pore sclerotized; P-2 ventral seta inserting near a minute medial extension; length ratio P-2/4 ca. 0.7; P-4 ventral sectors 1:1:1. Males: Idiosoma length/width 900/700, Cx-II posterior width 50 µm; posterior margin of Cx-II fused with pregenital sclerite; genital flap length 170 µm. Females: Idiosoma length/width 1400-1500/1000-1100, Cx-II posterior width 80; genital flap length 230-255 µm.

Preadult stages: Larva: (*sub nom. L. densa*): TUZOVSKIJ (2006a).

Similar species: *Lebertia dubia* is unique among *Mixolebertia* species in the combination of a lined integument, a relatively short medial suture to Cx-II (Cx-I/II median length ratio > 2.0), legs without swimming setae and the excretory pore with a sclerotized ring.

Habitat and Biology: Springs and streams in lowlands, at higher latitudes also in oligotrophic lakes (LUNDBLAD 1968). Life cycle unknown.

Distribution: Western Palaearctic, but excluding areas with a Mediterranean climate.

35. *Lebertia (Mixolebertia) excellens* LUNDBLAD, 1956 (*8-49 e*)

1956a *Lebertia tenuicollis excellens* LUNDBLAD, Ark. Zool. 10 (1): 118.

1975 *Lebertia excellens* BADER, Ergebn. wiss. Unters. schweizer. Nationalpark 14: 135.

Description: Integument with short lines, locally forming papillae; Cx-II medial suture shorter than that of Cx-I; legs without swimming setae, IV-L-6 with 4-7 ventral setae; excretory pore unsclerotized; P-2 ventral seta slender, ratio P-2/P-4 length 0.6-0.7, P-3 (length/height 2.3-2.8) and P-4 (length/height 3.4-3.9) slender and long, both basally narrow, P-4 maximum height in basal third, distally equally narrowed, ventral margin divided by seta insertions into sectors 3:2:3 or 2:2:3. Males: Idiosoma length/width 690-750/430 µm, genital bay embraced by large, posteromedially projecting extensions of Cx-IV which include V-1; genital flap length 120-130 µm, length/height P-2, 66-75/45-46; P-3, 80-90/34-35; P-4, 110-113/29-32. Females: Idiosoma length/width 800/650 µm; genital bay not embraced by extensions of Cx-IV, V-1 free; genital flap length 180 µm; length/height P-2, 86/50; P-3, 108/39; P-4, 125/33 µm.

Similar species: *Lebertia lativentris* is similar in the relatively long P-3, and the ventral setae of P-4 further apart, but differs in Cx-II medially as long as, or only slightly shorter than, Cx-I, and, in males, a longer genital field and the posterior margins of Cx-IV not extending beyond posterior margin of genital field. Females are similar to *L. depressostriata*, a species with the male unknown, of which *L. excellens* could be a junior synonym.

Habitat and Biology: Crenobiont. Preadult stages and life cycle unknown.

Distribution: Alps (mountain springs in Switzerland); Southern Italy.

36. *Lebertia (Mixolebertia) fontana* WALTER, 1912 (*8-51 d-g*)

1912b *Lebertia (Hexalebertia) fontana* WALTER, Zool. Anz. 40 (4-5): 146.

1922a *Lebertia (Hexalebertia) robusta* WALTER, Rev. Suisse Zool. 29 (7): 343 – GERECKE (2009).

1922a *Lebertia (Hexalebertia) extendens* WALTER, Rev. Suisse Zool. 29 (7): 353 – GERECKE (2009).

Description: Integument lined; sclerotized parts pale to dark red; coxal field extending laterally to flanks and including V-3, with sexual dimorphism; Cx-II medial suture shorter than that of Cx-I; posterior margin of Cx-II narrow; legs without swimming setae, IV-L-6 with 2-5 ventral setae; excretory pore with a sclerotized ring; P-2 ventral seta far from ventrodistal edge, P-3 dorsal setae equally spaced in proximal part of segment, dorsodistal setae close to each other, away from dorsodistal segment edge, P-2/P-4 length ratio 0.7-0.9; P-4 distally strongly narrowed, ventral margin proximally convex, distally concave, dorsal margin smoothly rounded, dorsodistal and ventral setae long and fine, ventral setae dividing segment into sectors 1:1:1, distoventral seta shifted to medial surface, peg-like distal seta slender and long; P-5 relatively long and slender. Males: Idiosoma length/width 690-750/550-700 µm; Cx-IV posteriorly extended by secondary sclerotization, embracing the genital flap and including V-1, but posteriorly leaving a space of membranous integument and pre- and postgenital sclerites not fused to ventral sclerotization; genital flap length 150-160, length/height P-2, 75-88/60-65; P-3, 80-90/41-65; P-4, 95-100/30-33 µm. Females: Idiosoma length/width 880-1200/690-900 µm; posterior margin of Cx-IV not extended by secondary sclerotization, not including V-1; genital bay "bell shaped", with straight, equally diverging lateral margins; genital flap length 205-240; length/height P-2, 85-93/68-78; P-3, 105-120/48-54; P-4, 130-150/39-43 µm.

Similar species: *Lebertia giardinai*, most closely related to *L. fontana*, can be distinguished with certainty in the male sex only, having the genital field completely surrounded by ventral sclerotization. Distinction of females is difficult between *L. cuneifera*, *L. aber-*

rata, *L. giardinai* and *L. fontana* – variability of characters used in the key may overlap, especially attribution of single specimens may be impossible.

Habitat and Biology: Crenobiont. Preadult stages and life cycle unknown.

Distribution: Alps (Austria, Germany, Italy, Switzerland), Carpathians.

37. *Lebertia (Mixolebertia) giardinai* Maglio, 1908 (*8-51 a-c*)

1908b *Lebertia (Hexalebertia) giardinai* Maglio, Rend. Ist. Lomb. Sci. Lett.: 674.

1922a *Lebertia (Hexalebertia) gracilipes* Walter, Rev. Suisse Zool. 29 (7): 346.

1962a *Lebertia (Hexalebertia) sefveoides* Schwoerbel, Rev. Suisse Zool. 69 (6): 204 – (?) K. O.Viets (1967), Gerecke (2009).

Description: Integument strongly lined; legs without swimming setae, IV-L-6 with 5-7 ventral setae; excretory pore with sclerotized ring; ventral seta of P-2 away from distoventral segment edge, P-3 dorsal setae equally separated in proximal part of segment, the two dorsodistal setae close to each other; P-4 slightly curved, with maximum height in proximal part, distally strongly narrowed, dorsal margin equally rounded, ventral margin proximally slightly convex, distally slightly concave, ventral sectors 3:2:2, distal peg seta well developed; P-2/P-4 length ratio ca. 0.7. Males: Idiosoma length/width 560-650/390-450 µm, coxae, pre- and postgenital sclerites fused to form an unique ventral shield completely surrounding genital field; Cx-I/II median length ratio 1.1-1.2, Cx-II posteriorly sharply pointed; genital flap length 128-150 µm; excretory pore directed posteriorly, not included in ventral sclerotization; length/height P-2, 61-63/45-50; P-3, 70-73/34-38; P-4, 91-93/28 µm. Females: Idiosoma length 800-850 µm; coxae and genitalia not fused to a ventral shield, postgenital area membranous; Cx-I/II median length ratio 1.5-1.7, Cx-II posterior width 30-40; genital flap length 190-200; length/height P-2, 85/58; P-3, 95/44; P-4, 120/38 µm.

Similar species: Male *Lebertia giardinai* are highly distinct due to the development of a shield completely embracing the genital field, but attribution of females to this species may be problematic (see under *L. fontana*).

Habitat and Biology: Crenobiont. Preadult stages and life cycle unknown.

Distribution: Central Alps; only known from Brenta (Italy) and the Engadin Valley (Switzerland).

38. *Lebertia (Mixolebertia) halberti* Koenike, 1902 (*8-47 g-h*)

1902b *Lebertia halberti* Koenike, Zool. Anz. 25 (679): 610.

1911d *Lebertia (Mixolebertia) schmidtii* Thor, Zool. Anz. 38 (18-19): 420 – Gerecke (2009).

Description: Idiosoma and appendages large in dimensions; integument smooth; Cx-II shortened (Cx-I/II median length ratio 2.0-2.5); legs with long swimming setae – numbers: II-L-5, 7-8; III-L-4, 8-11; III-L-5, 8-18; IV-L-4, 8-16; IV-L-5, 13-20; IV-L-6 with 7-14 ventral setae, terminal claws small; excretory pore sclerotized; P-2/P-4 length ratio 0.6-0.7, P-4 ventral sectors 1:1:1. Males: Idiosoma length/width 1400-1500/1300, Cx-II posterior width 49-60; genital flap length 215-256; length/height P-2, 103-118/73-78; P-3, 112-125/61-65; P-4, 173-188/50-53 µm. Females: Idiosoma length/width 1500-1600/1300, Cx-II posterior width 85; genital flap length 265; length/height P-2, 130/95; P-3, 133/75; P-4, 190/60 µm.

Preadult stages: Larva: Wainstein (*sub nom. L. schmidtii*: 1980).

Similar species: *Lebertia contracta*, the only other *Mixolebertia* species with such numerous ventral setae on IV-L-6 and with a distribution similar to *L. halberti*, has a relatively shorter median suture of Cx-II, some of the ventral setae on IV-L-6 fine and hair-like, an unsclerotized excretory pore, and fewer and shorter swimming setae. *Lebertia reducta* WAINSTEIN & TUZOVSKIJ, 1974a (known from a single specimen from Yaroslavl, Russia) differs only in P-2 lacking a ventral seta, the proximal dorsal seta of P-3 away from the two distal ones which are located in the distal third of the segment and in generally lower measurement values. As these character states are probably the result of malformation, or covered by general infraspecific variability, the species could be a junior synonym of *L. halberti* (GERECKE 2009).

Habitat and Biology: Standing waters. Life cycle unknown.

Distribution: Northern and Northeastern Europe.

39. *Lebertia (Mixolebertia) helocrenica* GERECKE, 2009 (*8-53 e-f*)

2009 *Lebertia (Mixolebertia) helocrenica* GERECKE, Abh. Senck. Naturf. Ges. 566: 121.

Description: Integument lined. Coxal field extended to lateral idiosoma surface, forming a nearly circular disk. Excretory pore sclerotized. Cx-I medially as long as, or slightly longer than, Cx-II. IV-L-5 with one rather short hair-like seta and 12-15 strong ventral setae, IV-L-6 with 4-7 ventral setae arranged in a group in the centre of segment. P-3 dorsal setae all located in proximal half of segment, mediodistal seta close to dorsodistal seta; P-4 straight and slender (length/height ratio 3.5-3.8), maximum height near insertion of proximoventral seta, ventral and dorsal margin equally converging towards tip of segment, distal margin rounded, with an elongate pointed peg-like seta, ventral seta insertions dividing segment into sectors 1:1:2. P-2/P-4 length ratio about 0.6. Males: Idiosoma length 630-670 µm; coxal field completely surrounding genital field, including pre- and postgenital sclerites, one of the lateroglandularia and in heavily sclerotized specimens, also the excretory pore; genital flap length 130; gnathosoma length 155; length/height P-2, 73-75/50-53; P-3, 75-83/35-38; P-4, 113-115/30-33 µm. Females: Idiosoma length 830-850 µm; coxal field forming right angles flanking the posterior edge of the genital field; genital flap length 160-170; gnathosoma length 175; length/height P-2, 83-84/55; P-3, 85-88/39-40; P-4, 125/35 µm.

Similar species: The only other species of the genus having a male coxal shield completely surrounding the genital field is *L. giardinai*. The latter species differs in its curved, relatively short and stouter P-4 (length/height ratio 3.2-3.5) with ventral setae at equal distance or the proximal sector the largest. Female *L. giardinai* and *L. fontana* differ from *L. helocrenica* also in the medial suture line of Cx-I being distinctly longer than that of Cx-II (length ratio 1.2-1.6).

Habitat and Biology: Crenobiont; in weakly seeping springs rich in organic debris. Preadult stages and life cycle unknown.

Distribution: Alps, Central Europe; only known from a mountain spring in northern Italy and from springs at low elevation in the northern prealps; rare.

40. *Lebertia (Mixolebertia) helvetica* THOR, 1906 (*8-48 d*)

1906d *Lebertia (Mixolebertia) helvetica* THOR, Zool. Anz. 30; 465.

1930a *Lebertia (Mixolebertia, Apolebertia) dubiaeformis* SOKOLOW, Zool. Jbch. Syst.: 178. – GERECKE (2009).

1962 *Lebertia (Mixolebertia) oudemansi helvetica* LUNDBLAD, Ark. Zool (2) 14 (1): 76.

Description: Integument with fine lineation; Cx-II short (Cx-I/II median length ratio 1.3-3.9), posteriorly large; legs without swimming setae; IV-L-6 with 6-7 ventral setae; excretory pore smooth; length ratio P-2/4, 0.6-0.8; P-4 ventral sectors 1:2:1. Males: Idiosoma length/width 1500/1200, Cx-II posterior width 90 µm; posterior edges of apodemes of Cx-III ending at the level of the anterior part of Ac-1; length/height P-2, 90-143/103; P-3, 130/68, P-4, 126-180/38-55 µm. Females: Idiosoma length/width 1600/1300, Cx-II posterior width 100 µm: posterior edges of apodemes of Cx-III ending anterior to the level of Ac-1; genital flap length 290; length/height P-2, 99-150/105; P-3, 143/73; P-4, 139-185/40-60 µm.

Preadult stages: Larva: WAINSTEIN (*sub nom. L. dubiaeformis*: 1980).

Similar species: *Lebertia oudemansi*, very similar from most points of view, differs in the shape of Cx-II (relatively shorter, Cx-I/II median length ratio > 4.0, posterior margin wider, > 100 µm), and posteromedial apodemes of Cx-III distinctly extending below the lateral margins of genital flaps, up to the level of the posterior part of Ac-1 in males, of the anterior part of Ac-1 in females.

Habitat and Biology: Lakes, occasionally in slowly-flowing streams connected to lakes, also in eutrophic waters. Rarely coexisting with the closely related *L. oudemansi*, in general with a preference for the littoral zone, while *L. oudemansi* is more often found in deeper waters (LUNDBLAD 1968). Life cycle unknown.

Distribution: Northern and Central Europe, in the southern part of the area only recorded from the prealpine lakes Bodensee (MUCKLE 1942) and Vierwaldstätter See (WALTER 1922a).

41. *Lebertia (Mixolebertia) holsatica* K.VIETS, 1920 (*8-36 a, 8-49 a-b*)

1920b *Lebertia (Hexalebertia) holsatica* K.VIETS, Arch. Hydrobiol. 12 (4): 807.

Description: Integument with fine zigzag lines in extensive parts of the dorsum forming a honeycomb-like pattern; coxal field with sexual dimorphism; Cx-I/II median length ratio 1.2-1.7; legs without swimming setae, IV-L-6 with two ventral setae; excretory pore smooth; palp slender; P-2 ventral seta near distal edge, P-3 dorsal setae equally spaced, P-2/P-4 length ratio ca. 0.5, P-4 straight and long (length/height ratio 3.5-4.3), ventral sectors 2:1:3, peg-like seta minute, pointed. Males: Idiosoma length/width 570-660/430-460, Cx-II posterior width 27-36 µm; coxal field laterally and posteriorly extended by secondary sclerotization, including V-3 and V-1, the latter on pointed posteromedial tips of Cx-IV embracing the genital flap, leaving a small area of postgenital integument; genital flap length 110; length/height P-2, 58-63/43-50; P-3, 73-75/35-40; P-4, 115/28-33 µm. Females: Idiosoma length/width 750-800/650-700, Cx-II posterior width 45-49 µm; coxal field not extended, neither V-3, nor V-1 included in secondary sclerotization; genital flap length 180-215; length/height P-2, 73-80/56-58; P-3, 85-98/40-43; P-4, 138-150/33-38 µm.

Preadult stages: Larva: MARTIN (2006).

Similar species: *Lebertia nitida* MOTAŞ & TANASACHI, 1963c, described from the Eastern Carpathians as a subspecies of *L. holsatica*, should differ in a lack of integumental sculpturing, a short median suture to Cx-I, the excretory pore surrounded by a sclerotized ring and a more slender genital field. However, this species is insufficiently defined and agrees with *L. semireticulata* in the shape of the coxae and palp (GERECKE 2009). Males of *L. haliki* K. VIETS, 1928c, a species also recorded from the Carpathians, and from Hungary (*L. barsica* SZALAY, 1937, syn. to *L. haliki*) are similar in the unsclerotized excretory pore, but differ in Cx-IV posteriorly extending beyond the posterior margin of the genital field, but not including V-1. Both sexes differ also in the shape of P-4 (maximum height in the basal third, distally strongly narrowed, ventral setae located side by side). A further similar species, *L. longipalpis* K. VIETS, 1936 from Macedonia, differs in some measurements (Cx-I median length < 160, Cx-II posterior width 30-32 µm) and a slightly shorter palp.

Habitat and Biology: Crenobiont; in Luxembourg preferably in habitats rich in macrophytes and fine detritus (GERECKE et al. 2005b), in Southern Europe often in weakly flowing rheohelocrenes at low altitudes; probably thermophilous. Larvae parasitize chironomid Diptera (no subfamily specification: MARTIN 2006).

Distribution: Central and southern Europe, isolated outposts in southern Scandinavia (LUNDBLAD 1968); not recorded from the Iberian peninsula, the Pyrenees (except for a locality at low elevation, MIGOT 1926b), and the Alps.

42. *Lebertia (Mixolebertia) lativentris* K. VIETS, 1922 (*8-49 f-g*)

1922d *Lebertia (Hexalebertia) lativentris* K. VIETS, Arch. Naturg. 88 (A 9): 64.

Description: Integument lined, most lines very short, in length not exceeding a glandularium diameter; legs without swimming setae, IV-L-6 with 3-7 ventral setae; genital bay not embraced by elongated Cx-IV; excretory pore smooth; P-2 ventral seta slender, P-3 "club-shaped", with a very narrow base, dorsal setae generally in proximal part of segment, one of them close to segment base, P-2/P-4 length ratio 0.65-0.70, P-4 straight, length/height ratio 3.0-3.6, with slightly convex dorsal and ventral margins, maximum height in basal third, ventral sectors 2:2:3 or 1:1:1, mediodistal peg seta small and stout, dorsodistal setae very long, P-5 extremely shortened. Males: Idiosoma length/width 700-800/500-750 µm, Cx-II medial suture relatively short (Cx-I/II medial length ratio 1.0-1.2), posterior width 20-40; genital flap length 145-170 µm, with numerous medial setae densely arranged in central and posterior part, scattered in anterior part, and 9 lateral setae; length/height P-2, 75-85/48-53; P-3, 95-105/38-43; P-4, 115-125/33-38 µm. Females: Idiosoma length/width 850-950/700-800 µm, Cx-II medial suture relatively long (sexual dimorphism, Cx-I/II median length ratio 1.1-1.6) posterior width 40-55 µm; genital flap length 190-200 µm, with 15 pairs of medial setae much further from each other than in males, and 3 pairs of lateral setae; palp more slender than in males, length/height P-2, 90-93/55-63; P-3, 110-115/40-48; P-4, 130-135/38-44 µm.

Preadult stages: Larva: MARTIN (2003).

Similar species: *Lebertia holsatica* and the SE European *L. longipalpis* K. VIETS, 1936b are similar in absence of swimming setae and having a smooth excretory pore, but differ in a relatively longer and more slender P-4 (P-2/P4 length ratio < 0.6, length/height ratio generally > 3.6), *L. holsatica* also in the reticulated integument (note that occasionally also in *L. lativentris*, regular arrangement of short lines may suggest reticulation).

Habitat and Biology: Crenobiont; preferably in weakly flowing (rheo)helocrenes (GERECKE & MARTIN 2006). Larvae parasitic on Chironomidae (Orthocladiinae); attachment site: thorax (MARTIN et al. 2010).

Distribution: Central Europe, a single record from southern Sweden (LUNDBLAD 1968).

43. *Lebertia (Mixolebertia) macilenta* K.Viets, 1926 (*8-49 b*)

1926b *Lebertia (Hexalebertia) macilenta* K.Viets, Zool. Anz. 65 (11-12): 312.

Description: Integument lined, in centre of dorsum wrinkled due to fusion of lines; Cx-I/II median length ratio 1.2-1.3, posterior width ca. 30 µm; legs without swimming setae; excretory pore with a sclerotized ring; palp slender; P-2/P-4 length ratio 0.55-0.65; P-4 straight, distally equally narrowed, peg seta small, ventral sectors 3:2:3; P-5 small and stout. Males: Idiosoma length/width 800/700 µm; Cx-IV posterior margin extending beyond, and slightly embracing, genital field, including V-3, but not V-1; IV-L-6 with 3 ventral setae; genital bay slender and longish; genital flap length 140; length/height P-2, 65/50; P-3, 80/41; P-4, 118/33 µm. Females: Idiosoma length/width 835-985/510-590 µm; Cx-IV including V-3, but not V-1; IV-L-6 with two ventral setae; genital flap length 180-190; length/height P-2, 78-85/60-65; P-3, 95-100/43-45; P-4, 123-130/35-38 µm.

Similar species: A straight, distally narrowed P-4 combined with a sclerotized excretory pore is typical for the *stigmatifera* species group. In the relatively narrow posterior margin of Cx-II, separated ventral setae of P-4, low number of ventral setae on IV-L-6, and measurements, *L. bracteata* is most similar. Both species are little known and need further investigation. Skin structure, the only distinctive character between the two species, could be caused by individual malformation in *L. macilenta*.

Habitat and Biology: Rhithrobiont? Preadult stages and life cycle unknown.

Distribution: Tatra, southern Alps.

44. *Lebertia (Mixolebertia) oudemansi* Koenike, 1898 (*8-48 c*)

1898a *Lebertia (Pilolebertia) oudemansi* Koenike, In: Oudemans & Koenike, Tijdschr. Entom., 's Gravenhage 40 (3): 240.

1926b *Lebertia (Mixolebertia, Apolebertia) berlesei* Thor, Arb. biol. Wolgastat. Saratow 9 (1-2): 17 – Gerecke (2009).

Description: Integument finely striated; Cx-II median suture very short (Cx-I/II median length ratio 4.0-4.8), posterior width > 100 µm; medial apodemes of Cx-III/IV strongly developed, those of Cx-III distinctly extending below lateral margins of genital flaps, posteriorly extended; III-L-4/5 and IV-L-5 each with one short distal seta, but without swimming setae; IV-L-6 bearing about 5 ventral setae; excretory pore unsclerotized (occasionally, with a fine, horseshoe-shaped sclerotization at posterior edge); P-3 rather slender; length ratio P-2/4 0.64; P-4 ventral sectors 1:1:1. Males: Idiosoma length/width 1200/1000 µm, posterior margin of Cx-II fused with pregenital sclerite; posterior margin of Cx-III extending to level of posterior part of Ac-1; genital flap length 240; length/height P-2, 120/80; P-3, 138/63; P-4, 180/51 µm. Females: Idiosoma length/width 1300/1200 µm; posterior margin of Cx-II not fused with pregenital sclerite; genital flap length 270, length/height P-2, 125/90; P-3, 130/63; P-4, 178/48 µm.

Similar species: *Lebertia helvetica* Thor, 1906d is closely related, but differs in the shape of Cx-II (with an absolutely and relatively longer median suture and differently developed posterior margin, see there).

Habitat and Biology: Preferably in deeper parts of oligotrophic lakes (Lundblad 1968), also recorded from high order streams. A slow-moving inhabitant of bottom sediments (Lundblad 1968). Preadult stages and life cycle unknown.

Distribution: Holarctic; northern Europe and northern prealpine lakes.

45. *Lebertia (Mixolebertia) sefvei* WALTER, 1911 (*8-52 a-c*)

1911a *Lebertia sefvei* WALTER, in: Naturwissenschaftliche Untersuchungen des Sarekgebirges in Schwedisch-Lappland, geleitet von Dr. AXEL HAMBERG 4, Zool. (5): 598.

Description: Integument lined; coxal field often rather dark brown in colour (also in slide mounted specimens); medial suture Cx-II short (Cx-I/II median length ratio > 1.5), posterior width > 40 µm; level of insertions of IV-L anterior to pregenital sclerite; Cx-IV including V-3; legs without swimming setae (one short hair-like terminal seta on IV-L-5); IV-L-6 with 4-8 ventral setae; excretory pore with a round or drop-shaped sclerotized ring; P-2/P-4 length ratio 0.59/0.65, P-2 ventral setae away from ventrodistal segment edge; P-4 very slender (length/height 3.6-4.3), with a fine lamella flanking the insertion of the dorsodistal setae, peg-like mediodistal seta small. Males: Idiosoma length/width 670-900/490-670 µm; coxal field with strongly developed lateral and posterior extensions, posterior margins of Cx-IV almost completely embracing genital field, leaving only a narrow strip of porose integument, including V-1, but not the crescent-shaped postgenital sclerite and the excretory pore; posterior margin of Cx-II often fused with pregenital sclerite; genital flap length 144-153; length/height P-2, 72/54-56; P-3, 85-85/38-43; P-4, 114-123/34 µm; P-4 ventral sectors 2:1:2. Females: Idiosoma length/width 950-1100/750-900 µm, coxal field without particular extensions, V-1 not included in coxal sclerite; genital flap length 193-225; length/height P-2, 81-92/58-67; P-3, 94-110/45-47; P-4, 132-143/34-36 µm; P-4 ventral sectors 2:2:3 (setae more apart than in males).

Preadult stages: Larva: MARTIN (2003).

Similar species: The extension of the secondary sclerotization of Cx-IV in males is probably age-dependent. Specimens with posteromedial margins of Cx-IV further apart (former subspecies *L. s. interposita*), or nearly in touch and leaving only a narrow membranous strip (former subspecies *L. s. circumclusa*), and all intermediate stages, may be found in the same population. Males of *L. giardinai*, similar in the presence of a nearly complete ventral shield, differ in IV-L inserting far anterior to the genital field, Cx-II with longer medial suture line and posterior edge narrow and pointed, and sclerotization posterior to the genital field being closed, also including the postgenital sclerite. Males of *L. oxa* Gerecke, 2009 agree in the posterior margins of Cx-IV completely embracing, but not fusing posterior to the genital field, but the facing margins are more rounded and touch at one point. Both sexes of *L. oxa* differ also in the narrower posterior margin of Cx-II, a relatively shorter P-2, the ventral setae on P-4 being further from each other and generally smaller body dimensions.

Habitat and Biology: Cold-stenotherm; rhithrobiont and crenophile in the northern part of the distribution area (LUNDBLAD 1968); crenobiont in the Alps – here in springs of various typology, but avoiding helocrenes with a weak flow (GERECKE & MARTIN 2006). Larvae parasitic on chironomid Diptera (Chironominae, Orthocladiinae; preferred attachment site thorax, MARTIN & STUR 2006; MARTIN et al. 2010).

Distribution: Europe; records from the Mediterranean area require confirmation (GERECKE 2009).

46. *Lebertia (Mixolebertia) semireticulata* K.VIETS, 1925 (*8-49 i-k*)

1925g *Lebertia (Hexalebertia) semireticulata* K.VIETS, Zool. Jbch. Syst. 50 (4-6): 487.

Description: Integument with honeycomb-like reticulation, in the meshes finely porose; preocular setae and Dgl-1 on a pair of oval platelets; coxal field extended, including V-1; legs without swimming setae; IV-L-5 with 8-9, IV-L-6 with 3 stout ventral setae and 2-3 fine, hair-like setae; genital field elongated; excretory pore sclerotized; distal setae of P-3 about as long as P-4, P-4 not continuously narrowed but with subparallel dorsal/ventral margins, P-2/P-4 length ratio 0.5-0.6, P-4 ventral sectors 2:1:2 or 3:1:3, P-5 stout. Males: Idiosoma length/width 750/600 µm, coxal field enlarged, both transversally and longitudinally; Cx-II medially long (Cx-I/II median length ratio 0.8-1.0), posteriorly narrow and acute-angled (width 10-30 µm); Cx-IV laterally embracing genital field with pointed, medially directed projections; area of postgenital membrane small; genital field with gonopore directed posteroventrally, genital flap length 120-130; length/height P-2, 60-65/45-49; P-3, 70-75/34-35; P-4, 113-120/30 µm. Females: Idiosoma length/width 800-900/600-650 µm, Cx-II medially shorter and posteriorly wider than in males (medial length ratio Cx-I/II 1.1-1.3, width 30-50 µm); posteromedial edges of Cx-IV blunt, not reaching posterior margin of genital field; genital flap length 170-185, length/height P-2, 71-80/55-60; P-3, 80-100/38-43; P-4, 120-133/30-36 µm.

Similar species: In slide-mounted *Lebertia*-specimens, the relative location of preocularia and Dgl-1 is often not visible and therefore badly documented. This character merits particular attention in future studies. From other species of the *stigmatifera* group, characterized by a straight P-4 with a small mediodistal peg seta, it differs in the integument structure and a more elongated P-4 (38-39 % total palp length, in other species < 38 %). *Lebertia nitida* MOTAŞ & TANASACHI, 1963c was incompletely described from Romania as a subspecies of *L. holsatica*. It should be characterized by complete lack of integumental sculpturing, a character not found in other species of the subgenus, but possibly resulting from individual malformation. From other points of view it is similar to *L. semireticulata* and could be its junior synonym (GERECKE 2009). Specimens from Greece differ in slightly smaller dimensions and have the posterior margin of Cx-II narrower (GERECKE 2009).

Habitat and Biology: Weakly seeping rheohelocrenes. Preadult stages and life cycle unknown.

Distribution: Central European mountains, northern prealps and Alps, Greece; obviously rare.

47. *Lebertia (Mixolebertia) separata* LUNDBLAD, 1930 (*8-53 d*)

1930c *Lebertia (Hexalebertia) stigmatifera var. separata* LUNDBLAD, Dansk. Vidensk. Selsk. biol. Meddel. 8 (7): 32.

2006 *Lebertia (Hexalebertia) separata* GERECKE, Lauterbornia 57: 79.

Description: Integument lined; median length ratio Cx-I/II 0.9-1.4; three long and slender lateral setae on flanks of Cx-III; legs without swimming setae; IV-L-6 with 7-11 ventral setae; V-1 not fused to Cx-IV; excretory pore with a sclerotized ring; P-2/P-4 length ratio 0.62-0.65, P-4 slender, distally equally narrowed, peg seta small, ventral setae close together, dividing segment into sectors 3:1:3. Males: Idiosoma length/width 750-800/620-680; Cx-II posterior width 25-32; genital flap length 145-160 µm, with 13-15 medial setae arranged more densely in posterior than in anterior part, and 7 lateral setae; length/height P-2, 88-93/55-60; P-3, 95-105/43-45; P-4, 135-143/35-38 µm. Females: Idiosoma length/width 750-850/600-650; Cx-II posterior width 25-40; genital flap length 190 µm, with 13 medial setae, mostly in posterior part, and 7 setae at posterior and lateral margins; length/height P-2, 85-93/55-58; P-3, 95-105/43-45; P-4, 138-143/33-38 µm.

Similar species: The only, but stable differences to the sibling species *L. stigmatifera* (posterior margin of Cx-IV not reaching posterior margin of genital field, V-1 completely separate in the membranous integument) is found in males only; females of the two species are indistinguishable.

Habitat and Biology: Crenobiont. Preadult stages and life cycle unknown.

Distribution: Central and southeastern Europe.

48. *Lebertia (Mixolebertia) sinuata* K.VIETS, 1930 (*8-48 e*)

1930d *Lebertia (Mixolebertia) sinuata* K.VIETS, Arch. Hydrobiol. 21 (2): 366.

Description: Integument lined; Cx-I/II median length ratio 1.8-2.1; Cx-II posterior width 45-50 µm; legs without swimming setae; IV-L-1 with 9 dorsal and dorsodistal setae; IV-L-6 with 9-11 ventral setae; genital bay without sexual dimorphism; excretory pore sclerotized; palp slender (e.g. length/height ratio P-3, 2.2-2.3; P-4, 3.7-3.9); P-2/P-4 length ratio 0.5-0.7, ventral seta P-2 away from distal segment edge, ventral setae P-4 dividing segment into sectors 2:1:2. Males: Idiosoma length/width 1000/800, genital flap length 200 µm, with 32 densely arranged medial setae and 10 lateral setae; length/height P-2, 83/73; P-3, 115/50; P-4, 155/40 µm. Females: Idiosoma length/width 1100/900, genital flap length 230 µm, with 20 medial and 7 lateral setae; length/height P-2, 110/75; P-3, 118/53; P-4, 160/44 µm.

Similar species: *Lebertia sinuata* is similar to *L. dubia* in the lined integument, the short medial suture of Cx-II, absence of swimming setae, a sclerotized excretory pore and distally equally narrowed P-4 with a strong, pointed mediodistal peg seta. *Lebertia dubia* differs in having a stouter palp (length/height P-3, 1.9-2.0; P-4, 2.9-3.3), the ventral setae of P-4 located closer to each other than to the proximal/distal segment edges respectively, and a higher number of ventral setae on IV-L-6.

Habitat and Biology: Rhithrobiont? Preadult stages and life cycle unknown.

Distribution: Spanish Pyrenees, only known from the type locality.

49. *Lebertia (Mixolebertia) stigmatifera* THOR, 1900 (*8-53 a-b*)

1900a *Lebertia (Pseudolebertia) stigmatifera* THOR, Nyt Mag. Naturv. Oslo 38 (3): 275.

Description: Integument sculptured with distinct lines, some short, others more than 100 µm in length; Cx-II medially short (medial length ratio Cx-I/II 1.1-1.8, generally 1.2-1.4), posteriorly forming a pointed tip; legs without swimming setae; IV-L-6 bearing 4-10 strong ventral setae and several fine hair-like setae; excretory pore with sclerotized ring; P-2 dorsal margin forming a subrectangular edge, ventral margin concave, the strong seta away from distal edge; P-3 setae long and fine, paired dorsal and dorsodistal setae close to each other, proximomedial seta close to segment base, P-2/P-4 length ratio 0.61-0.65; P-4 slender, maximum height (often at a slight inflation of ventral margin) in basal third, distally strongly narrowed, ventral sectors 3:1:3, the distal seta bent upwards, peg-like mediodistal seta minute; P-5 small and slender. Males: Idiosoma length/width 600-800/600-700 µm; posteromedial edges of Cx-IV acute-angled, but not wholly embracing genital bay; genital flap length 140-170; length/height P-2, 85-90/55-63; P-3, 90-103/40-43; P-4, 130-143/34-38 µm. Females: Idiosoma length/width 650-1000/500-800, genital flap length 160-230; length/height P-2, 68-103/48-63; P-3, 75-120/38-49; P-4, 110-158/33-41 µm.

Preadult stages: Larva: LUNDBLAD (1924c), MARTIN (2000).

Similar species: See discussion of *L. bracteata*.

Habitat and Biology: Cold stenotherm, in waters with both low and high conductivity; in the north both in springs and streams (in the latter with a preference for summer-cold pools: LUNDBLAD 1968, MARTIN & SPETH 1996), in central, western and southern Europe a crenobiont (GERECKE & MARTIN 2006, the new data given here). Larvae parasitic on chironomid Diptera (Tanytarsini, attachment site: cervix, MARTIN 2000 – non-parasitic larvae with direct development to deutonymph reported by LUNDBLAD 1924c).

Distribution: Northern, Western and Central Europe, Pyrenees, Apennines and Balkans; apparently not in a warm Mediterranean climate.

Subgenus *Pilolebertia* Thor, 1900

1900a *Pilolebertia* Thor, Nyt Mag. Naturv. Oslo 38 (3): 268.

Type species: *Lebertia insignis* Neuman, 1880b, by original designation.

Diagnosis: Integument smooth or with fine striation; coxal field not expanding to lateral and posteroventral idiosoma, anterior tips of Cx-I not bent ventrally; legs with numerous swimming setae (minimum numbers: II-L-5, 2; III-L-4, 2; III-L-5, 3; IV-L-4, 2; IV-L-5, 5); palp generally slender; P-3 bearing five medial setae, one of them in ventrodistal position; P-4 distally narrowed, dorsal setae concentrated to a longitudinal pit in distal quarter of segment.

Discussion: The subgenus is probably a monophyletic unit, known from all continents where *Lebertia* is present. As some species with numerous swimming setae (name-giving for *Pilolebertia*) are surely taxonomically distant, belonging to *Lebertia s. str.*, and not related to each other, adaptations to life in stagnant water probably evolved several times within the genus. On the other hand, two species endemic to Madeira, *Lebertia maderigena* and *L. madericola*, combine a P-4 setation typical for *Pilolebertia* (very similar to *L. (Pilolebertia) inaequalis*) with characters generally found in *Mixolebertia* (reduction of swimming setae, papillosity of the integument, in *L. madericola* also presence of 6 medial setae on P-3). Following Gerecke (2009) both species could have evolved from a *Pilolebertia*-like ancestor colonizing this island lacking suitable habitats for standing-water-dwellers.

Key to species

The following (sub)species are excluded as *species incertae* (Gledhill & K. O. Viets 1976; Gerecke 2009): *Lebertia curvipalpis* Halbert, 1944; *L. minuticornis* K. Viets, 1949b; *L. plauta* Halbert, 1944; *L. saxonica* Thor, 1911c.

1 Ventral setal insertions on P-4 dividing segment into a long proximal, and two short distal sectors; proximal setal pore located in the centre or distal half of segment (*8-54 a, c, e*). 2

\- Ventral setal insertions on P-4 dividing segment into three sectors +/- equal in length; proximal setal pore located in the proximal half of segment (*8-54 h, k*). 4

2 (1) Mediodistal seta on P-3 away from distal segment margin, the three distal setal insertions forming a triangle; ventral setal insertions on P-4 close together, with distal seta closer to proximal seta than to distal edge of segment (*8-54 a*). Palp smaller in dimensions (e. g. P-2 length < 80, P-3 length < 100 µm). . . . . *Lebertia pilosa* (page 143)

\- Mediodistal seta close to distal segment margin; the three distal seta insertions forming a line; ventral seta insertions on P-4 well separated, with distal setae closer to distal edge of segment than to proximal seta (*8-54 c, e*). Palp larger in dimensions (e. g. P-2 length > 80, P-3 length > 100 µm). 3

3 (2) Ventrodistal seta on P-2 fine, exceeding total length of P-2+3; P-2 often stouter (length/height 1.4-1.6); mediodistal seta on P-3 closer to dorsodistal seta (*8-54 c*). II-L-5 with six to seven posterior swimming setae. . . . . *Lebertia longiseta* (page 143)

\- Ventrodistal seta on P-2 thicker and shorter, not exceeding total length of P-2+3, P-2 often more slender (length/height 1.5-1.8); mediodistal seta on P-3 near centre of the segment (*8-54 e*). II-L-5 with two to six posterior swimming setae. . . . *Lebertia inaequalis* (page 142)

4 (1) Mediodistal seta on P-3 close to ventrodistal seta (*8-54 k*). Posteromedial margin of Cx-I ending in a pointed tip. Palp small (e. g. length P-2 ♂ < 120, ♀ < 130; P-4 ♂ < 150, ♀ < 160 µm). . . . . *Lebertia insignis* (page 142)

\- Mediodistal seta on P-3 away from ventrodistal seta (*8-54 h*). Posteromedial margin of Cx-I more enlarged (width 30-50 µm). Palp larger (e. g. length P-2 ♂ > 120, ♀ > 130; P-4 ♂ in most cases > 150, ♀ > 160 µm). . . . . *Lebertia porosa* (page 144)

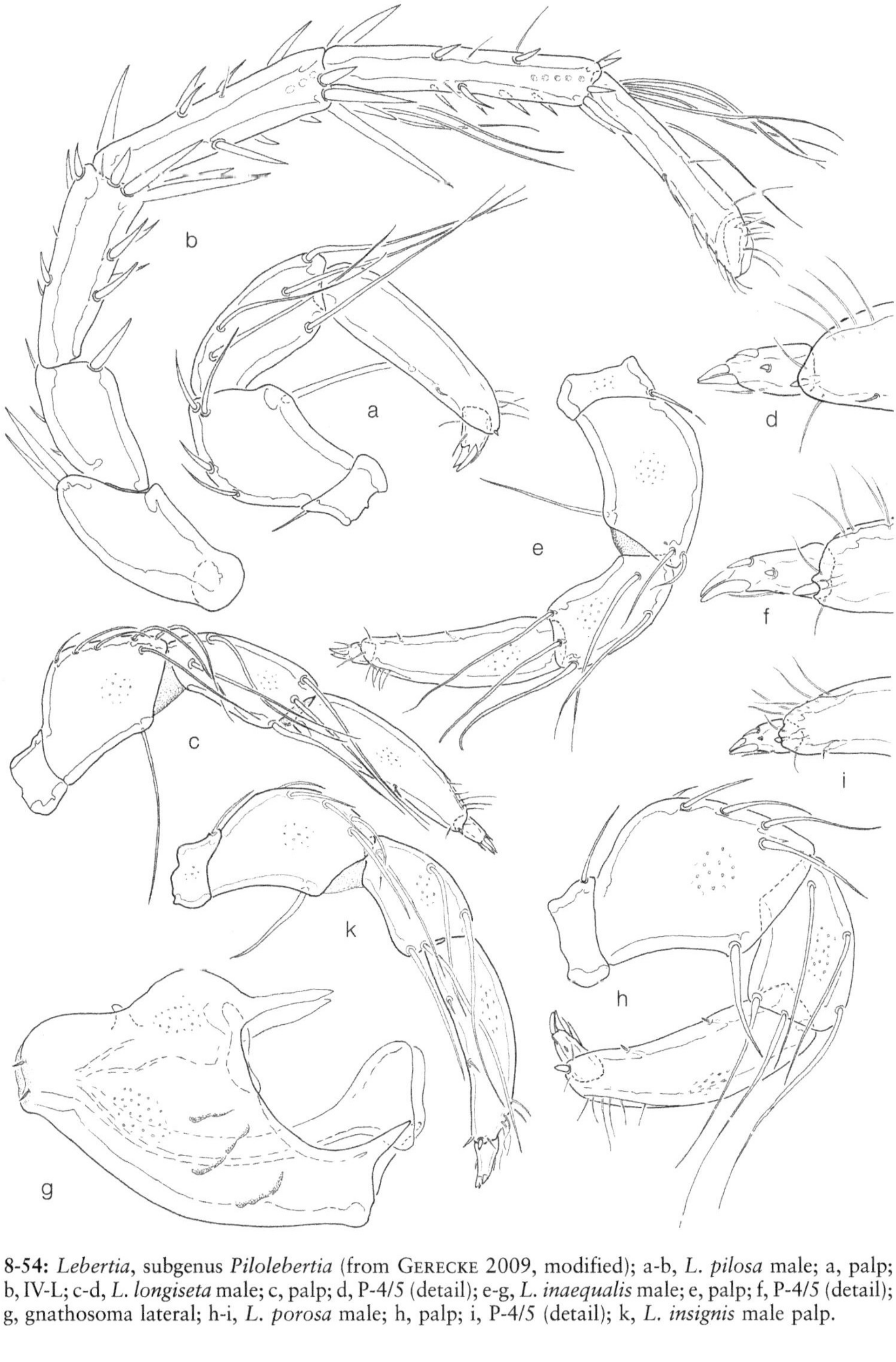

8-54: *Lebertia*, subgenus *Pilolebertia* (from Gerecke 2009, modified); a-b, *L. pilosa* male; a, palp; b, IV-L; c-d, *L. longiseta* male; c, palp; d, P-4/5 (detail); e-g, *L. inaequalis* male; e, palp; f, P-4/5 (detail); g, gnathosoma lateral; h-i, *L. porosa* male; h, palp; i, P-4/5 (detail); k, *L. insignis* male palp.

50. *Lebertia (Pilolebertia) inaequalis* (Koch, 1837) (*8-54 e-g*)

1837 *Hygrobates inaequalis* Koch, Deutschlands Crustaceen, Myriapoden und Arachniden 11: 20.

1903 *Lebertia quadripora* Koenike, Zool. Anz. 26 (702): 536 – Gerecke (2009).

1955 *Lebertia (Pilolebertia) wackeri* Bader, Verh. naturforsch. Ges. Basel 66 (1): 62 – Gerecke (2009).

1955 *Lebertia (Pilolebertia) neocomensis* Bader, Verh. naturforsch. Ges. Basel 66 (1): 69 – Gerecke (2009).

Description: Integument smooth; Cx-II medially short (Cx-I/II median length ratio generally > 1.3), posterior width 20-40 µm; excretory pore not sclerotized, swimming setae numbers II-L-5, 2-6; III-L-4, 2-10; III-L-5, 6-11; IV-L-4, 2-8; IV-L-5, 5-11; IV-L-6 with 2-5 fine ventral setae; gnathosoma with slight, but distinct indentation near ventral edge of mouth opening, in lateral view also visible in undissected specimens; P-2 relatively long, ventrodistal seta stout and shorter than length P-2; P-2/P-4 length ratio 0.7-0.9. P-3 mediodistal seta halfway between ventro- and dorsodistal setae or slightly towards dorsodistal seta; P-4 ventral sectors 2:1:1 or 3:2:1, mediodistal peg-like seta small (length 2-3 µm, that part projecting from insertion cavity approximately as long as wide). Males: Idiosoma length 800-1200; genital flap length 155-220; length/height P-2, 98-130/63-84; P-3, 85-120/40-48; P-4, 133-180/40-48 µm; Females: Idiosoma length 1000-1300; genital flap length 150-200; length/height P-2, 103-125/65-80; P-3, 103-123/41-53; P-4, 130-158/40-45 µm.

Preadult stages: Larva: Wainstein (1980), Martin (2000).

Similar species: *Lebertia algeriensis* Lundblad, 1942d, considered synonymous with *L. inaequalis* by former authors, replaces this species in the Western Mediterranean area. It differs in the larger, more elongate peg-like seta of P-4 (L 4-8 µm, external part distinctly longer than wide), the ventral margin of gnathosoma equally rounded, without an indentation near anterior margin, and segments P-5 and IV-L-4-6 being more slender and Cx-I/II being medially almost equal in length.

Habitat and Biology: Pools of higher order streams, macrophyte belt and wave-belt of larger standing waters (under oligotrophic, as well as dys- and eutrophic conditions, Lundblad 1968, Smit & Van der Hammen 2000). Larvae parasitic on chironomid Diptera (Tanytarsini, attachment site: thorax, Martin 2000).

Distribution: Extended parts of the Palaearctic, but infrequent in the far North and not recorded from some parts of eastern Europe (Lundblad 1968). Several published records could refer to the similar *L. longiseta* or *L. algeriensis* (Gerecke 2009).

51. *Lebertia (Pilolebertia) insignis* Neuman, 1880 (*8-54 k*)

1880b *Lebertia insignis* Neuman, Svensk Vetensk-Akad. Handlingar Stockholm (N. F.), 1879 17 (3): 68.

Description: Integument smooth; Cx-II medially short (Cx-I/II median length ratio 1.2-1.3), posteriorly narrow, pointed (width 15-20 µm); Cx-IV posterior margin truncated; swimming setae numbers: II-L-5, 2; III-L-4, 7-8; III-L-5, 14-15; IV-L-4, 8-10; IV-L-5, 12; on III-/IV-L-4 restricted to distal third, on III-/IV-L-5 to distal half; IV-L-6 with 3-6 ventral setae; excretory pore smooth; P-3 with mediodistal seta close to ventrodistal seta, P-4 distally narrowed, P-2/P-4 length ratio ca. 0.8, P-4 ventral sectors 1:1:1. Males: Idiosoma length/width 1100/800; genital flap length 185; length/height P-2, 113/70; P-3, 100/60; P-4, 145/41 µm. Females: Idiosoma length/width 1000/700; genital flap length 190; length/height P-2, 123/80; P-3, 110/63; P-4, 153/48 µm.

Preadult stages: Larva: THOR (1912b), WAINSTEIN (1980).

Similar species: Individuals are regularly found which bridge the gap between *L. insignis* and *L. porosa*, a species similar in general dimensions and palp shape, but with mediodistal seta on P-3 away from ventrodistal seta. Such intermediate specimens were formerly classified as *L. saxonica*. The separation of *L. insignis* and *L. porosa* needs further investigation. (GERECKE 2009).

Habitat and Biology: Mostly in pools of streams, rarely in larger standing waters, probably preferring relatively warm water (LUNDBLAD 1968). Life cycle unknown.

Distribution: Central, Western and Northern Europe; rare in the Mediterranean area (only records: Spain, Oviedo, LUNDBLAD 1956a; Hercegovina, SCHWOERBEL 1963).

52. *Lebertia (Pilolebertia) longiseta* BADER, 1955 (*8-54 c-d*)

1955 *Lebertia (Pilolebertia) longiseta* BADER, Verh. naturforsch. Ges. Basel 66 (1): 64.

Description: Integument smooth, finely porose; Cx-II as long as, or slightly shorter than, Cx-I (median length ratio 0.9-1.3), posteriorly 20-30 µm wide; leg swimming setae: II-L-5, 4-7; III-L-4, 3-6; III-L-5, 7-9; IV-L-4, 3-6; IV-L-5, 6-8; IV-L-6 with 2-3 ventral setae; excretory pore smooth; P-2 with a long, slender ventral seta, P-3 relatively long and stout (length/height 2.3), P-2/P-4 length ratio ca. 0.8, P-4 ventral sectors 4:2:1. Males: Idiosoma length/width 900/800; genital flap length 165-178 µm, with 20 pairs of medial setae arranged mostly regularly and 8 lateral pairs of setae; length/height P-2, 92-99/62-69; P-3, 100-109/44-45; P-4, 115-127/35-38 µm. Females: Idiosoma length/width 800-1000/700-900; genital flap length 185-190 µm, with 15 pairs of medial setae arranged regularly and 3-4 pairs of lateral setae; length/height P-2, 96-108/63-69; P-3, 107-115/48-50; P-4, 125-135/38-40 µm.

Similar species: *Lebertia inaequalis* is very similar to *L. longiseta* in dimensions and the particular shape of the palp (P-4 both ventral setae distally, mediodistal peg seta very small), but differs in having a shorter and stouter ventral seta on P-2. Furthermore, P-2 tends to be more slender and on P-3 the mediodistal setae are closer to the dorsodistal ones, and the number of swimming setae on II-L-5 is generally lower (see also K. O. VIETS 1976b).

Habitat and Biology: Standing waters. Preadult stages and life cycle unknown.

Distribution: Northern Alps and prealps.

53. *Lebertia (Pilolebertia) pilosa* MAGLIO, 1924 (*8-54 a-b*)

1924 *Lebertia (Neolebertia) pilosa* MAGLIO, in: MONTI, R. (ed.): La Limnologia del Lario. Roma: 87.

1958 *Lebertia (Pilolebertia) pilosa* K. O. VIETS, Mem. Ist. Ital. Idrobiol. 10: 58.

Description: Integument smooth, locally finely striated; Cx-II short (medial length ratio Cx-I/II 1.1-1.6), posteriorly 20-30 µm wide; leg swimming setae: II-L-5, 2-4; III-L-4, 2, III-L-5, 6-8, IV-L-4, 3-4, IV-L-5, 5-6; IV-L-6 with 1-3 ventral setae; excretory pore unsclerotized; palp slender, P-2 with a strong ventrodistal seta; P-3 drop-shaped, with a narrowed base, dorsal setae inserting proximally, mediodistal seta away from segment margin, forming a triangle with the other distal setae, P-2/P-4 length ratio ca. 0.7, P-4 slender and straight, equally narrowed distally and with a very small mediodistal peg seta, ventral sectors 3:1:1 or 4:1:2; P-5 relatively long and slender. Males: Idiosoma length/width 750/600; genital flap length 160-170, length/height P-2, 68-75/48-53; P-3, 83-93/35-38; P-4, 98-

110/26-28 µm. Females: Idiosoma length/width 900/750, genital flap length 160-170; length/height P-2, 80-88/55-60; P-3, 95-108/40-43; P-4, 113-130/30-33 µm.

Similar species: *Lebertia pilosa* is highly distinct in the particular shape of its palp, smaller dimensions and lower numbers of swimming setae. A very similar, possibly synonymous species is *L. kuzini* WAINSTEIN & TUZOVSKIJ, 1974a, described from Russia. It should differ from *L. pilosa* in the reduced number of only 4 dorsal hair-like setae on P-4 (in *L. pilosa* 5-6); and a stronger sclerotization of the sclerites surrounding the glandularia.

Habitat and Biology: In Scandinavia mostly in lakes, but colonizing lowland rivers in other parts of the area (LUNDBLAD 1968) – rare, found in small populations only. Preadult stages and life cycle unknown.

Distribution: Europe; scattered records.

54. *Lebertia (Pilolebertia) porosa* THOR, 1900a (*8-54 h-i*)

1900a *Lebertia porosa* THOR, Nyt Mag. Naturv. Oslo 38 (3): 273.

1936b *Lebertia stackelbergi saxicola* SOKOLOW, Arch. Hydrobiol. 30 (3): 495 – GERECKE (2009).

1940a *Lebertia caucasica cubanica* SOKOLOW, In: SERNOW & STACKELBERG, Faune de l'URSS, Arachnides 5 (2): 204 – GERECKE (2009).

Description: Integument smooth or with very fine striation; Cx-II medially generally shorter than Cx-I, but variable (median length ratio 1.0-2.0), posteriorly 30-50 µm wide, in exceptional cases narrower; swimming seta numbers: I-L-4/5, 0, but several fine distal setae; II-L-5, 4-7; III-L-4, 5-10, III-L-5, 10-14 (7-9); IV-L-4, 5-12; IV-L-5, 8-14; IV-L-6 with 3-6 ventral setae; genital flaps in their anterior quarter nearly devoid of medial setae, posteriorly with dense cover of hair-like setae; anterior of gnathosoma in lateral view equally rounded, without an indentation near distal edge; palp rather large, P-2 ventral seta away from ventrodistal edge; P-2/P-4 length ratio generally about 0.8, up to 0.9, P-4 ventral sectors 1:1:1 or 3:3:2, peg-like mediodistal setae relatively large. Males: Idiosoma length/width 1000-1300/800-1000; genital flap length 200-250 µm, with 30 or more pairs of medial setae, scattered in anterior part, very dense, locally in double rows in posterior part; length/height P-2, 121-163/78-108; P-3, 99-130/61-78; P-4, 148-193/47-60 µm. Females: Idiosoma length/width 1200-1500/1000-1200; genital flap length 210-260 µm, with 16-20 pairs of medial setae, scattered in anterior third, equally spaced posteriorly; length/height P-2, 148-163/93-103; P-3, 95-135/70-78; P-4, 173-203/58-60 µm.

Preadult stages: Larva: THOR (1912b), WAINSTEIN (1980), MARTIN (2000).

Similar species: See discussion under *L. insignis*.

Biology: Adults from March to November in Central Europe but year-round in the Mediterranean area (GERECKE 2009).

Habitat and Biology: Eurytopic and eurythermous, in standing waters and pools of streams – also in eutrophic and slightly polluted waters (LUNDBLAD 1968, GERECKE & SCHWOERBEL 1992 under "*Pilolebertia* sp.").

Distribution: Holarctic.

Family Oxidae K. VIETS, 1926

1926l Oxidae VIETS, Zool. Anz. 69: 197.

Diagnosis: Idiosoma usually elongated, often laterally compressed. Lateral eyes not in capsules, lying below integument. Coxal plates fused and in most genera greatly expanded, encompassing a large part of the idiosoma except a narrow dorsal strip. Suture lines of coxal plates generally visible only laterally; a median suture line rarely present. Coxoglandularia 1 lying near tip of first coxal plates, accompanied by a hyaline membrane and variously modified setae. All legs located anteriorly, II-IV-L with swimming setae; I-II-L with claws bearing a dorsal clawlet, claws of IV-L reduced, IV-L-6 bearing a long terminal seta. Genital field anteriorly embraced or surrounded by Cx-IV, genital flaps typically covering three pairs of Ac. Palp with either four of five segments, P-2 and P-3 without ventral projections or setae. Cheliceral claw stout.

Remarks: Oxidae is probably a sister taxon of Lebertiidae from which it differs in having a reduced ventral palp setation. The description of a species of *O. halophilus* E. ANGELIER, 1947 from Southern France, with a short ventrodistal seta on P-2, seems to contradict this interpretation. In view of unsatisfactory figures and scanty information about this taxon (combined with the loss of the holotype and only known specimen), DI SABATINO et al. (2009) consider it a *species incerta*, with the palp misshaped or incorrectly described. In addition to *Oxus*, the only genus recorded from the treated area (for synonymy see remarks on this genus), the family includes only *Africoxus* COOK, 1966, including two subgenera recorded exclusively from the Ethiopian region.

Distribution: Worldwide except Antarctica.

Genus *Oxus* KRAMER, 1877

1877 *Oxus* KRAMER, Arch. Naturg. 43: 239.

1891b *Frontipoda* KOENIKE, Zool. Anz. 14: 19 - BESCH (1964).

Type species: *Oxus strigatus* (MÜLLER, 1776) (*sub nom. O. oblongus* KRAMER, 1877), designated by KRAMER (1879a).

Diagnosis: Idiosoma as wide as high or laterally compressed. Coxal plates may or may not extend onto dorsal surface. Degree of sexual dimorphism of the coxal plates variable. Soft integument with or without dorsalia, excretory pore with or without associated sclerotization. Medial suture line of coxal field generally absent (present in subgenus *Flabellifrontipoda*, which is not in the treated area). Palp five-segmented.

Remarks: The species treated here were traditionally attributed to three genera, i. e. *Oxus* KRAMER, 1877, *Gnaphiscus* KOENIKE, 1898b, and *Frontipoda* KOENIKE, 1891b which appeared very distinct within Europe. However, with the description of many more species from other areas of the world, differences became blurred. As the degree of lateral compression was found highly diverse in Australian species (COOK 1986) and also the number of glandularia in the expanded coxal field provides only an artificial separation, DI SABATINO et al. (2009) decided to follow BESCH (1964), the first author to synonymize *Frontipoda* with *Oxus*. Consequently, *Flabellifrontipoda* LUNDBLAD, 1947, described as a genus from the Southern Hemisphere and reduced to the rank of a subgenus of *Frontipoda* by SMIT (2002), is a subgenus of *Oxus*. LUNDBLAD (1956a) and several authors since also proposed synonymization of *Gnaphiscus* with *Frontipoda* or *Oxus* (see K. O. VIETS 1987), but here COOK's (1974) line is followed, ranking this taxon as a subgenus of *Oxus*. Members of *Africoxus*, the only known other oxid genus differ in having a four-segmented palp.

Preadult stages: Larva: PRASAD & COOK (1972), WAINSTEIN (1980), I. M. SMITH (1982); deutonymph: TUZOVSKIJ (1990). SMITH & OLIVER (1986) report on larvae of North American species parasitizing chironomid Diptera (Chironominae, Orthocladiinae), but life cycle and parasitism are unknown for all species from the treated area.

Distribution: Worldwide except Antarctica.

Key to species

Excluded as *species incertae* (DI SABATINO et al. 2009): *Oxus lineatus* WALTER, 1912b, *O. halophilus* E. ANGELIER, 1947 and *O. integer* (THOR, 1901b).

1 Coxal field nearly completely enclosing idiosoma (*8-55 a-b*), not leaving a membranous postgenital area; excretory pore on a large plate (*8-55 b, e-f*). 2

- Coxal field less extensive, leaving a membranous postgenital area; excretory pore not sclerotized (*8-56 a, e, k*). 3

2 (1) Excretory pore plate with a long, pointed extension (*8-55 b*), P-2 longer than P-4 (*8-55 c*). IV-L-6 with one long and two short terminal setae. . . . . *Oxus musculus* (page 149)

- Excretory pore plate without or with a short, blunt posterior extension (*8-55 e-f*), P-2 shorter than P-4 (*8-55 d*); IV-L-6 with two long and one short terminal setae. . . . . *Oxus carpenteri* (page 148)

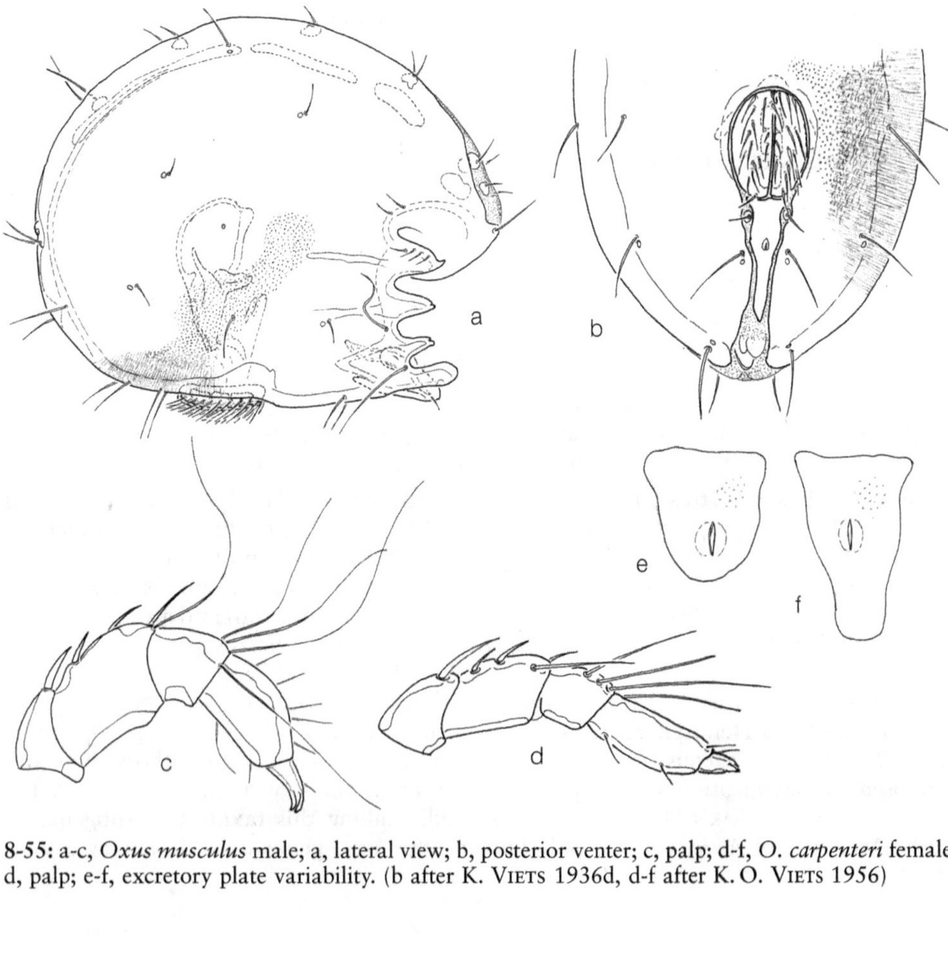

8-55: a-c, *Oxus musculus* male; a, lateral view; b, posterior venter; c, palp; d-f, *O. carpenteri* female; d, palp; e-f, excretory plate variability. (b after K. VIETS 1936d, d-f after K. O. VIETS 1956)

3 (1) Tips of Cx-I with long, curved setae which are longer than width of anterior part of Cx-I (*8-56 b, d*). 4

\- Tips of Cx-I with short, often blade-like setae (*8-56 f-g, i*). 5

4 (3) Setae at tips of Cx-I very long and slender (length > 60 µm, *8-56 d*); in both sexes genital field not enclosed by coxal field. . . . . *Oxus longisetus* (page 149)

\- Setae at tips of Cx-I shorter, (length < 50 µm), one of them thickened (*8-56 b*); male genital field enclosed by coxal field (*8-56 a*). . . . . *Oxus setosus* (page 151)

5 (3) Acetabula by far shorter than their basal sclerites (*8-56 h*). . *Oxus nodigerus* (page 150)

\- Acetabula only slightly shorter than their basal sclerites (*8-56 e, k,* see also *8-56 c*). . 6

6 (5) Setae at tips of Cx-I short and stout (*8-56 f-g*). 7

\- Setae at tips of Cx-I longer and more slender (*8-56 i*). 8

7 (6) Excretory pore lying anterior to associated glandularia, anterior margin of genital bay straight (*8-56 k*). . . . . *Oxus ovalis* (page 150)

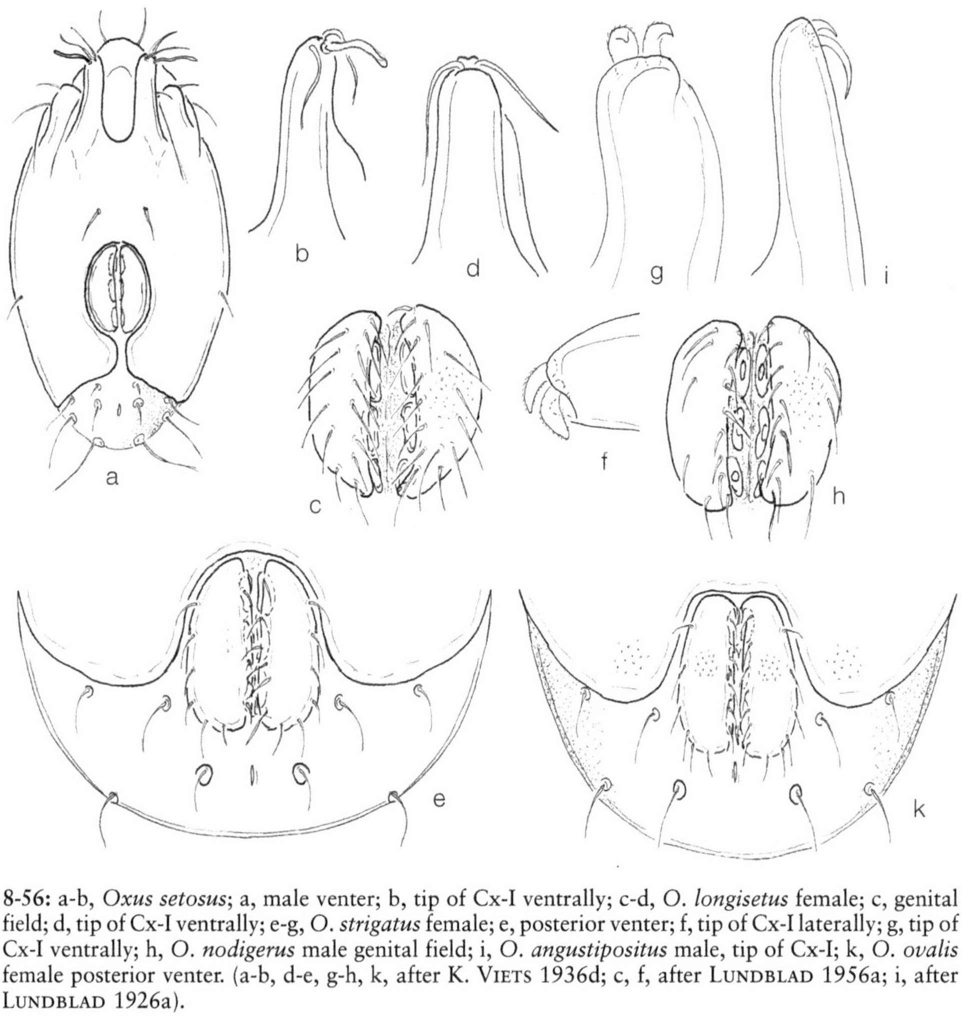

8-56: a-b, *Oxus setosus*; a, male venter; b, tip of Cx-I ventrally; c-d, *O. longisetus* female; c, genital field; d, tip of Cx-I ventrally; e-g, *O. strigatus* female; e, posterior venter; f, tip of Cx-I laterally; g, tip of Cx-I ventrally; h, *O. nodigerus* male genital field; i, *O. angustipositus* male, tip of Cx-I; k, *O. ovalis* female posterior venter. (a-b, d-e, g-h, k, after K. Viets 1936d; c, f, after Lundblad 1956a; i, after Lundblad 1926a).

-	Excretory pore lying between associated setae, anterior margin of genital bay concave (*8-56 e*).	*Oxus strigatus* (page 150)
8 (6)	Genital bay wide, large species (idiosoma length > 1000 µm).	*Oxus tenuisetis* (page 151)
-	Genital bay narrow, small species (idiosoma length < 1000 µm).	*Oxus angustipositus* (page 148)

Subgenus *Oxus* KRAMER, 1877

Diagnosis: Without sexual dimorphism in the extension of the coxal field (if genital field is embraced by coxal plates, this character state is found in both sexes). Species with extended coxal field have the excretory pore on a sclerotized platelet.

Remark: Probably a paraphyletic group, including species from the whole distribution area of the genus.

1. *Oxus (Oxus) angustipositus* K. VIETS, 1908 (*8-56 i*)

1908b *Oxus angustipositus* K. VIETS, Zool. Anz. 33: 51.

Description: Tips of Cx-I with two relatively long, curved, blade-like setae; margins of these setae serrated. Posterior margin of Cx-IV directed perpendicularly to lateral idiosoma margin. Genital bay narrow. Excretory pore located between associated glandularia. Distal seta of IV-L-6 smooth, length between 84 and 98 % of IV-L-6. Males: Idiosoma length 660-724, width 390-500 µm. Females: Idiosoma length 810-875, width 480 µm.

Preadult stages: Larva: WAINSTEIN (1980). Deutonymph: TUZOVSKIJ (1978c, 1990).

Biology: Hatching of eggs occurs within 14-15 days of deposition, after which the larva moves within a gelatinous cocoon for 4-6 days. According to NOCENTINI (1960a) larvae forego a parasitic phase. Another 9-10 days are required for transformation to the deutonymph. Thus the whole development from egg to deutonymph takes about a month. NOCENTINI (1960a) found maximum numbers in August-September, with minimum numbers in March-April. Deutonymphs were frequently found in August-September. Deutonymphs hibernate, and have disappeared by spring.

Habitat: Mainly in the littoral of lakes, occasionally in ponds and streams. In Sweden both in eutrophic and oligotrophic lakes (LUNDBLAD 1968).

Distribution: Western Palaearctic. Widespread in Europe. Common in Central and Southern Sweden (LUNDBLAD 1968), but apparently rare in other parts of Europe.

2. *Oxus (Oxus) carpenteri* (HALBERT, 1911) (*8-55 d-f*)

1911 *Frontipoda Carpenteri* HALBERT, Proc. Irish Acad., Dublin 31: 22.

Description: Tips of Cx-I each with two plain hair-like setae. Excretory plate rounded posteriorly, without or with a short, blunt posterior extension (K. O. VIETS 1956). P-1 slightly longer than P-2, the latter segment distinctly shorter than P-4. Males: Idiosoma length 725-850, height 585-700, genital flap length 150-165 µm. Females: Idiosoma length 755-925, height 710-775, genital flap length 185-200 µm.

Similar species: The shape of the posterior margin of the excretory plate is highly variable and in extreme cases may resemble the condition found in *O. musculus* (see below). A clear difference between both species is found in the proportions and setation of the palp (in *O. musculus*, setae generally longer, P-4 shorter than P-2) and shape of distal setae on Cx-I (in *O. musculus* stouter, pectinate). In addition, *O. carpenteri* differs from *O. mus-*

culus in the shape of the hyaline distal projection of Cx-I (enlarged, straight or slightly bent and distally truncate, in *O. musculus* short and rounded, TUZOVSKIJ 2006b).

Preadult stages: Larva unknown. Deutonymph: HALBERT (1911), TUZOVSKIJ (2006b).

Habitat and Biology: Lakes, ponds and ditches, occasionally in streams. Life cycle unknown.

Distribution: Western Palaearctic, reported from a limited number of countries: Ireland, England, Sweden, Finland, Germany, France, Switzerland, Italy and The Netherlands; rare. In Sweden occurring far north (LUNDBLAD 1962).

3. *Oxus (Oxus) longisetus* (BERLESE, 1885) (*8-56 c-d*)

1885 *Pseudomarica longiseta* BERLESE, Acari, Myriopoda et Scorpiones 3 (23): 10.

1899 *Oxus koenikei* THOR, Arch. Math. Naturv. 21: 34 – DI SABATINO et al. (2009).

Description: Idiosoma brownish-yellowish. Tips of Cx-I each with two very long, curved setae, which are longer than the width of Cx-I. Coxal field extending posteriorly almost to posterior idiosoma margin. Terminal seta of IV-leg-6 shorter than half of this leg segment. Males: Idiosoma length 700-750, width 433-520 µm; length of terminal seta 1/4 length of IV-L-6. Females: Idiosoma length 1030-1300, width 800-900 µm; length of terminal seta 1/3 length of IV-L-6.

Preadult stages: Larva: WAINSTEIN (1980). Deutonymph: ÖZKAN, ERMAN & BOYACI (1995).

Similar species: LUNDBLAD (1969) placed *O. pictus* DADAY, 1898a from Sri Lanka into synonymy with *O. longisetus*. However, according to COOK (1967), the long distal setae of Cx-I are not curved in *O. pictus*. See also discussion of *O. setosus*.

Habitat and Biology: Standing waters, e.g. ditches, lakes. Life cycle unknown.

Distribution: Palaearctic. Widespread in Europe, but rare in the north. In Sweden only a few records from the southern part of the country (LUNDBLAD 1962). Questionable records from Burma, India and Sri Lanka (LUNDBLAD 1969).

4. *Oxus (Oxus) musculus* (MÜLLER, 1776) (*8-55 a-c*)

1776a *Frontipoda musculus* MÜLLER, Zoologiae Danicae prodromus, seu animalium Daniae et Norvegiae indigenarum characteres, nomina, et synonyma imprimis popularium: 191.

Description: Idiosoma reddish, occasionally blue-green. Tips of Cx-I each with two blade-like setae with fine pectination. P-1 distinctly shorter than P-2, the latter segment longer than P-4; P-4 stocky. Excretory plate ending in a long, pointed posterior extension variable in length. Males: Idiosoma length 850-900, width 610, height 720 µm. Females: Idiosoma length 875-1050, width 500-570, height 720-770 µm.

Similar species: See discussion of *O. carpenteri*.

Preadult stages: Larva: PIERSIG (1896-1899), TUZOVSKIJ (2002). Deutonymph: TUZOVSKIJ (1990).

Habitat and Biology: Standing waters, e.g. lakes, ponds and ditches. According to LUNDBLAD (1968) mainly in oligotrophic waters. In The Netherlands collected from mid-April until mid-October, with a peak in spring (SMIT & VAN DER HAMMEN 2000). PIERSIG (1896-1899) assumed a parasitic life style of the larva.

Distribution: Palaearctic. Widespread in Europe, in Sweden occurring far north (LUNDBLAD 1962).

5. *Oxus (Oxus) nodigerus* KOENIKE, 1898 (*8-56 h*)

1898b *Oxus nodigerus* KOENIKE, Zool. Anz. 272.

Description: Idiosoma brownish to reddish, occasionally greenish-yellow. Tips of Cx-I each with two short, bowed, blunt blade-like setae; these with fine pectination. Genital bay deep, widened posteriorly. Acetabula small, lying on elongated basal sclerites. Terminal seta of IV-leg-6 longer than 1/2 length of segment. Excretory pore located between associated glandularia. Males: Idiosoma length 1005-1034, width 591-586 µm. Females: Idiosoma length 1170-1327, width 700-905 µm.

Preadult stages: Larva unknown. Deutonymph: LUNDBLAD (1930a).

Habitat and Biology: Standing, mostly acid, waters: ponds, moorland pools and lime-poor coastal dune waters (SMIT & VAN DER HAMMEN 2000). Life cycle unknown.

Distribution: Palaearctic. Widespread, but not recorded from Southern Europe.

6. *Oxus (Oxus) ovalis* (MÜLLER, 1776) (*8-56 k*)

1776a *Hydrachna ovalis* MÜLLER, Zoologiae Danicae prodromus, seu animalium Daniae et Norvegiae indigenarum characteres, nomina, et synonyma imprimis popularium: 190.

Description: Idiosoma brownish-yellow. Tips of Cx-I each with two short blade-like setae with fine pectination; one of these setae more bowed than the other. Genital bay with straight anterior margin. Terminal seta of IV-L-6 in length about 80-100 % of leg segment. Excretory pore anterior to associated glandularia. Males: Idiosoma length 775-895, width 565, height 530 µm; genital bay posteriorly widened. Females: Idiosoma length 800-1075, width 530, height 490 µm; genital bay posteriorly strongly widened.

Similar species: *Oxus ovalis* is most close to *O. strigatus*, and some authors, e. g. LUNDBLAD (1962), consider the two species conspecific. However, differences in the shape of the genital bay (especially in the female), configuration of the excretory pore and its associated glandularia, and size warrant separate status for these two species.

Preadult stages: Larva: WAINSTEIN (1980). Deutonymph: TUZOVSKIJ (1990).

Habitat and Biology: Standing waters, e. g. lakes, ponds and ditches; according to LUNDBLAD (1968), mainly in oligotrophic waters, but also in eutrophic and dystrophic waters. According to SMIT & VAN DER HAMMEN (2000) in freshwater with a low nutrient content. Life cycle unknown.

Distribution. Western Palaearctic. Widespread in Europe, in Sweden only in the central and southern part of the country (LUNDBLAD 1968).

7. *Oxus (Oxus) strigatus* (MÜLLER, 1776) (*8-56 e-g*)

1776a *Hydrachna strigata* MÜLLER, Zoologiae Danicae prodromus, seu animalium Daniae et Norvegiae indigenarum characteres, nomina, et synonyma imprimis popularium: 191.

Description: Idiosoma brownish-yellow. Tips of Cx-I each with two short, blade-like setae with fine pectination; one of these more bowed than the other. Genital bay with concave anterior margin. Excretory pore between the associated glandularia. Terminal seta of IV-leg-6 as long as segment. Males: Idiosoma length 595-700, width 450, height 400 µm. Females: Idiosoma length 720-915, width 450, height 400 µm.

Similar species: Following the original description, *Oxus lineatus* WALTER, 1912b from Austria should be similar to *O. strigatus*. As *O. lineatus* was not figured, with the holotype in poor condition and no other specimens recorded since, it is considered a *species incerta* by DI SABATINO et al. (2009). See also discussion of *O. ovalis*.

Preadult stages: Larva and deutonymph: PIERSIG (1896-1899).

Habitat and Biology: Standing waters, e. g. ponds, ditches and coastal dune lakes. Also found in canalized lowland streams (SMIT & VAN DER HAMMEN 2000). Life cycle unknown.

Distribution: Western Palaearctic. Widespread in Europe.

8. *Oxus (Oxus) tenuisetis* PIERSIG, 1898

1898c *Oxus tenuisetis* PIERSIG, Zool. Anz. 21: 525.

Description: Coxal field short, anteriorly not extending beyond anterior idiosoma margin, posteriorly just a little beyond middle of venter. Setae of tips of Cx-I strongly bowed, blade-like, with serrated margins. Genital bay wide. Terminal seta of IV-L-6 serrate, in length 60-85 % of segment. Males: Idiosoma length 1100-1350 µm; Females: Idiosoma length 1350-1640, width 825-1000, height 1120 µm.

Habitat: Standing waters, mainly in ponds, occasionally in streams. Preadult stages and life cycle unknown.

Distribution: Western Palaearctic; reported from a limited number of countries: Germany, Sweden, Poland, Romania, Russia and the former Czechoslovakia, rare. No records from Southern Europe.

Subgenus *Gnaphiscus* KOENIKE, 1898

1898b *Gnaphiscus* KOENIKE, Zool. Anz. 21: 267.

Type species: *Gnaphiscus setosus* KOENIKE, 1898b, by monotypy.

Diagnosis: Coxal field with pronounced sexual dimorphism; in males, genital field enclosed by the coxae, except for a narrow median strip, in females, coxae embracing only the anterior part of genital field.

Remark: Possibly a monophyletic taxon, including four species recorded from the Northern Hemisphere.

9. *Oxus (Gnaphiscus) setosus* (KOENIKE, 1898) (*8-56 a-b*)

1898b *Gnaphiscus setosus* KOENIKE, Zool. Anz. 21: 267.

1913 *Gnaphiscus ekmani* THOR, Zool. Anz. 41: 166 – TUZOVSKIJ (2001b).

Description: Idiosoma brownish-yellow to blue-green. Tips of Cx-I each with two laterally-bowed setae. P-4 stocky. Males: Idiosoma length 600, width 320, height 400 µm; IV-leg-6 terminal seta slightly shorter than segment length; Females: Idiosoma length 700-900, width 400-650, height 450 µm; IV-leg-6 terminal seta distinctly shorter than (but > 50 %) segment length.

Similar species: The male is well characterized by the shape of the coxae enclosing the genital field. The female is similar to *O. longisetus*, which also has long setae on the tip of the first coxal plates. However, these setae are much longer in the latter, and more bowed. *Oxus ekmani* (THOR 1913a) from Sweden was introduced for specimens with a somewhat shorter coxal field, but this does not warrant a separate specific status (LUNDBLAD 1962, 1968).

Preadult stages: Larva: TUZOVSKIJ (2001b) – *sub nom. Oxus ekmani* (THOR, 1913a); deutonymph: TUZOVSKIJ (1990).

Habitat and Biology: Lakes and streams. According to LUNDBLAD (1968) living near the bottom of lakes, up to a depth of 34-36 m, only occasionally in the littoral. However, in The Netherlands only found in man-made springs and canalized lowland streams (SMIT & VAN DER HAMMEN 2000).

Distribution: Palaearctic. Widespread in Europe. Life cycle unknown.

Family Rutripalpidae SOKOLOW, 1934

1934c Rutripalpidae SOKOLOW, Zool. Anz. 106: 325.

Diagnosis: Lateral eyes with two lenses in a small eye capsule, no frontal eye present. Five pairs of dorsocentralia well sclerotized. Coxal plates in paired anterior and posterior groups; Cgl-1 in the integument between Cx-II and Cx-III. Legs stout, central segments with heavy distal setae, without swimming setae. Leg claws I-III with ventral blade, claws IV stronger, lacking ventral blade. Excretory pore with a small semicircular anterior sclerite. Gnathosomal base and chelicerae short and stocky, cheliceral basal segment proximally with a strong, pointed dorsal projection, claw short and hook-like. All palp segments without ventral projections; P-2 strongly thickened, P-3 and P-4 very short, P-5 flattened and enlarged, in dorsal view with a lateral indentation, with four strong, nail-like claws arranged in two groups. Males: Six pairs of dorsolateralia occasionally sclerotized as small muscle attachments. Genital field consisting of an angled, movable penis-like tube formed by fused genital plates; tip of this tube when in normal position, the organ is embedded in a pit of membranous integument, directed posteriorly, in erect position directed anteriorly. In anterior part of the organ paired, long , posteriorly directed setae, three pairs of Ac on distal part of "penis" tube. Acetabula-bearing area flanked by a collar diverging posteriorly and forming a pair of tips on level of Ac-3; this collar and its surroundings with numerous spine-like short setae. Between Ac, no gonopore slit visible but a posteriorly-directed tube originating medially from Ac-3. Ejaculatory complex small, proximal arms and proximal horns reduced. Genital field flanked by a pair of large sclerites. Palp very stout, P-5 in dorsal view with a lateral indentation, nail-like claws arranged in two groups. Females: Dorsolateralia and ventralia not sclerotized. Genital field with a pair of flaps flanking three pairs of Ac and the posterior half of the long gonopore; pregenital sclerite prominent, pent- or hexangular. Palp slightly more slender than in males, P-5 in dorsal view without lateral indentation, nail-like claws arranged in one group.

Remarks: A rather isolated, monotypic taxon within Lebertioidea. Larval morphology suggests a relationship to Bandakiopsidae PANESAR, 2004, with the only genus *Bandakiopsis* SMITH, 1979, recorded from North America and the Himalayas. However, in comparison with *Bandakiopsis* larvae, the larva of *Rutripalpus* is characterized by a high number of plesiomorphies, suggesting an early separation of the two clades. Also the deutonymphal provisional genital organ, with paired halfmoon-shaped sclerites embracing Ac (unknown in *Bandakiopsis*), is a plesiomorphic condition, found in Teutoniidae, Lebertiidae and Sperchontidae (GERECKE & TUZOVSKIJ 2001).

Genus *Rutripalpus* SOKOLOW, 1934

1934c *Rutripalpus* SOKOLOW, Zool. Anz. 106: 325.

Type species: *Rutripalpus limicola* SOKOLOW, 1934c, by monotypy.

Diagnosis and Remarks: Only genus, with the characters of the family. In addition to the European *R. limicola*, only one further species known, *R. canadiensis* SMITH, 1991.

Rutripalpus limicola Sokolow, 1934 (*8-35 c-d; 8-57 a-e*)

1934c *Rutripalpus limicola* Sokolow, Zool. Anz. 106: 325.

Description: Colour red. Integument with fine lines formed by the interspaces of flat, pillow-shaped papillae. Males: Idiosoma length/width 670-750/550-600 µm; second pair of dorsocentralia may fuse medially to one single platelet; length/height P-1, 36/72; P-2, 136/94; P-3, 52/67; P-4, 76/51; P-5, 67/29 µm. Females: Idiosoma length/width 780-820/600-630 µm; dorsocentralia smaller than in males, never fused; length/height P-1, 34/72; P-2, 154/96; P-3, 61/72; P-4 96/40; P-5, 78/22 µm.

Preadult stages: Larva and deutonymph: Gerecke & Tuzovskij (2001).

Habitat and Biology: Crenobiont. In weakly seeping helocrenes. Larvae parasitic on Ptychopteridae (Diptera). Preparasitic attendance on host larvae, parasitic-phoretic behaviour on adults. The whole male genital organ can be moved due to a snap mechanism (protrusion induced by increased haemolymph pressure, accompanied by expansion of the otherwise folded integument posterior to the genital field, retraction probably by muscle contraction (Gerecke & Tuzovskij 2001, Gerecke 2006).

Distribution: Central-Northeastern Europe, very rare. Only known from a few sites in NW Russia, S Germany and Montenegro (Gerecke & Tuzovskij 2001, Gerecke 2006).

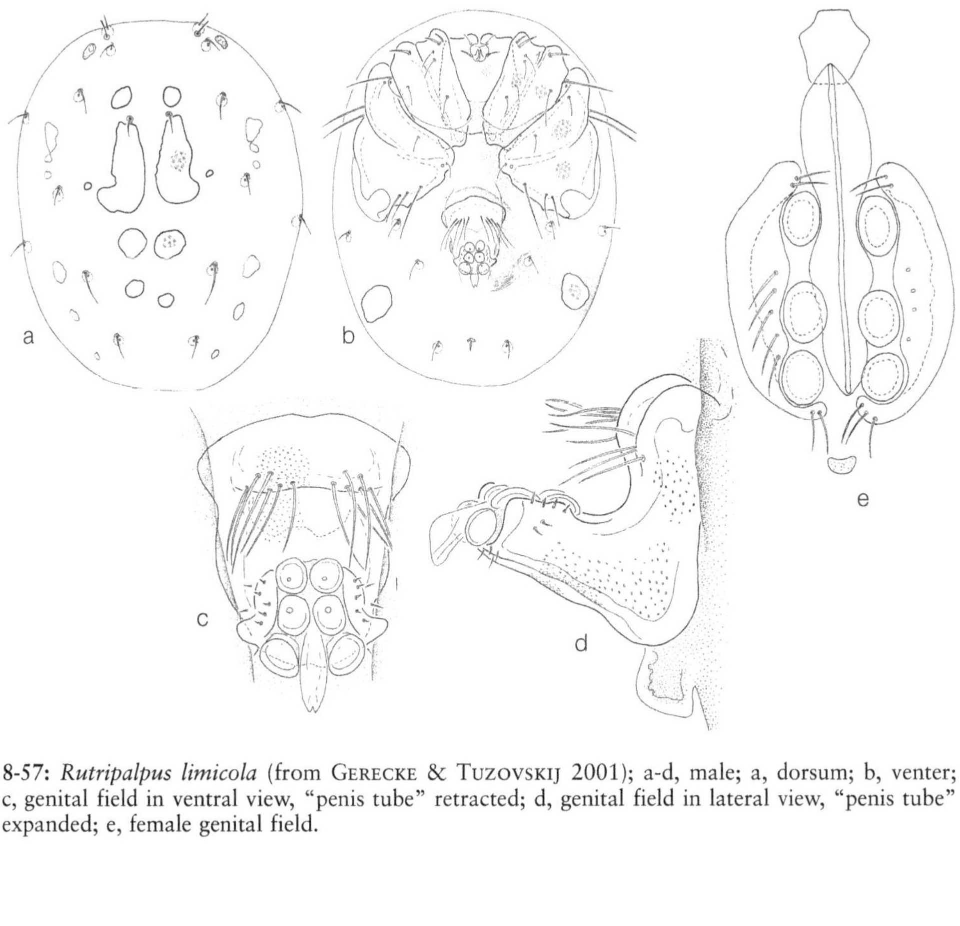

8-57: *Rutripalpus limicola* (from Gerecke & Tuzovskij 2001); a-d, male; a, dorsum; b, venter; c, genital field in ventral view, "penis tube" retracted; d, genital field in lateral view, "penis tube" expanded; e, female genital field.

Family Sperchontidae THOR, 1900

1900 Sperchontidae THOR, Nyt. Mag. Naturv. 38: 2.

Diagnosis: Idiosoma sclerotization various, from none to extensive, with development of a complete dorsal shield. If muscle attachments sclerotized, these generally more extended and often fused with neighbouring plates in males, but smaller and unfused in females. Lateral eyes encapsulated, frontal eye generally absent but occasionally visible in males as a small pore in the anterior dorsal shield. Coxae typically in four groups, but occasionally fused on their respective sides. Legs with well-developed claws bearing claw blades and clawlets, without swimming setae. Genital field with well-developed flaps bearing a row of medial setae, but without setation on flap surface, more slender and elongate in females than in males. Three pairs of Ac between gonopore and flaps. Pregenital sclerite weakly developed or absent in males, crescent-shaped, with a medial extension in females. Anterior to genital field a round sclerotized platelet present in males, absent in females. Palp five-segmented, neither chelate nor uncate; P-2 typically with a ventral projection.

Remarks: On the basis of larval morphology and behaviour, COOK (1974) interpreted this family as a transitional group between lower, hydryphantid-like, and higher, lebertioid-like water mites, and HARVEY (1998) proposed to rank sperchontids in their own superfamily Sperchontoidea. However, many questions about relationships of this probably monophyletic family are open and more data on larval morphology, also in related families are needed for a thorough discussion. In addition to Sperchontinae (see below), only the monotypic subfamily Apeltosperchontinae COOK, 1974 (known from Chile and New Zealand) is presently accepted.

Subfamily Sperchontinae THOR, 1900

Diagnosis: Coxae normally in four groups – if fused, suture lines Cx-I/II, Cx-II/III and Cx-III/IV on their respective sides complete. Cx-I without a sclerotized line crossing from suture Cx-I/II to lateral gnathosomal bay margin. Tips of Cx-I with several setae, but without a particularly large seta arising from a prominent setal base. P-3 without a long heavy ventrodistal seta, P-4 without a heavy mediodistal seta.

Remarks: A rather homogenous assemblage of taxa, possibly in paraphyletic relationship with the highly distinct Apeltosperchontinae (COOK 1974).

Distribution: Worldwide except for Australia and Antarctica.

Key to genera

1 Glandularia not particularly raised, their surface smooth (*8-58 a*). Ventral setae of P-4 peg-like, separated from each other; cheliceral basal segment without an anteriorly-directed, hook-like dorsal extension (*8-58 b*). . . . . *Sperchon* (page 154)

\- Glandularia raised, with a papillate surface (*8-58 c*). Ventral setae of P-4 not peg-like, in European species located close to each other on a prominent projection in proximal part of segment; cheliceral basal segment with an anteriorly-directed, hook-like dorsal extension (*8-58 d*). . . . . *Sperchonopsis* (page 154)

Genus *Sperchon* KRAMER, 1877

1877 *Sperchon* KRAMER, Arch. Naturg. 43 (1): 240.

1985b *Charoelia* BADER, Rev. Suisse Zool. 92: 461 – DI SABATINO et al. (2009).

Type species: *Sperchon squamosus* KRAMER, 1879, by original designation.

Diagnosis: Glandularia not strongly raised above integument surface, their surface smooth or porose, not papillate or warty. P-4 with two ventral peg-like setae (reduced in some extra-European species).

Remarks: Probably a monophyletic clade; some species in palp morphology similar to *Sperchonopsis* are obviously close relatives of typical *Sperchon* species, having secondarily reduced peg setae of P-4. In addition to the two subgenera treated here, two further subgenera are known: *Acadiosperchon* HABEEB, 1959 (Eastern North America) and *Palpisperchon* LUNDBLAD, 1941 (Western North America, Burma, Russia).

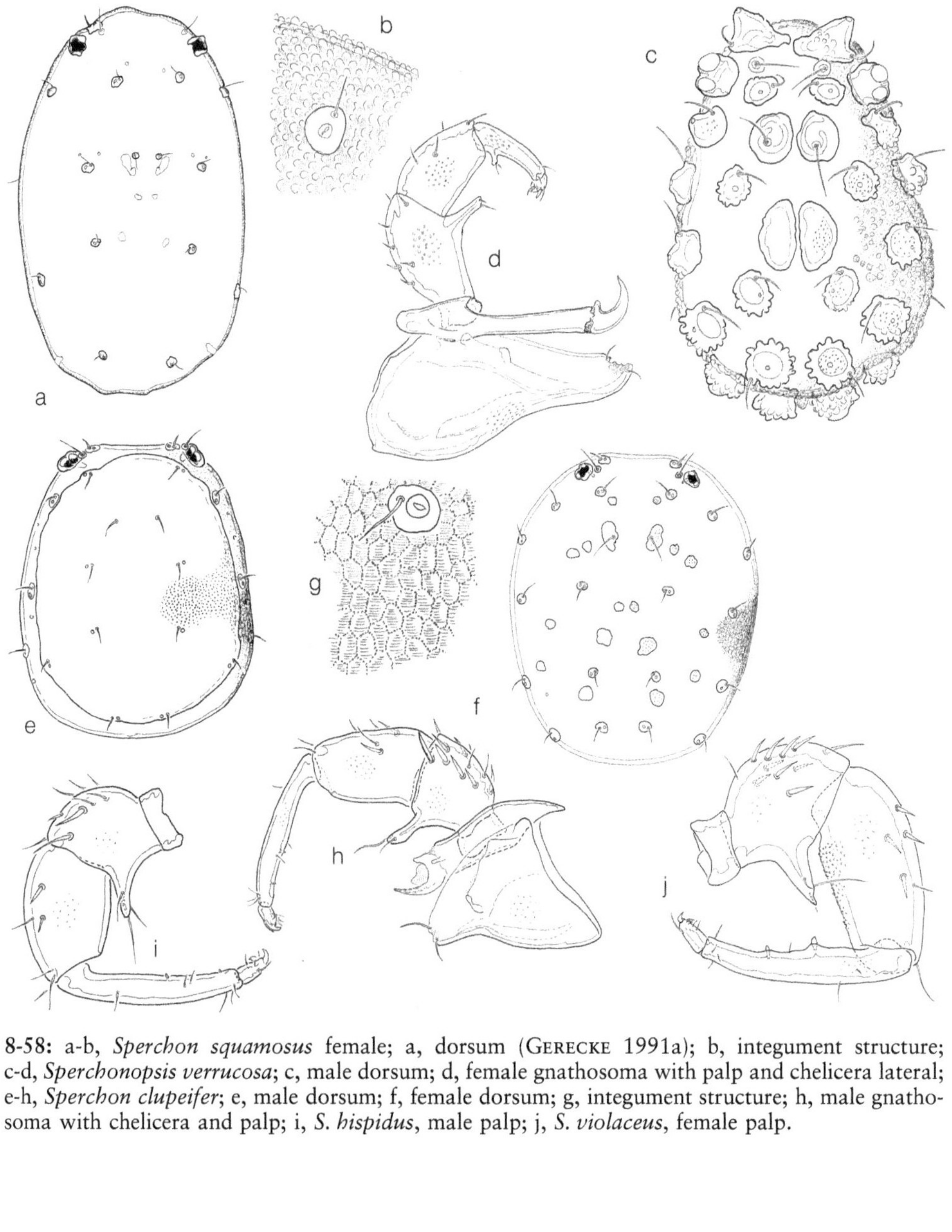

8-58: a-b, *Sperchon squamosus* female; a, dorsum (GERECKE 1991a); b, integument structure; c-d, *Sperchonopsis verrucosa*; c, male dorsum; d, female gnathosoma with palp and chelicera lateral; e-h, *Sperchon clupeifer*; e, male dorsum; f, female dorsum; g, integument structure; h, male gnathosoma with chelicera and palp; i, *S. hispidus*, male palp; j, *S. violaceus*, female palp.

In addition to the indications given for the single species treated here (see below), descriptions of *Sperchon* larvae and important indications on their host range are given by Prasad & Cook (1972), I. M. Smith (1982), Smith & Oliver (1976c) and Wainstein (1980). A thorough treatment and discussion of deutonymphal characters is presented by Tuzovskij (1990).

Distribution: Worldwide except for Australis and Antarctica.

Key to subgenera and species

Excluded from the following key as *species incertae* (Di Sabatino et al. 2009): *Sperchon carpaticus* Biesiadka, 1972, *S. hystrix* Komárek, 1921; *S. montanus* Thon, 1901d and *S. ornatus* Halbert, 1944.

1 Membranous integument – at least in localized areas of the venter – reticulate (*8-58 g*). Dorsum and posterior venter with sclerotized muscle attachment plates (*8-58 f*) which may fuse to form larger shields in males (*8-58 e*): subgenus *Hispidosperchon*. 2

\- Membranous integument rugose or papillate (*8-58 b*), nowhere reticulate. Muscle attachment plates in European species often unsclerotized, rarely one to three pairs of small dorsal plates (*8-58 a*): subgenus *Sperchon* s.str. 11

2 (1) Well developed sexual dimorphism: Males (pregenital sclerite hardly developed or absent) palp stout, with inflated P-2 and P-3 distal margins forming strongly developed sheaths flanking base of subsequent segments (*8-59 a*), idiosoma with a large dorsal shield (*8-59 d*). Females (pregenital sclerite well developed, with a posteromedial extension) palp slender, ventral margin of P-3 covered by fine denticulation (*8-58 j*); idiosoma membranous, with small muscle insertion platelets (as in *8-58 f*): *denticulatus* species group. 3

\- Poorly developed sexual dimorphism. In both sexes, palp neither inflated, nor with a denticulate ventral margin of P-3. Males with, or (more frequently), without dorsal shield. 6

3 (2) Sclerotized parts purple to violet in colour. Males: postgenital plate broadly oval, length/width > 0.7 (*8-59 f*); tip of anterior ramus of ejaculatory complex with an U-shaped cleft flanked by a pair of laterally rounded, flange-like extensions (*8-59 g*). . . . . *Sperchon violaceus* (page 166)

\- Sclerotized parts uncoloured, coriaceous (species distinguishable in males only). . . . 4

4 (3) Male palp relatively slender, P-4 dorsal length/proximal constriction height (= least proximal height) > 7.0 (*8-59 h*); postgenital plate very small and roundish, length/width > 1.0 (*8-59 i*); tip of anterior ramus of ejaculatory complex with a broad V-shaped cleft flanked by a pair of angular flange-like extensions, base of cleft with three small projections (*8-59 j*). . . . . *Sperchon hibernicus* (page 163)

\- Male palp stout, P-4 dorsal length/proximal constriction height < 7.0 (*8-59 a*); postgenital plate and tip of anterior ramus of ejaculatory complex various. 5

5 (4) Male postgenital plate large, anteriorly rounded, posteriorly more-or-less straight, length/width < 1.0 (*8-59 b*); tip of anterior ramus of ejaculatory complex shallowly indented with laterally rounded, posteriorly slightly pointed extensions (*8-59 c*). . . . . *Sperchon denticulatus* (page 163)

\- Male postgenital plate small, generally oval, length/width > 1.0 (*8-59 l*); tip of anterior ramus of ejaculatory complex with an U-shaped cleft flanked by a pair of sharply angled lateral extensions (*8-59 m*). . . . . *Sperchon vaginosus* (page 166)

6 (2) Cx-III with a glandularium near medial margin (as in *8-60 g*). Palps in a normal position, held parallel to each other. 7

\- Cx-III without a glandularium (as in *8-61 g*). Palps rotated to an oblique horizontal position, parts of their medial surface visible in dorsal view (*8-36 c*). 8

7 (6) Excretory pore surrounded by a sclerotized ring (*8-59 n*); IV-L-3-5 with densely plumose dorsal setae (*8-59 o*). In both sexes, dorsum with several isolated plates; palps stouter (length/height P-4, 4.5-5.4) (*8-58 i*). . . . . *Sperchon hispidus* (page 164)

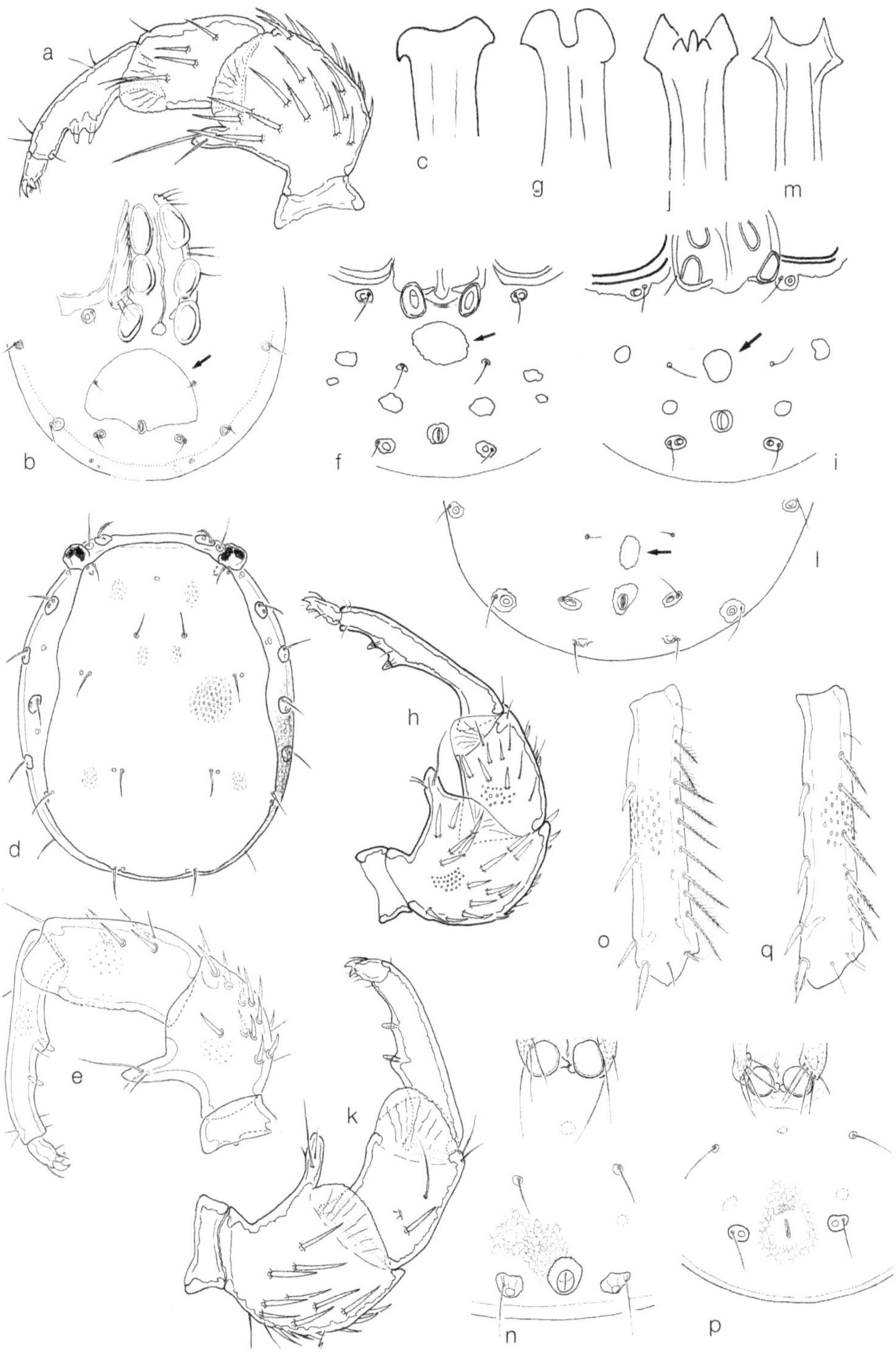

8-59: *Sperchon*; a-c, *S. denticulatus*; a, male palp; b, male venter (posterior part); c, tip of ejaculatory complex; d-g, *S. violaceus*; d, male dorsum; e, male palp; f, male venter (posterior part); g, tip of ejaculatory complex; h-j, *S. hibernicus*; h., male palp; i, male venter (posterior part); j, tip of ejaculatory complex; k-m, *S. vaginosus*; k, male palp; l. male venter; m, tip of ejaculatory complex; n-o, *S. hispidus*; n, excretory pore area; o, IV-L-4; p-q, *S. clupeifer*; p, excretory pore area; q, IV-L-4 (n-q: Gerecke 1991a) (arrows indicate postgenital plate).

- Excretory pore smooth (*8-59 p*); IV-L-3-5 with less plumose dorsal setae (*8-59 q*). Males with a dorsal shield, females with several isolated plates; palps more slender (length/height P-4, 5.4-6.5) (*8-58 h*). . . . . *Sperchon clupeifer* (page 162)

8 (6) P-3 without ventral setae (*8-60 a*). . . . . *Sperchon compactilis* (page 162)

- P-3 with ventral setae (*8-60 b, d*). 9

9 (8) P-3 with four, rarely three ventral setae (*8-60 b*); integument sculpture formed exclusively by fine denticles (as in *8-58 g*). . . . . *Sperchon setiger* (page 165)

- P-3 with up to two ventral setae (*8-60 d*); integument various. 10

10 (9) Integument sculpture made up of a network of fine denticles combined with additional larger papillae (*8-60 c*). . . . . *Sperchon papillosus* (page 165)

- Integument sculpture made up of a simple network of fine denticles, without larger papillae (as in *8-58 g*). . . . . *Sperchon insignis* (page 164)

11 (1) Gnathosoma length > 270 µm, distinctly longer than wide, rostrum nearly as long as gnathosomal base; P-2 dorsoventrally high, P-5 very short (< 30 µm) (*8-61 a-b, d, f*). 12

- Gnathosoma length < 270 µm, rostrum shorter than gnathosomal base; P-2 not dorsoventrally high, P-5 absolutely and relatively longer (> 30 µm) (e.g., *8- 62 b, d-e*). 16

12 (11) Cx-III with a medial glandularium (*8-60 g*). P-2 without a neck-like narrow proximal section (*8-60 h*). . . . . *Sperchon longirostris* (page 168)

- Cx-III without a glandularium (*8-61 g*). P-2 with a neck-like narrow proximal section (*8-61 a-b, d, f*). 13

13 (12) P-2 extremely high, with a dorsal hump (as high as or higher than long); P-4 ventral peg-like setae directed proximally (*8-61 a-b*). 14

- P-2 less high (longer than high); P-4 ventral peg-like setae directed ventrally (*8-61 d, f*). 15

14 (13) P-2 as long as high, its distoventral projection short, forming a broad-based triangle; P-4 distoventral peg-seta located ventrally, closer to distal segment margin than to proximoventral peg seta (*8-61 a*). . . . . *Sperchon resupinus* (page 169)

- P-2 higher than long, its distoventral projection narrow, finger-shaped; P-4 distoventral peg-seta located medially, closer to proximoventral peg seta than to distal segment margin (*8-61 b*). . . . . *Sperchon turgidus* (page 171)

15 (13) Gnathosoma in dorsoventral view with a narrowing at the rostral base (*8-61 e*). Idiosoma elongated (*8-61 g*). . . . . *Sperchon longissimus* (page 169)

- Gnathosoma without a narrowing at rostral base (*8-61 c*). Idiosoma roundish. . . . . *Sperchon squamosus* (page 170)

16 (11) Cx-I medially fused, completely surrounding gnathosomal base (*8-62 a*); gnathosomal rostrum obtuse or nearly absent. 17

- Cx-I medially separated by a fine strip of membranous integument (as in *8-61 g*); gnathosomal rostrum more elongate and pointed. 18

17 (16) Excretory pore surrounded by a sclerotized ring; palp very slender (length/height P-4 > 6.0); P-2 with a long distoventral projection (*8-62 b*); idiosoma sclerites generally deep red or brownish in colour. *Sperchon brevirostris* (page 167)

- Excretory pore smooth; palp stouter (length/height P-4 < 6.0); P-2 with a short distoventral projection (*8-62 d*); idiosoma sclerites generally pale in colour. . . . . *Sperchon mutilus* (page 169)

18 (16) Excretory pore surrounded by a sclerotized ring. . . . . *Sperchon glandulosus* (page 168)

- Excretory pore not surrounded by a sclerotized ring. *Sperchon thienemanni* (page 170)

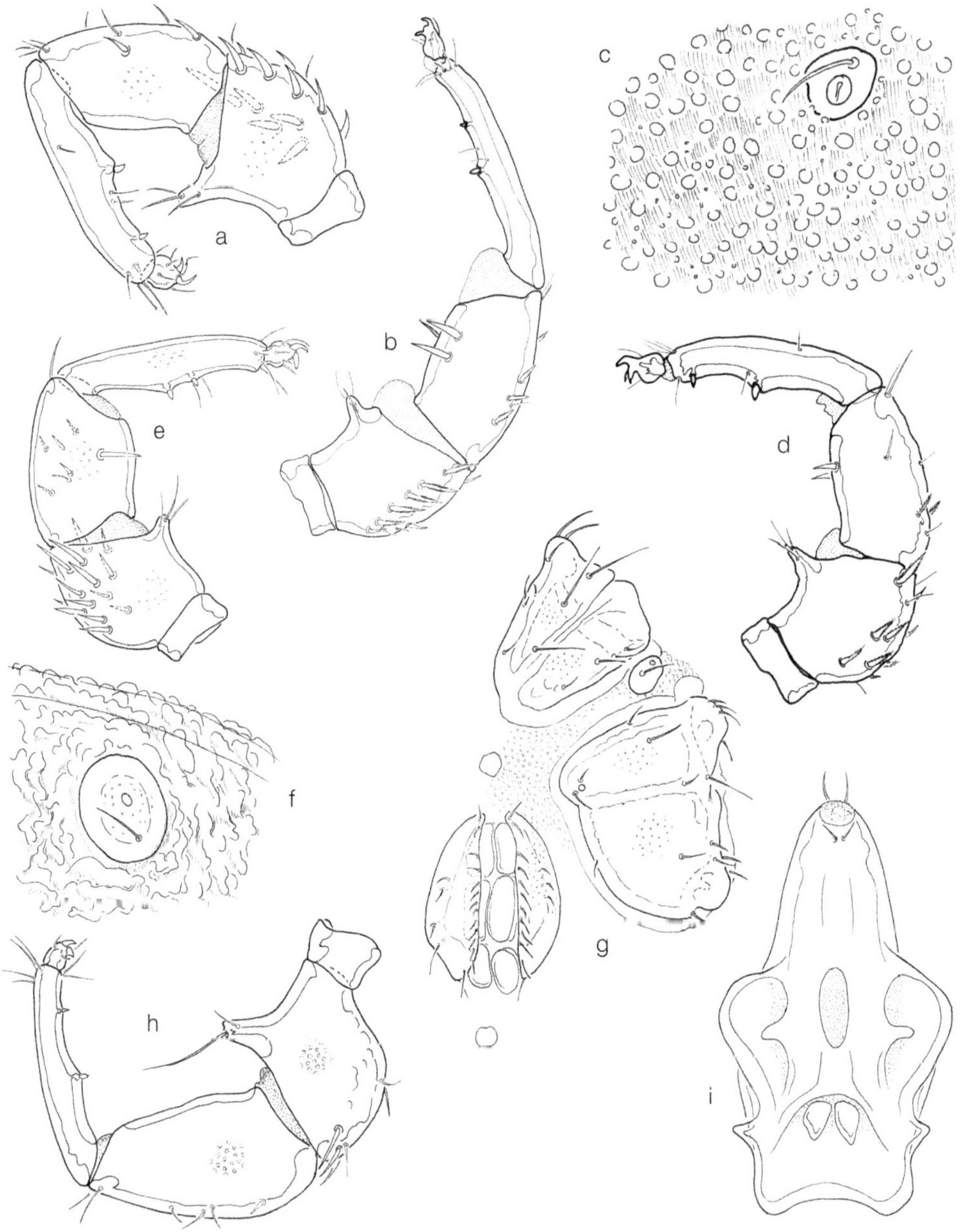

8-60: *Sperchon*; a, *S. compactilis*, male palp; b, *S. setiger*, male palp (GERECKE 1991a); c-d, *S. papillosus*; c, integument structure; d, male palp (GERECKE 1991a); e, *S. insignis*, male palp; f, *S. thienemanni* male, integument structure; g-i, *S. longirostris* male, g, genital field and left coxae; h, palp i, gnathosoma in dorsal view.

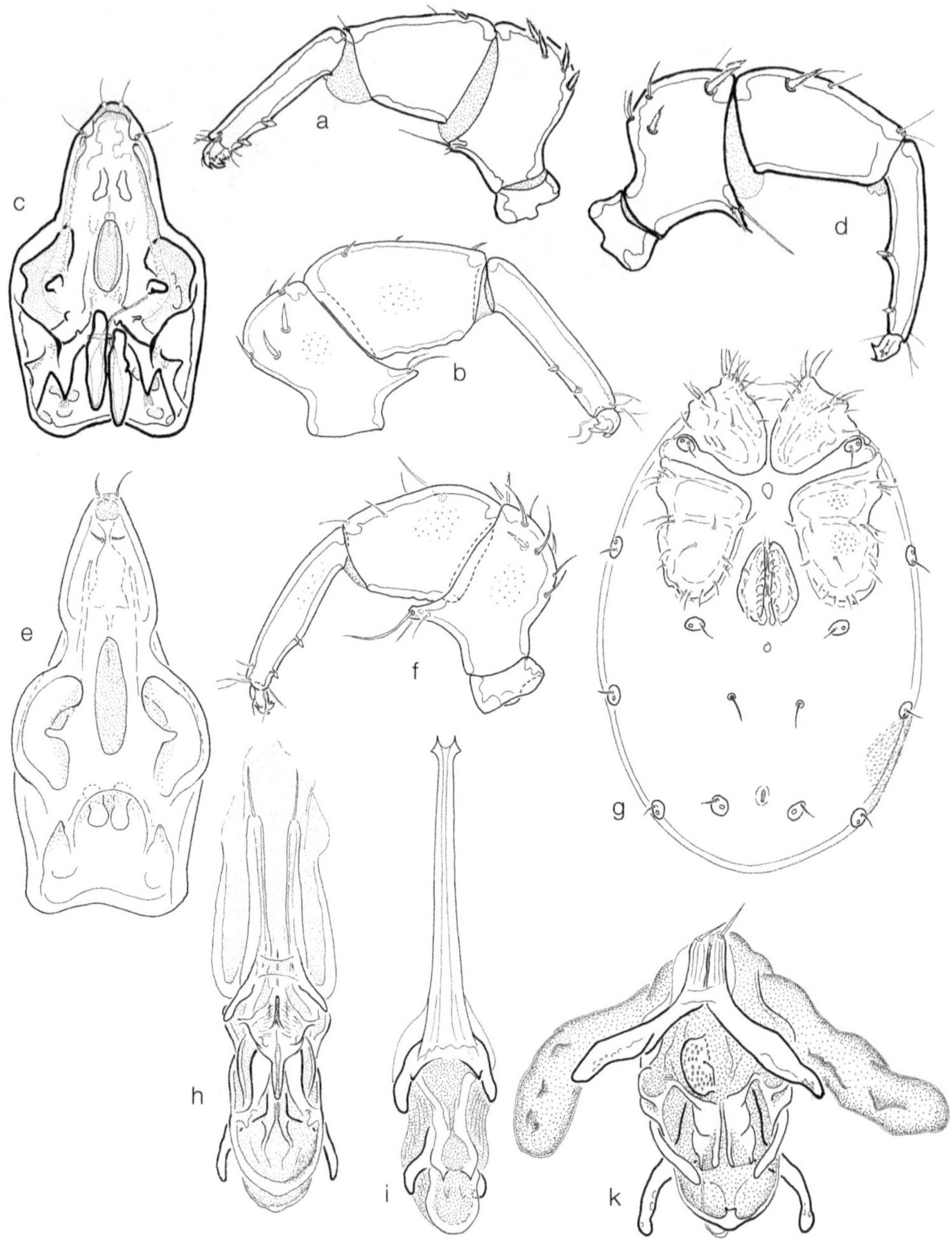

8-61: *Sperchon*; a, *S. resupinus,* male palp (Gerecke 1991a); b, *S. turgidus*, female palp; c-d, *S. squamosus* (Gerecke 1991a); c, gnathosoma, dorsal view; d, palp; e-g, *S. longissimus*; e, male gnathosoma, dorsal view; f, palp; g, venter; h-k, male ejaculatory complex, anterior view (Gerecke 1991a); h, *S. hispidus*; i, *S. vaginosus*; k, *S. squamosus*.

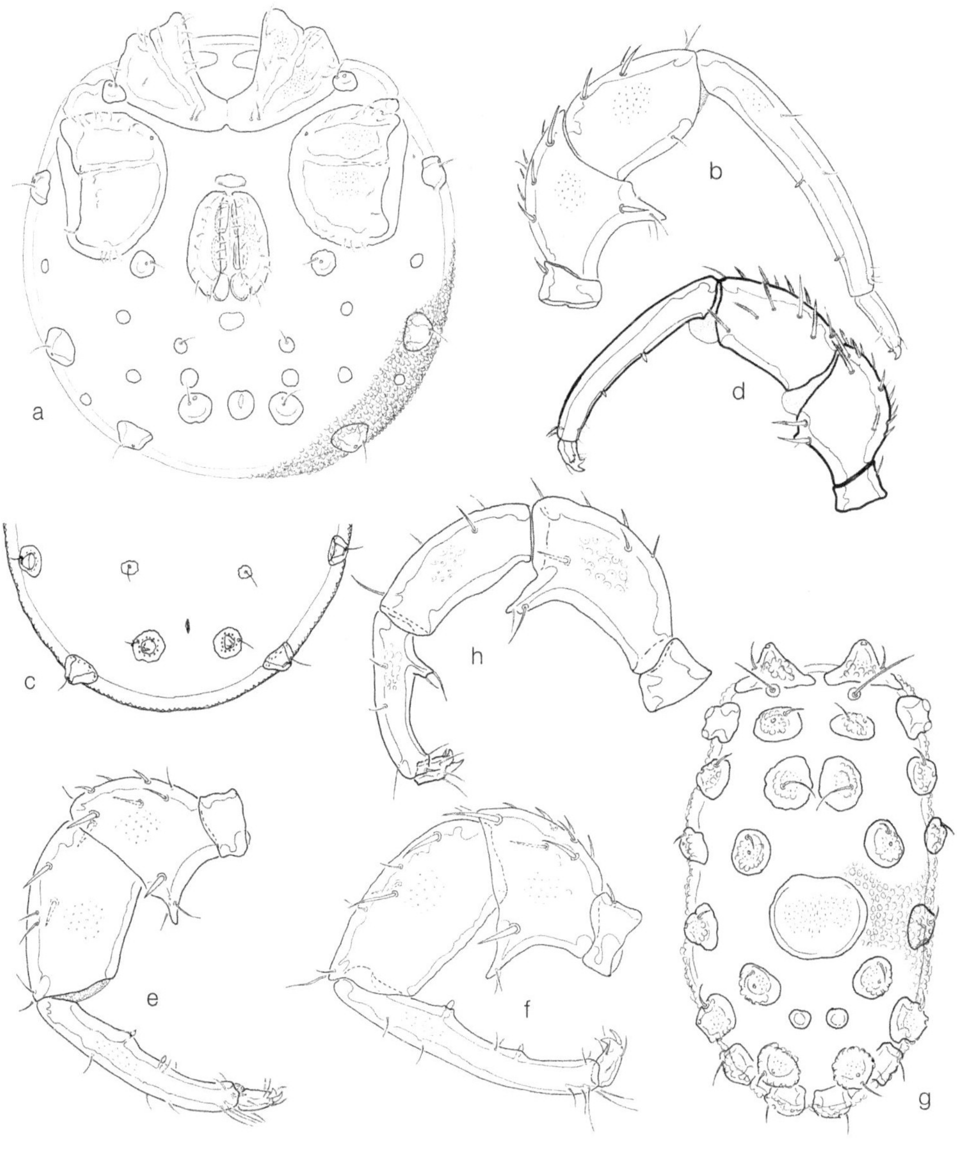

8-62: a-b, *Sperchon brevirostris* male; a, venter; b, palp; c-d, *Sperchon mutilus* male (GERECKE 1991a); c, excretory pore area; d, palp; e, *Sperchon glandulosus*, male palp; f, *Sperchon thienemanni*, male palp; g-h, *Sperchonopsis procera* male, g, dorsum; h, palp.

Subgenus *Hispidosperchon* THOR, 1901

1901b *Hispidosperchon* THOR, Arch. Math. Naturvid. 23: 24.

1926l *Mixosperchon* K. VIETS, Zool. Anz. 69: 190 – DI SABATINO et al. (2009).

Type species: *Sperchon hispidus* KOENIKE, 1895, by original designation.

Diagnosis: Integument areolate due to honeycomb-like arrangement of fine denticles. Muscle attachments often sclerotized, forming plates that tend to fuse in males of many species. P-3 may be finely denticulate or bear setae on the ventral surface, but is without projections.

Remarks: Possibly a monophyletic group, distinguished by the characteristic integument structure. In some species, on the edges of the "honeycombs" may be larger papillae, and these may be conspicuously enlarged as an individual variation, obfuscating the characteristic mesh-like pattern. However, remnants of the integument structure remain visible at least on the lateral or posteroventral idiosoma.

1. *Sperchon (Hispidosperchon) clupeifer* PIERSIG, 1896 (*8-58 e-h, 8-59 p-q*)

1896a *Sperchon clupeifer* PIERSIG, Zool. Anz. 19: 438.

Description: Colour yellow to reddish brown. Excretory pore unsclerotized. Palp slender, P-4 with very small ventral peg setae. Males: Dorsum with a large porose shield. Idiosoma length/width 540-800/450-600; genital flap length 120-145 µm; a faintly sclerotized, somewhat triangular plate on each side of excretory pore; length P-2, 90-110; P-3, 116-150; P-4, 160-195 µm. Females: Dorsum with five to seven paired dorsal muscle attachment plates. Idiosoma length/width 700-1100/600-900; length genital flap 150-166; P-2, 130-157; P-3, 180-220; P-4, 208-260 µm.

Similar species: See discussion under *S. hispidus*.

Preadult stages: Larva: ULLRICH (1976); TUZOVSKIJ (1977b); WAINSTEIN (1980); MARTIN (2000).

Habitat and Biology: Low to high order streams (also in potamal reaches: BADER 1988b); able to withstand organic pollution (VAN DER HAMMEN & SMIT 1996), but absent from strongly polluted sites (GERECKE & SCHWOERBEL 1992). Larvae parasitic on chironomid Diptera (MARTIN 2000).

Distribution: Western Palaearctic. Recorded from all over Europe except for Iceland and Scandinavian tundra's; known also from North Africa and Asia Minor.

2. *Sperchon (Hispidosperchon) compactilis* KOENIKE, 1911 (*8-60 a*)

1911a *Sperchon compactilis* KOENIKE, Zool. Anz. 37: 323.

1956a *Sperchon papillosus compactilis* LUNDBLAD, Ark. Zool. 10: 69.

Description: Colour yellow-brown; dorsal integument papillae occasionally enlarged, namely in males obscuring the areolate ground pattern which however remains visible posteroventrally. P-3 without ventral setae. Males: Idiosoma length/width 550-700/450-600; length genital flap 110-135; gnathosoma 153-170; P-2, 83-94; P-3, 98-119; P-4, 127-148 µm. Females: Idiosoma length/width 800-1100/750-950; length genital flap 140-160; P-2, 114-135; P-3, 142-160; P-4, 180-195 µm.

Similar species: Frequently found together with *S. papillosus* but by LUNDBLAD (1956a) considered a subspecies of the latter. Homogeneous *S. compactilis*-populations are also known. *Sperchon papillosus* and the closely related *S. algeriensis* LUNDBLAD, 1942 differ mostly in the presence of at least one ventral seta on P-3.

Habitat and Biology: Shaded low-order streams. Preadult stages and life cycle unknown.

Distribution: Central and SW Europe, Maghreb, Caucasus, Asia Minor.

3. *Sperchon (Hispidosperchon) denticulatus* KOENIKE, 1895 (*8-59 a-c*)

1895d *Sperchon denticulatus* KOENIKE, Rev. Suisse de Zoologie 4: 422.

Description: Males: Colour reddish. Idiosoma length/width 600-900/450-700 µm; dorsal shield present; genital flap length 180-240 µm; postgenital plate large, length/width 150-258/210-300, ratio length/width 0.7-0.9, semicircular, may include excretory pore within posterior boundary; ejaculatory complex with an elongate, rod-shaped anterior ramus, the tip shallowly indented with laterally rounded, posteriorly slightly pointed, extensions; gnathosoma length 275-325 µm, in lateral view with an anteroventral extension forming a mouth-like outline; length chelicera 330-390, claw 96-126 µm; palp stout, length P-2, 156-190; P-3, 120-150; P-4, 156-205 µm; P-2 ventral projection close to distal segment margin, anterior length ca. 50 µm, lateral face of segment with numerous (ca. 15) stout setae, medial face with ca. 14 stout setae close to dorsal margin; P-2 dorsally longer than P-3; P-3 ventrally slightly concave, medial and lateral faces with numerous stout setae; P-4 as long as or longer than P-2, ventral setae close together (distance 15-25 µm), their bases touching; P-4 proximoventrally slightly concave forming a slight constriction, ratio dorsal length/constriction height 4.4-4.9; I-IV-L-5 with prominent distolateral extensions. Females: See 'Similar species'.

Similar species: Together with *S. vaginosus*, *S. violaceus*, and *S. hibernicus*, *S. denticulatus* forms a monophyletic clade. In addition to the distinct sexual dimorphism in palp morphology, males share the particular shape of the ejaculatory complex in the form of an extremely elongate projection of the anterior ramus (LUNDBLAD 1956a, GERECKE 1991a). Tip morphology of the anterior ramus is thus a useful character for separation of males at species level. Because of confusion with other species in the *denticulatus*-group and the uncertain identity of females formerly described as *S. denticulatus*, apart from *S. violaceus*, distinguished by its name-giving colouration of sclerotized parts, these species can be distinguished in the male only.

Preadult stages: Larva: SPARING (1959); ULLRICH (1976). Identity of larvae in these studies questionable.

Habitat and Biology: Middle-order streams rich in vegetation; halotolerant: extending to estuarine sectors and found in semihaline inland waters (HEUSS 1966, GERECKE 1991a). Larvae parasitic on chironomid Diptera and Trichoptera (ULLRICH 1978, attribution to *S. denticulatus* uncertain).

Distribution: West Palaearctic. Although records show *S. denticulatus* to be widely distributed in Europe, many are questionable regarding species determination; not recorded from northeastern parts including most of Scandinavia.

4. *Sperchon (Hispidosperchon) hibernicus* HALBERT, 1944 (*8-59 h-j*)

1944 *Sperchon hibernicus* HALBERT, Proc. R. Irish. Acad. 50: 51.

1954c *Sperchon denticulatus hibernicus* E. ANGELIER, Vie et Milieu 5: 88.

1957 *Sperchon monstruosus* BADER, Abh. naturw. Ver. Bremen 35 : 32 – DI SABATINO et al. (2009).

Description: Colour brownish. Males: Idiosoma length/width 660-870/540-720; dorsal shield length/width 610-820/490-640; genital flap length 195-250 µm; postgenital plate small, round to oval, length/width 35-70/30-50 µm, length/width ratio 1.2-1.4; ejaculatory complex with rod-shaped anterior ramus, the tip with a broad V-shaped cleft flanked by a pair of angular, flange-like extensions, the base of the cleft with three small projections, the central one the longest; gnathosoma length/width 260-340/210-265; length chelicera 320-400 µm, cheliceral claw 85-115 µm; palp large; length P-1, 48-60, P-2, 198-270; P-3, 156-216; P-4, 240-330, P-5, 48-54 µm; P-2 ventral projection pointed or rounded, anterior length 36-48 µm, lateral face with 8-18 stout setae, medial face with 9-14 stout setae arranged close to dorsal margin; P-2 longer than P-3, both with a few stout setae on inner and outer faces; P-4 longer than P-2, ventral peg setae well separated (distance 60-90 µm); P-4 proximoventrally concave, forming a dorsoventral constriction, ratio dorsal length/constriction height 7.2-9.4; distolateral extensions of I-IV-L-5 if present then generally short. Females: See *S. denticulatus*.

Similar species: See discussion under *S. denticulatus*.

Habitat and Biology: Small streams to rivers. Preadult stages and life cycle unknown.

Distribution: British Isles including the Orkney Isles, Central and South Europe to Bulgaria.

5. *Sperchon (Hispidosperchon) hispidus* KOENIKE, 1895 (*8-58 i, 8-59 n-o, 8-61 h*)

1895e *Sperchon hispidus* KOENIKE, Zool. Anz. 18: 385.

1988b *Sperchon sandozi* BADER, Verhandl. Naturf. Ges. Basel 98: 60 – DI SABATINO et al. (2009)

Description: Colour brownish-yellow; dorsum with seven paired and one unpaired muscle attachment plates. III/IV-L- 3-5 with numerous pinnate dorsal setae, arranged densely so that ramifications of neighbouring setae may touch. Males: Idiosoma length/width 580-680/530-670; length genital flap 135; P-2, 90-120; P-3, 120-150; P-4, 155-180 µm. Females: Idiosoma length/width 570-1000/470-800; length genital flap 130-150; gnathosoma 200-260; P-2, 100-180; P-3, 145-244; P-4, 180-270 µm.

Similar species: While males of *S. clupeifer* are clearly distinguished by their dorsal shield, females of both species have often been confused. In fact, pinnate dorsal setae on IV-L-3-5, a character in former times considered diagnostic for *S. hispidus*, are also found in *S. clupeifer* (however, less well developed and less densely arranged – on female IV-L-4 of *S. hispidus* > 6, of *S. clupeifer* < 6). *Sperchon clupeifer* is best distinguished by the unsclerotized excretory pore and a more slender palp (length/height P-4 > 5.5).

Habitat and Biology: Middle and high order streams. Preadult stages and life cycle unknown.

Distribution: Palaearctic. All over Europe except for England, Iceland and Northern Scandinavia; Maghreb, Asia Minor.

6. *Sperchon (Hispidosperchon) insignis* WALTER, 1906 (*8-60 e*)

1906b *Sperchon setiger insignis* WALTER, Zool. Anz. 30: 574.

Description: Colour light reddish-brown. Males: Idiosoma length/width 580-780/450-600; length genital flap 120-145; gnathosoma 160-200; P-2, 90-120; P-3, 105-136; P-4, 145-180 µm. Females: Idiosoma length/width 670-2000/570-1500; genital flap length 140; length P-2, 130; P-3, 160; P-4, 220 µm.

Similar species: Very similar to *S. setiger*, from which it differs exclusively in P-3 being smaller in size and, at the adult stage, bearing only one, not two or more ventral seta(e), at the deutonymphal stage bearing no seta (*S. setiger* one). Many authors doubt the justification of the separation from the latter species, but larvae can be separated from *S. setiger* by morphometric characters of the legs and dorsal plate (MARTIN 2006).

Preadult stages: Larva: MARTIN (2006).

Habitat and Biology: Springs, low- and middle order streams. Larvae parasitic on simuliid Diptera (MARTIN & STUR 2006).

Distribution: Europe; scattered records.

7. *Sperchon (Hispidosperchon) papillosus* THOR, 1901 (*8-60 c-d*)

1901c *Sperchon papillosus* THOR, Zool. Anz. 24: 65.

Description: Colour reddish brown; dorsal integument papillae can be enlarged, namely in males obfuscating the areolate ground pattern which however always remains visible posteroventrally. P-4 with stout, prominent ventral peg setae. Males: Idiosoma length/width 510-750/450-600; length genital flap 115-130; gnathosoma 160-170; P-2, 87-98; P-3, 105-122; P-4, 129-154 µm. Females: Idiosoma length/width 550-1200/450-1000; length genital flap 130-156; gnathosoma length/width 200-220/145-150; length P-2, 110-138; P-3, 138-174; P-4, 160-190 µm.

Similar species: *Sperchon algeriensis* LUNDBLAD 1942, occasionally considered a subspecies of *S. papillosus* and recorded from a wide area in the Central and Western Mediterranean region, differs from *S. papillosus* in having an absolutely and relatively longer P-3 (in males > 160, in females > 220 µm, 36.5-38 % total palp length, in *S. papillosus* 33-36,5 %) bearing two pairs of ventral setae.

Habitat and Biology: Low and middle-order streams. Preadult stages and life cycle unknown.

Distribution: Europe, also in the far north; Asia Minor, widely distributed.

8. *Sperchon (Hispidosperchon) setiger* THOR, 1898 (*8-36 c, 8-60 b*)

1898d *Sperchon setiger* THOR, Arch. Math. Naturv. 20 (13): 6.

1954c *Sperchon meridianus* E. ANGELIER, Vie et Milieu 5: 90 – GERECKE (1991a).

Description: Colour pale orange or red, with greenish sclerotized parts. Dorsum with numerous small paired muscle attachment plates. Gnathosoma with short rostrum (< 1/4 total length). Palp slender, P-3 basally narrow; P-4 ventral peg setae stout, generally located close to each other. Males: Idiosoma length/width 600-900/540-700; length genital flap 130-160; P-2, 100-144; P-3, 128-176; P-4, 150-230 µm. Females: Idiosoma length/width 700-2000/600-1700; length genital flap 150-190; P-2, 140-230; P-3, 190-290; P-4, 220-370 µm.

Similar species: See discussion under *S. insignis*.

Preadult stages: Larva: BESSELING (1939); ULLRICH (1976); BIESIADKA & CICHOCKA (1984); WAINSTEIN (1980); MARTIN (2000, 2006).

Habitat and Biology: Middle-order streams; adults with a preference for parts with reduced flow, often also in pool habitats. In summer-warm lowland streams under Mediterranean climate often the only representative of the lotic fauna, together with lenitic species of the genera *Limnesia*, *Arrenurus*, or *Hydrodroma* (GERECKE 1991a). Larvae generally parasitic on simuliid Diptera (Besseling 1939, Ullrich 1978, 1978a, Gledhill et al. 1982, Semushin 1981), exceptionally on chironomid Diptera (Martin 2000); preparasitic attendance on pupae, parasite load on adult simuliids generally low; host species with pupae bearing long and fine respiratory appendages are preferred to those with short and stout appendages (Gledhill et al. 1982); occasionally, haemolymph ingestion starts on simuliid pupae, and the host insects may be killed at this stage (Renz et al. 2005).

Distribution: Western Palaearctic; widely distributed in Europe, Maghreb.

9. *Sperchon (Hispidosperchon) vaginosus* Thor, 1902 (*8-59 k-m, 8-61 i*)

1902d *Sperchon vaginosus* Thor, Zool. Anz. 26: 151.

1957 *Sperchon hibernicus* Bader, *nec* Halbert, 1944, Abh. Naturw. Ver. Bremen 35: 39 – Di Sabatino et al. (2009).

Description: Males: Colour reddish; idiosoma length/width 620-900/530-700; dorsal shield length/width 600-830/400-620; genital flap length 190-230 µm; postgenital plate small, generally oval in the longitudinal plane, length/width 42-50/30-36 µm, ratio length/width 1.4-1.8; ejaculatory complex with an elongate, rod-shaped anterior ramus, the tip with a broad U-shaped cleft flanked by a pair of sharply-angled lateral extensions; gnathosoma length/width 215-275/150-180; length chelicera 245-335, claw 70-100 µm; palp stout, length P-1, 30-42; P-2, 129-174; P-3, 108-147; P-4, 153-198; P-5, 30-40 µm; P-2 ventral projection anterior length 45-70 µm, lateral face with 7-13 stout setae, medial face with 4-5 setae close to dorsal margin; P-2 dorsally longer than P-3; P-3 ventrally slightly concave, both outer and inner lateral faces with a few stout setae; P-4 longer than P-2; distance between ventral peg setae 21-33 µm; P-4 proximoventrally slightly concave forming a slight constriction, ratio dorsal length/constriction height 4.4-5.0; I-IV-L-5 with distolateral extensions. Females: See under *S. denticulatus*.

Similar species: See under *S. denticulatus*.

Habitat and Biology: Small streams to rivers. Preadult stages and life cycle unknown, possibly larvae parasitic on chironomid Diptera as previously considered conspecific with *S. denticulatus*.

Distribution: Scattered records for Europe (some as *S. denticulatus*). Great Britain to Sicily.

10. *Sperchon (Hispidosperchon) violaceus* Walter, 1944 (*8-58 j, 8-59 d-g*)

1944 *Sperchon violaceus* Walter, Int. Rev. Hydrobiol. Hydrogr. 43: 291.

Description: Colour deep red, sclerotized parts violet. Palp rather slender. Males: Idiosoma length/width 650-900/560-650; dorsal shield length/width 670-730/490-550, occasionally with a frontal eye pore; genital flap length 180-240 µm; postegenital plate a broad oval, length/width 70-80/90-110 µm, ratio length/width 0.7-0.8; ejaculatory complex with an elongate rod-shaped anterior ramus, the tip having an U-shaped cleft flanked by a pair of laterally rounded, flange-like extensions; gnathosoma length/width 200-245/145-175; length chelicera 260-295, claw 60-70 µm; palp slender, length P-2, 120-145; P-3, 135-200; P-4, 170-250 µm; P-2 ventral projection away from ventrodistal corner, anterior length 48-54 µm, lateral surface with 6-10 stout setae, medial surface with 5-8 stout setae close to dorsal margin; P-3 longer than P-2, ventrally denticulate, more-or-less straight but with a proximal concavity, both outer and inner surfaces with a few stout setae; P-4 longer than P-3, ventral setae well separated (distance 36-42 µm); P-4 slightly bowed distally, with only a very slight proximal constriction, being more-or-less equal in height from proximal ventral peg setae to segment base, ratio dorsal length/constriction height 6.4-6.8; distolateral extensions of I-IV-L-3/4 longer than those of I-IV-L-5. Females: Idiosoma length/width 780-1200/630-900; genital flap length 200-230; gnathosoma length/width 230-300/170-200; length P-2, 145-210; P-3, 210-300; P-4, 230-330 µm; P-3 ventrally denticulate; P-2 and P-3 with small distal hyaline extensions; P-4 with ventral peg setae well separated (distance 50-70 µm).

Similar species: See under *S. denticulatus*.

Preadult stages: Larva: MARTIN (2003).

Habitat and Biology: Rhithrobiont, crenophilous; in montane and alpine low-order streams and springs with extended mineral substrata. Host spectrum of larva unknown. Due to larval morphology and the absence as a parasitc on host insects from emergence traps a reduced parasitism is postulated (GERECKE & MARTIN 2006)

Distribution: Alps, Pyrenees, Tatra, Carpathians. A "variety" recorded from Japan (IMAMURA 1953b). Following BADER (1975), all records published before 1944 from Alpine sites above 800 m asl. under the name *S. denticulatus* refer, in reality, to *S. violaceus*.

Subgenus *Sperchon* KRAMER, 1877

Diagnosis: Integument papillate to rugose, without regular patterns of areolate surface structures. Dorsalia in European species poorly developed or absent. P-3 without ventral denticulation or setation; ventral peg-like setae on P-4 not enlarged and not located close together.

Remark: Probably a paraphyletic clade defined by plesiomorphies. As the subgenus includes closely related species with and without a medial gland pore on Cx-III, this character may not warrant subgenus definition.

Distribution: Worldwide except for Australian region and Antarctica.

11. *Sperchon (Sperchon) brevirostris* KOENIKE, 1895 (*8-62 a-b*)

1895f *Sperchon brevirostris* KOENIKE, Rev. Suisse de Zoologie 3: 416.

Description: Colour brownish red, sclerotized parts purple. Integument with wrinkled lineation, glandularia rather large. Leg claws with ventral clawlets and large claw blade. Excretory pore on a sclerotized ring in size similar to neighbouring glandularia. Gnathosoma with very stout rostrum, palp with very slender P-4 and P-5, ventral peg-like setae of P-4 fine and insignificant. Males: Idiosoma length/width 720-1000/580-900; length genital flap 180-230; gnathosoma 165-210; P-2, 145-195; P-3, 140-210; P-4, 235-400 µm. Females: Idiosoma length/width 850-1500/730-1250; length genital flap 190-200; gnathosoma 155-200; P-2, 140-170; P-3, 140-160; P-4, 275-335 µm.

Preadult stages: Larva: LUNDBLAD (1927c, 1930a); SPARING (1959).

Habitat and Biology: Low- and middle-order streams, at the northern border of the distribution area, also in lake littoral, avoiding springs. Life cycle unknown. Colonization of remote islands (Azores, Faroes, Bear Island/Svalbard) suggests effective phoretic dispersal.

Distribution: Europe except for areas with a Mediterranean climate; islands of the Atlantic and northern seas. In the Alps up to 2000 m a. s. l. (BADER 1975).

12. *Sperchon (Sperchon) glandulosus* Koenike, 1886 (*8-62 e*)

1886b *Sperchon glandulosus* Koenike, Zeitschr. f. wiss. Zool. 43: 279.

Description: Colour red to brown, sclerotized parts yellowish or greenish; leg claws with ventral clawlet and small claw blade. Excretory pore with a sclerotized ring. According to Lundblad (1956a), gnathosoma distally with a more or less elevated ventral keel. P-4 ventral peg setae strongly developed and projecting, dividing segment in three parts equal in size, all with a slightly concave ventral margin. Males: Idiosoma length/width 800-900/700-800; length genital flap 180-200 µm; gnathosoma 200, P-2, 115-135; P-3, 160-180; P-4, 210-225 µm. Females: Idiosoma length/width 1050-1300/850-1100; genital flap length 195-200 µm; a small sclerotized platelet posterior to genital field; length gnathosoma 235; P-2, 155-160; P-3, 205-210; P-4, 260-270 µm.

Similar species: For a long time confused with *S. thienemanni* which differs in the unsclerotized excretory pore (Bader 1974). Most records published from spring habitats under the name of *S. glandulosus*, in fact refer to *S. thienemanni*. The situation is yet more complicated as some authors interpreted the two taxa in the opposite sense (e. g., K. Viets 1936d, at subspecies level: *Sperchon glandulosus glandulosus* as a crenobiont, *S. glandulosus thienemanni* as a rhithrobiont in lowland waters)

Preadult stages: Larva: Motaş (1961); Ullrich (1976); Smith (1982); Biesiadka & Cichocka (1984), in all cases species attribution uncertain.

Habitat and Biology: Low- and middle-order streams. In Scandinavia also in the lake littoral. Larvae parasitic on chironomid Diptera (Motaş 1961; Efford 1963; Ullrich 1976, 1978; Smith 1982 – all attributions uncertain).

Distribution: Central and Northern Europe (incl. Faroe Islands and Iceland); distribution in East, West and South Europe unclear due to former confusion with *S. thienemanni*.

13. *Sperchon (Sperchon) longirostris* Koenike, 1895 (*8-60 g-i*)

1895f *Sperchon longirostris*, Koenike, Revue Suisse de Zoologie 3: 420.

Description: Colour brownish; claws with ventral clawlet and small claw blade; genital flaps each with up to 15 medial setae; a small roundish platelet posterior to genital field. P-2 proximally narrow. Males: Idiosoma length/width 630-1000/460-800; genital flap length 160-200; gnathosoma length/width 270-310/130-165; length P-2, 114-135; P-3, 145-185; P-4, 150-180 µm. Females: Idiosoma length/width 650-1100/520-800; genital flap length 185-215 µm; gnathosoma length/width 270-330/140-175; length P-2, 115-145; P-3, 165-200; P-4, 180-190 µm; P-3 a little stouter than in males.

Similar species: Often confused with *S. squamosus*. In addition to the characters given in the key, the latter species differs also in the absence of a claw blade on leg claws and in having an additional pair of longer, apparently unsclerotized, tubular sacs slightly anterior to the distal arms of the anterior ramus of the ejaculatory complex (Lundblad 1956a). *Sperchon vesnae* Pešić, 2002 from Montenegro is similar in the shape of the ejaculatory complex (without large anterolateral sacs) and the absence of a narrowed proximal sector of P-2, but differs in the absence of a medial gland pore on Cx-III and a stouter gnathosomal rostrum.

Habitat and Biology: Low order streams, mostly in montane areas. Life cycle and preadult stages unknown.

Distribution: Central and Western Europe (including Ireland); rare, many records uncertain (K. Viets 1936d). A questionable record from North America (Crowell 1952).

14. *Sperchon (Sperchon) longissimus* K. Viets, 1920 (*8-61 e-g*)

1920b *Sperchon longissimus* Koenike, Arch. Hydrobiol. 12: 805.

Description: Colour dark orange to brown, with reddish frontal and excretory pore areas; integument regularly papillate, with two pairs of small dorsal platelets; leg claws with ventral clawlet and narrow claw blade. A small, round platelet anterior to genital field; genital flaps each with about 10 setae. P-2 proximally narrow, P-4 ventral setae very small. Males: Idiosoma length/width 845-975/515-600; length genital flap 125; gnathosoma 225; P-2, 85-95; P-3, 90-100; P-4, 103-166 µm. Females: Idiosoma length/width 1000-1250/550-730; length genital flap 190-205; gnathosoma 330; P-2, 125-130; P-3, 132-142; P-4, 150-152 µm.

Similar species: *Sperchon vesnae* Pešić, 2002 from Montenegro differs in the absence of a narrowed proximal sector of P-2 and a stouter gnathosomal rostrum without narrowed basal part of rostrum.

Preadult stages: Larva: Martin (2006). Deutonymph: K. Viets (1923a).

Habitat and Biology: Crenobiont. Preferably in helocrenes of lowlands and lower mountain ranges; not recorded from Alpine areas. Larvae parasitic on chironomid Diptera (Martin & Stur 2006).

Distribution: England, Central and Southeastern Europe; Asia Minor.

15. *Sperchon (Sperchon) mutilus* Koenike, 1895 (*8-36 d-e,8-62 c-d*)

1895f *Sperchon mutilus* Koenike, Revue Suisse de Zoologie 3: 425.

1985b *Charoelia schloethi* Bader, Rev. Suisse Zool. 92: 461 – Di Sabatino et al. (2009).

Description: Colour orange to deep red, often a little milky. Males: Idiosoma length/width 730-1200/600-860; genital flap length 175-215; gnathosoma length/width 180-230/140-170; length P-2, 125-180; P-3, 150-190; P-4, 225-305 µm. Females: Idiosoma length/width 900-1400/650-1100; genital flap length 168-265; gnathosoma length/width 180-270/150-200; length P-2, 140-220; P-3, 150-245; P-4, 230-375 µm.

Preadult stages: Larva: Martin (2003).

Habitat and Biology: Crenobiont. Preferably in cold rheo(helo)crenes with sand and gravel substrata and rich moss vegetation (Gerecke & Martin 2006). Larvae parasitic on chironomid Diptera (Tanytarsini: Martin et al. 2010); attachment site: thorax.

Distribution: Central, East and South Europe, montane to Alpine; in the southern part of the distribution area rare and restricted to upper mountain reaches.

16. *Sperchon (Sperchon) resupinus* K. Viets, 1922 (*8-61 a*)

1922d *Sperchon resupinus* K. Viets, Arch. Naturgesch. 88: 58.

Description: Colour brown. Integument regularly papillate. Leg claws with small ventral clawlet, without claw blade. P-2 as high as long; P-4 with both ventral peg setae in distal half, directed proximally, distoventral peg seta located ventrally and closer to distal segment margin than to proximoventral seta; P-5 very small. Males: Idiosoma length/width 680-915/510-700; length genital flap 155-160; gnathosoma 280-310; P-2, 115-130; P-3, 125-130; P-4, 135-140 µm. Females: Idiosoma length/width 620-1300/560-1050; length genital flap 180-210; gnathosoma 320-360; P-2, 130-160; P-3, 135-165; P-4, 150-170 µm.

Preadult stages: Larva: MARTIN (2003).

Habitat and Biology: Crenobiont. Preferably in helocrenes, also in habitats exposed to sunlight and organic input (GERECKE 1991a, GERECKE & MARTIN 2006). Larvae parasitic on chironomid Diptera (MARTIN et al. 2010).

Distribution: Central and South Europe and southern Scandinavia.

17. *Sperchon (Sperchon) squamosus* KRAMER, 1879 (*8-58 a-b, 8-61 k*)

1879a *Sperchon squamosus* KRAMER, Arch. Naturgesch. 45: 2.

Description: Colour greyish to dark brown; integument regularly papillate, with two pairs of small, bean-shaped dorsal platelets; claws with insignificant ventral clawlet, without claw blade. Males: Idiosoma length/width 700-1000/600-860; length genital flap 140-190; gnathosoma 275-375; P-2, 130-159; P-3, 145-220; P-4, 165-270 µm. Females: Idiosoma length/width 750-1300/570-1100; length genital flap 185-253, gnathosoma 290-410; P-2, 155-178; P-3, 135-198; P-4, 155-220 µm.

Similar species: See under *S. longirostris*.

Preadult stages: Larva: K. VIETS (1923a); LUNDBLAD (1927c); MOTAŞ (1961); ULLRICH (1976), MARTIN (2006). Deutonymph: K. VIETS (1936d).

Habitat and Biology: Rhithrobiont, crenophilous; often in helocrenes. In the northern part of the distribution area also in lake littoral, in the southern part found exclusively in spring habitats (GERECKE 1991a). Larvae parasitic on chironomid Diptera (ULLRICH 1976, 1978, CICOLANI et al. 2001, MARTIN & STUR 2006).

Distribution: Western Palaearctic; preferably in lowlands and lower mountain ranges.

18. *Sperchon (Sperchon) thienemanni* KOENIKE, 1907 (*8-62 f*)

1907b *Sperchon thienemanni* KOENIKE, Abh. naturw. Ver. Bremen 19: 133.

1911a *Sperchon rugosus* KOENIKE, Zool. Anz. 37: 324 – BADER (1974).

Description: Colour orange to dark red; excretory pore unsclerotized. Males: Idiosoma length/width 600-1100/500-900; genital flap length 160-210; gnathosoma 210-240, P-2, 125-160; P-3, 180-230; P-4, 220-290 µm. Females: Idiosoma length/width 570-1250/500-950; length genital flap 165-205; gnathosoma 215-260, P-2, 120-200; P-3, 180-240; P-4, 180-295 µm (data from a population in the study area, BADER 1975 – distinctly smaller dimensions observed in the southern part of the distribution area, GERECKE 1991a).

Similar species: See under *S. glandulosus*.

Preadult stages: Larva: LUNDBLAD (1927c); ULLRICH (1976); MARTIN (2000).

Habitat and Biology: Crenophilous. Often large populations in springs of various typology, but with a preference for rheohelocrenes rich in organic material (GERECKE & MARTIN 2006); in Central and Northern Europe frequently recorded from low order streams. Larvae parasitic on chironomid Diptera (ULLRICH 1976, MARTIN 2000, CICOLANI et al. 2001, MARTIN & STUR 2006), adult life span can extend to at least two years and females are able to oviposit over at least two years (MARTIN 1997).

Distribution: Europe – in many parts of the continent, detailed distribution unclear due to former confusion with *S. glandulosus*.

19. *Sperchon (Sperchon) turgidus* K. VIETS, 1914 (*8-61 b*)

1914c *Sperchon turgidus* K. VIETS, Abh. naturw. Ver. Bremen 33: 336.

Description: Colour unknown. Integument regularly papillate. Leg claws with a small ventral clawlet and a narrow remnant of claw blade. Genital field with small Ac (size of interspaces equal to acetabula diameter). P-2 higher than long; P-4 with both ventral peg-setae in anterior half of segment, distal one located medially, closer to proximoventral peg-seta than to distal segment margin; P-5 extremely shortened. Males: Idiosoma length/width 750/550; length genital flap 145-155; gnathosoma 270; P-2, 110; P-3, 137; P-4, 132 µm. Females: Idiosoma length/width 960/650; genital flap length 170-210; gnathosoma length/width 280-300/120-125; length P-2, 103; P-3, 139; P-4, 90 µm.

Habitat and Biology: Low order streams, but also dredged from bottom sediments in the potamal sector of River Rhine (LANGE 1990). Life cycle and preadult stages unknown, but observation of loss of parasitism in the laboratory (B. P. SMITH 1998).

Distribution: Central and Northern Europe; extremely rare, only known from scattered single records.

Genus *Sperchonopsis* PIERSIG, 1896

1896c *Sperchonopsis* PIERSIG, Sitz. ber. Naturf. Ges. Leipzig 22/23: 52.

Type species: *Sperchon verrucosus* (PROTZ, 1896), by original designation.

Diagnosis: Glandularia enlarged, with a papillate or warty surface. Dorsum smooth or with one to two pairs of platelets. Cx-III without glandularia. Genital area in males without a pregenital sclerite but with a small round sclerotized patch between genital area and posterior margin of Cx-I+II, in females with a pregenital sclerite but without a sclerotized patch anterior to genital area; chelicera often with a greatly enlarged claw; P-3 without a ventral projection; P-4 of European species without peg-like ventral setae.

Remarks: A homogenous group, but including one species with a P-4 setation bridging the gap to *Sperchon*.

Descriptions of *Sperchonopsis* larvae from outside the area treated here and indications of their host range are given by PRASAD & COOK (1972); LESAGE & HARRISON (1980), SEMUSHIN (1981), I. M. SMITH (1982) and SMITH & OLIVER (1976c); a thorough treatment and discussion of deutonymphal characters is presented by TUZOVSKIJ (1990).

Distribution: Holarctic, Neotropical and Oriental regions.

Key to species

1 Colour brown to grey, sclerotized parts dark; mid dorsum with a pair of elongate plates. Idiosoma stout (length/width ratio < 1.5, *8-58 c*). . . . *Sperchonopsis verrucosa* (page 172)

\- Colour yellow to orange, sclerotized parts pale; mid dorsum with an unpaired rounded plate. Idiosoma more slender (length/width ratio > 1.5, *8-62 g*). . . . . *Sperchonopsis procera* (page 172)

1. *Sperchonopsis verrucosa* (Protz, 1896) (*8-58 c-d*)

1896a *Sperchon verrucosa* Protz, Zool. Anz. 19: 23.

1940a *Sperchonopsis verrucosa reducta* Sokolow, Fauna SSSR 6 (2):190 – Özkan (1989).

Description: Colour brownish, sometimes with bluish sclerotized parts. Integument strongly rugose. Tips of Cx-I projecting and covered by numerous setae. Gnathosoma with rostrum slightly shorter than gnathosomal base. Males: Idiosoma length/width 530-750/400-500; length genital flap 110-160; gnathosoma 250-260; P-2, 105-145; P-3, 90-130; P-4, 90-125 µm. Females: Idiosoma length/width 750-1000/500-700; length genital flap 135-240; gnathosoma 250-390; P-2, 110-190; P-3, 100-175; P-4, 100-154 µm.

Preadult stages: Larva: Martin (2000). Deutonymph: e.g., Imamura (1954), Tuzovskij (1990).

Similar species: *Sperchonopsis akguli* Oezkan, 1989 from Turkey differs in extremely enlarged glandular plates, leaving only narrow membranous interspaces between, or in touch with, neighbouring plates.

Habitat and Biology: Rhithrobiont. In low order streams, often on muddy, fine sediment; occasionally in the hyporheic, rarely in springs, but frequently in spring brooks. Larvae parasitic on chironomid Diptera (Martin 2000, Martin & Stur 2006); deutonymphs and adults feeding on chironomid larvae (Martin 2005). In the laboratory, females survived up to 3.5 years and oviposited during three consecutive years (Martin 1998).

Distribution: Holarctic, Oriental (Nearctic records need revision). From lowlands to above 2000 m a. s. l. (Bader 1975).

2. *Sperchonopsis procera* Láska, 1965 (*8-62 g-h*)

1965 *Sperchonopsis verrucosa procera* Láska, Annot. Zool. Bot. 15: 1.

1975 *Sperchonopsis phreatica* Biesiadka, Fragmenta Faunistica 20: 76 – Di Sabatino et al. (2009).

2009 *Sperchonopsis procera* Láska, 1965; Di Sabatino et al. (2009).

Description: Colour brownish yellow. Idiosoma slender, integument strongly papillate. Muscle attachment plates sclerotized, rounded, in the centre of dorsum one pair of dorsalia fused to form a larger, rounded plate. Gnathosoma with rostrum as long as, or slightly longer than, gnathosomal base. P-2 with a long ventrodistal projection bearing 2-3 setae, P-4 with a ventral projection close to segment base, bearing 2 setae. Males: Idiosoma length/width 700-730/420-450; length genital flap 143; gnathosoma 234; P-2, 126; P-3, 126; P-4, 111 µm. Females: Idiosoma length/width 880-920/530-535; length genital flap 126-140; gnathosoma 233-312; P-2, 138-162; P-3, 138-150; P-4, 124-140 µm.

Habitat and Biology: Middle order streams, hyporheobiont. Preadult stages and life cycle unknown.

Distribution: Eastern Europe, only known from sites in Slovakia, Poland and Montenegro.

Family Teutoniidae KOENIKE, 1910

1910b Teutoniidae KOENIKE, Abh. naturw. Ver. Bremen 20: 144.

Diagnosis: Integument thin, finely striate, without sclerotized muscle attachments. Lateral eye lenses free under the integument, not encapsulated. Coxae in three or four groups, finely porose, more strongly sclerotized along suture lines and apodemes, in particular at medial margin of Cx-III. Cx-IV large, subrectangular. A pair of glandularia associated with Cx-IV, located anteromedially, partly or completely surrounded by coxal plate, or on an anterior extension deeply indenting posterior margin of Cx-III. Legs with swimming setae. I-III-L claws with ventral clawlet and claw blade, claws of IV-L reduced to small, seta-like structures. Anterior part of genital field in a bay formed by Cx-IV, with three elongate Ac and large flaps. Excretory pore smooth. Gnathosoma with short rostrum. P-2 with one ventral seta.

Remarks: The two genera included in this family are morphologically rather distant. Both differ from members of neighbouring families in the "sperchontid type" provisional genital field of the deutonymph (with two pairs of Ac surrounded by a sclerotized ring). While this character state is possibly symplesiomorphic, the character set developed as adaptation to fast swimming (leg setation, coxal apodemes, reduction of IV-L-claws) probably indicates a closer relationship of *Teutonia* and *Limnolegeria* (GERECKE 1988).

Key to genera

1 Glandularia of Cx-IV anteromedial, partly or completely surrounded by a membranous area within the coxal plate (*8-35 e, 8-63 c, e*); palp slender, P-2 with a ventrodistal projection, ventral seta away from ventral margin and relatively short; P-4 ventrally with a very small, blunt peg-like seta (*8-63 d, f*). . . . . *Teutonia* (page 174)

\- Glandularia of Cx-IV placed on an anterior protrusion indenting posterior margin of Cx-III (*8-35 f, 8-63 a*); palp stocky, P-2 with a long ventral seta, but without projection; P-4 without peg-like seta (*8-63 b*). . . . . *Limnolegeria* (page 173)

Genus *Limnolegeria* MOTAŞ, 1928

1928a *Limnolegeria* MOTAŞ, C. R. Acad. Sci. Paris 186: 176.

Type species: *Limnolegeria longiseta* MOTAŞ, 1928a, by monotypy.

Diagnosis: Glandularia with very long, slender setae. Coxae strongly sclerotized, with pores in hexagonal groups, older specimens with a border of secondary sclerotization at Cx-III+IV. Cx-I+II fused to a compact anterior coxal plate with rounded medioposterior margin and short posterior apodemes, glandularia C2 fused to this plate. A pair of glandularia on an anterior protrusion of Cx-IV deeply indenting posterior margin of Cx-III. Lateral to this protrusion, a finger-shaped posteriorly-directed apodeme at suture Cx-III/IV. Flattened swimming setae in dense rows posteriorly on distal parts of II-III-L-3-5, anteriorly and posteriorly on IV-L-3-5. Genital flaps with a few, scattered setae laterally and medially, in males posteromedially slightly longer and more numerous than in females. Pregenital sclerites distinctly smaller than postgenital sclerites, both well developed. Palp with thick and short segments 1-4; P-2 without a ventrodistal projection, ventral seta long and strong. P-3 with a long ventrodistal sword seta, P-4 ventrodistally and dorsally with numerous fine setae, but without a peg-like seta. P-5 with two strong, but weakly curved claws. Cheliceral basal segment stout.

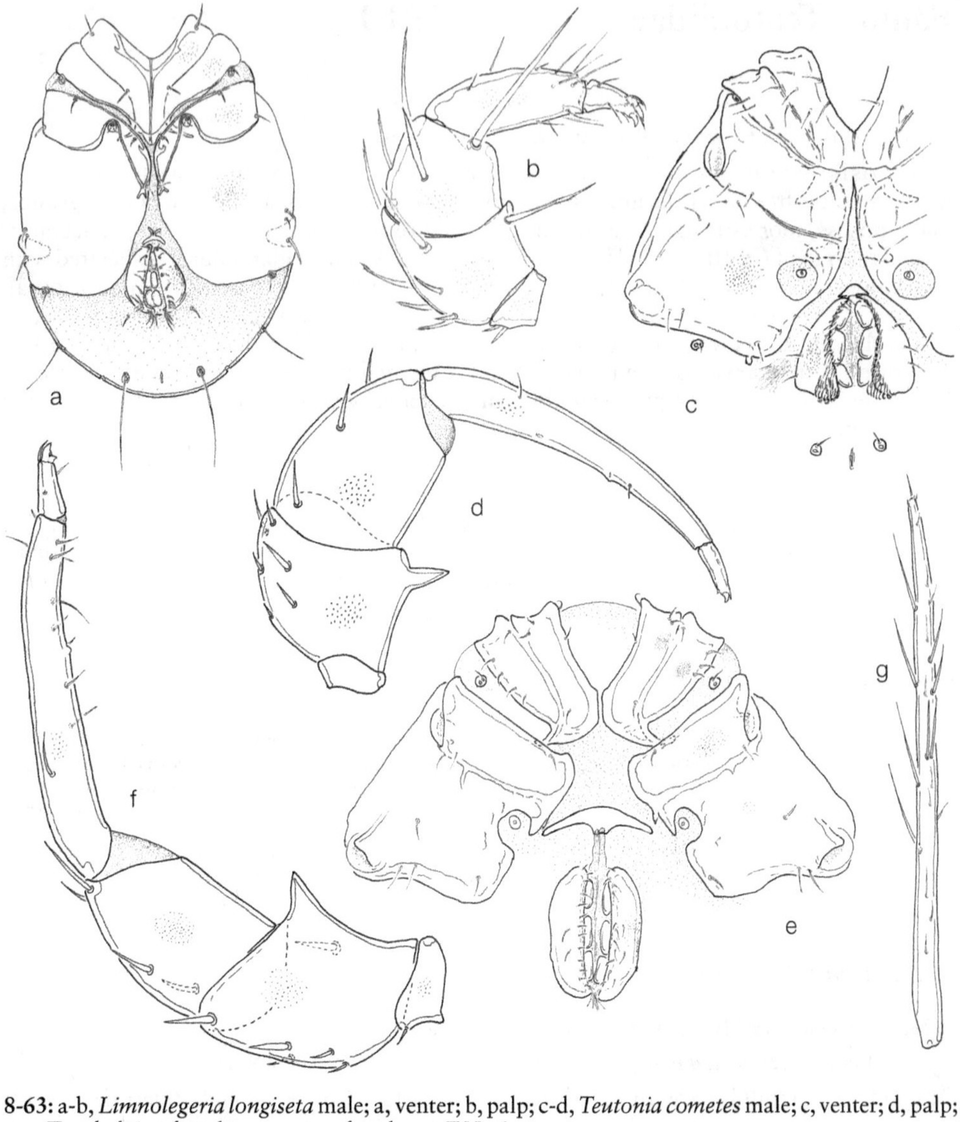

8-63: a-b, *Limnolegeria longiseta* male; a, venter; b, palp; c-d, *Teutonia cometes* male; c, venter; d, palp; e-g, *T. subalpina* female; e, venter; f, palp; g, IV-L-6.

Remarks: For a long time, this monotypic genus was considered a representative of the Anisitsiellidae (Cook 1974, see Gerecke 1988).

Limnolegeria longiseta Motaş, 1928 (*8-35 f; 8-63 a-b*)

1928a *Limnolegeria longiseta* Motaş, C. R. Acad. Sci. Paris 186: 176.

Description: Colour orange (probably in juveniles) to deep red, sclerotization bluegreen or violet. Males: Idiosoma length/width 700-900/550-700 µm; genital field oval in shape, pregenital sclerite compact (length/width ratio 1.7-2.5); length P-1, 18-27; P-2, 60-82; P-3,

46-63; P-4, 75-96; P-5, 36-51 µm. Females: Idiosoma length/width 900-1400/650-1100 µm; genital field subtrapezoidal (with posteriorly diverging lateral flap margins), pregenital sclerite enlarged, crescent-shaped (length/width ratio 2.6-4.2); length P-1, 22-34; P-2, 90-105; P-3, 64-75; P-4, 102-111; P-5, 42-48 µm.

Preadult stages: Larva unknown. Deutonymph: GERECKE (1988).

Habitat and Biology: Described from a karst lake. All other records from pools of shaded low-order streams (GERECKE 1988). Probably the fastest swimming species in the European fauna. Life cycle unknown.

Distribution: SE France, Central Italy, Sicily (GERECKE 1991a, unpubl.).

Genus *Teutonia* KOENIKE, 1889

1889a *Teutonia* KOENIKE, Zool. Anz. 12: 14.

Type species: *Hygrobates cometes* KOCH, 1837, by original designation (KOENIKE 1889a).

Diagnosis: Glandularia small, with fine and short setae. Coxae generally weakly sclerotized; Cx-I+II medially separate or fused by a weak bridge, their posterior apodemes strongly developed, "moustache-shaped", with laterally-bent tips; glandularia C2 free in the membranous interspace between Cx-II and Cx-III. A pair of glandularia located medially on Cx-IV, partly to completely surrounded by coxal plate. Swimming setae in dense rows on posterior surface of distal parts of III-IV-L-3-5. Genital flaps with a few fine setae on the flap surface and numerous densely-arranged medial setae, in females in one row, in males posteriorly extending to multiple rows and posterior to Ac-3 densely covering a large, but poorly sclerotized posteromedial projection of the flaps. Pregenital sclerite in females distinctly larger, in males smaller, than postgenital sclerites, the latter weakly sclerotized. Palp slender, in particular P-4 and P-5. P-2 ventral seta away from ventral margin; P-3 without ventral seta; P-4 with a small, peg-like ventral seta and numerous fine, hair-like setae, most of them ventral; P-5 with stout, hook-like claws. Cheliceral basal segment long and curved, claws short.

Remark: Species of this well-defined genus are recorded from the Holarctic, arranged in the two subgenera *Teutonia s. str.*, and *Subteutonia* HABEEB, 1958a (Canada, monotypic). No information available on life cycles of European species, larvae of a North American species parasitic on Diptera (Chironomidae: SMITH 1982, SMITH & OLIVER 1986).

Subgenus *Teutonia* KOENIKE, 1889

Diagnosis: P-2 with a distinct ventral projection. Setae on P-3 and ventral peg-seta on P-4 relatively short.

Key to species

1 Female gonopore all over its length flanked by Ac and genital flaps (*8-63 c*). In both sexes, ventral projection of P-2 on a narrow base, peg-like (*8-63 d*). . . . . *Teutonia cometes* (page 176)

\- Female gonopore very long, only the posterior 2/3 flanked by Ac and genital flaps (as given in *8-63 e* for the male). In both sexes, ventral projection of P-2 on a large base, triangular (*8-63 f*). . . . . *Teutonia subalpina* (page 176)

1. *Teutonia cometes* (Koch, 1837) (*8-35 e; 8-63 c-d*)

1837 *Hygrobates cometes* Koch, Deutschlands Crustaceen, Myriapoden und Arachniden 10: 23.

1889a *Teutonia primaria* Koenike, Zool. Anz. 12: 103.

1933e *Teutonia cometes* K. Viets, Zool. Anz. 104: 272.

Description: Colour yellow-orange to brown, coxae and appendages bluish-green. Integument finely striate. Coxal field with a few, sparse setae. Leg claws with claw blade and ventral clawlet; swimming setae restricted to III-/IV-L-3-4, numbers: III-L-3, 3-4; III-L-4, 9-13; III-L-5, 10-12; IV-L-3 3-5; III-L-4, 9-12; III-L-5, 12-18. P-2 generally with a narrow-based ventrodistal projection (but occasionally appearing broad-based in juveniles) and one stronger setae inserting in ventral part; P-3 with smooth, nearly straight or slightly convex ventral margin, P-4 slender, bearing numerous fine hair-like setae, mainly in the anteroventral part. P-2/P-4 length ratio 0.43-0.47. Males: Idiosoma length/width 750-1000/ 560-800 µm. All coxae medially touching, generally fused. Glandularium on Cx-IV generally completely surrounded by coxal sclerotization. Genital flap length 130-170 µm, medially with a dense cover of fine, anteriorly-directed setae (clearly visible only in opened condition); pregenital sclerite narrow (width 45-60 µm), postgenital sclerite as wide as genital field (105-135 µm). Length/height P-1, 22-27/45-52; P-2, 108-134/108-116; P-3, 136-148/90-110; P-4, 250-300/38-42; P-5 47-50/13-14 µm; P-4 length/height ratio 6.5-7.2. Females: Idiosoma length/width 850-1500/700-1200 µm. Coxae medially touching or separated by narrow strips of membranous integument. Glandularium on Cx-IV generally partly surrounded by coxal sclerite, medially in contact with the perigenital membrane. Genital flap length 170-200 µm, with a few, regularly arranged medial setae; pregenital sclerite large, crescent-shaped (width 150-160 µm), postgenital sclerite as in males (width 110-135 µm). Length/height P-1, 23-32/50-57; P-2, 145.153/135-139; P-3, 175-184/121-135; P-4, 310-360/43-45; P-5 54-57/16 µm; P-4 more slender than in males (length/height ratio 7.2-8.0).

Similar species: Probably all species of the genus characterized by rather weak sclerotization of coxae, genital plates and appendages. Consequently, legs and palps are frequently distorted or folded, and the verification of character states may become difficult. In larger populations, specimens may also be found having the ventral projection of P-2 similar to the description given for *T. subalpina* (see there).

Preadult stages: Larva: Piersig (1896-99, 1901b); Sparing (1959). Deutonymph: Piersig (1896-99, 1901b); K. Viets (1936d); Tuzovskij (1990).

Biology: Oviposition on macrophyte leaves; larval development eventually direct to deutonymphal stage within the egg clutch; duration of tritonymphal stage (attached to water plants) about 8 days (Piersig 1896-1899). Juvenile adults appearing in late autumn, overwintering in lakes at the adult stage (Steiner 1911).

Habitat: Cold montane lakes (preferably in shallow littoral waters: Lundblad 1968); pools of shaded streams.

Distribution: Palaearctic.

2. *Teutonia subalpina* Thor, 1897 (*8-63 e-g*)

1897b *Teutonia subalpina* Thor, Arch. Math. Naturvid. Kristiania 19 (6): 21.

Description: From most points of view similar to *T. cometes* or not described in detail. P-2 with broad-based ventrodistal projection. Length/height ratio P-2, 1.5; P-4, 6.8; length ratio P-2/P-4, 0.54. Males: Idiosoma length 900 µm, gonopore not longer than genital flaps. Females: Idiosoma length 1200-1500 µm, gonopore anteriorly distinctly extending over margin of genital flaps.

Similar species: Very little known (figured here from a single specimen from the Senckenberg Museum, Frankfurt). Based on published characters, only females can be clearly distinguished from *T. cometes*. Additional studies on the variability of the female genital field are needed before the taxonomic status of *T. subalpina* can be cleared, possibly a taxon based on unsclerotized, freshly hatched, misshapen adults of *T. cometes*.

Preadult stages: Larva unknown. Deutonymph: Sokolow (1940a).

Habitat and Biology: Oligo- and dystrophic lakes, preferably in shallow littoral reaches, not in running waters.

Distribution: In Europe restricted to the far north (alpine and subalpine coniferous forests of Norway and Sweden), Siberia, possibly Holarctic (syn.? *T. lunata* Marshall, 1924c, North America; Lundblad 1968)

Family Torrenticolidae Piersig, 1902

1902a Atractideidae Thor, Zool. Anz. 25: 408.

1902 Torrenticolidae Piersig, Zool. Centrbl. 53: 849.

Diagnosis: Idiosoma completely sclerotized, with dorsal and ventral shields separated by a dorsal furrow. Dorsal shield formed by a large plate and one or two pairs of anterior platelets completely separate or variously fused with the main plate (*8-64 b; 8-68 b*) – in *Testudacarus* Walter, 1928c, a genus not recorded from the study area, numerous other setae-bearing platelets surround the lateral and posterior margin of the dorsal plate. Coxal plates fused with postcoxal sclerotization to form a ventral shield. A Y-shaped suture anterior to genital field separating Cx-I from Cx-II+III, and Cx-II+III medially (*8-64 a; 8-68 a; 8-74 a*). Suture line Cx-II/III medially obliterated, suture Cx-III/IV distinct. Legs without swimming setae, IV-L with well developed claws. Three or six pairs of Ac in the membranous area between gonopore and genital flaps. Gnathosoma with rostrum various in length, but clearly set off from base. Palp four- or five-segmented.

Discussion: Wiles (1997b) suggests that Torrenticolidae evolved from an anisitsiellid-like ancestor and branched to the subfamilies Testudacarinae Cook, 1974 and Torrenticolinae Piersig, 1902. In the Palaearctic, only members of Torrenticolinae are present.

In the treatment of the Torrenticolidae, the following terms are used: medial suture Cx-II+III = distance between posterior edge of Cx-I and anterior margin of genital field; line of primary sclerotization = posterior edge of Cx-IV; postgenital area = sclerotized area extending from posterior margin of genital field to posterior edge of body; dorsal shield = sclerotized area covering the entire dorsal idiosoma, formed by a large dorsal plate and two to four small anterior platelets; ventral shield = sclerotized area of ventral idiosoma formed by fused coxae (primary sclerotization) and sclerotized postcoxal integument (secondary sclerotization).

Subfamily Torrenticolinae Piersig, 1902

1902 Torrenticolinae Piersig, Zool. Centrbl. 53: 849.

1941c Neoatractidinae Lundblad, Zool. Bidr. 20: 364.

Diagnosis: Dorsal shield without lateral platelets (*8-64 b*). A gland near the anterior edge of Cx-III present or absent. Genital field with six pairs of Ac. Insertion of IV-L-sockets without condyles and without ridge extending anteriorly. P-4 ventrally with two to three hair-like setae on tubercles (*8-64 d, f, h*).

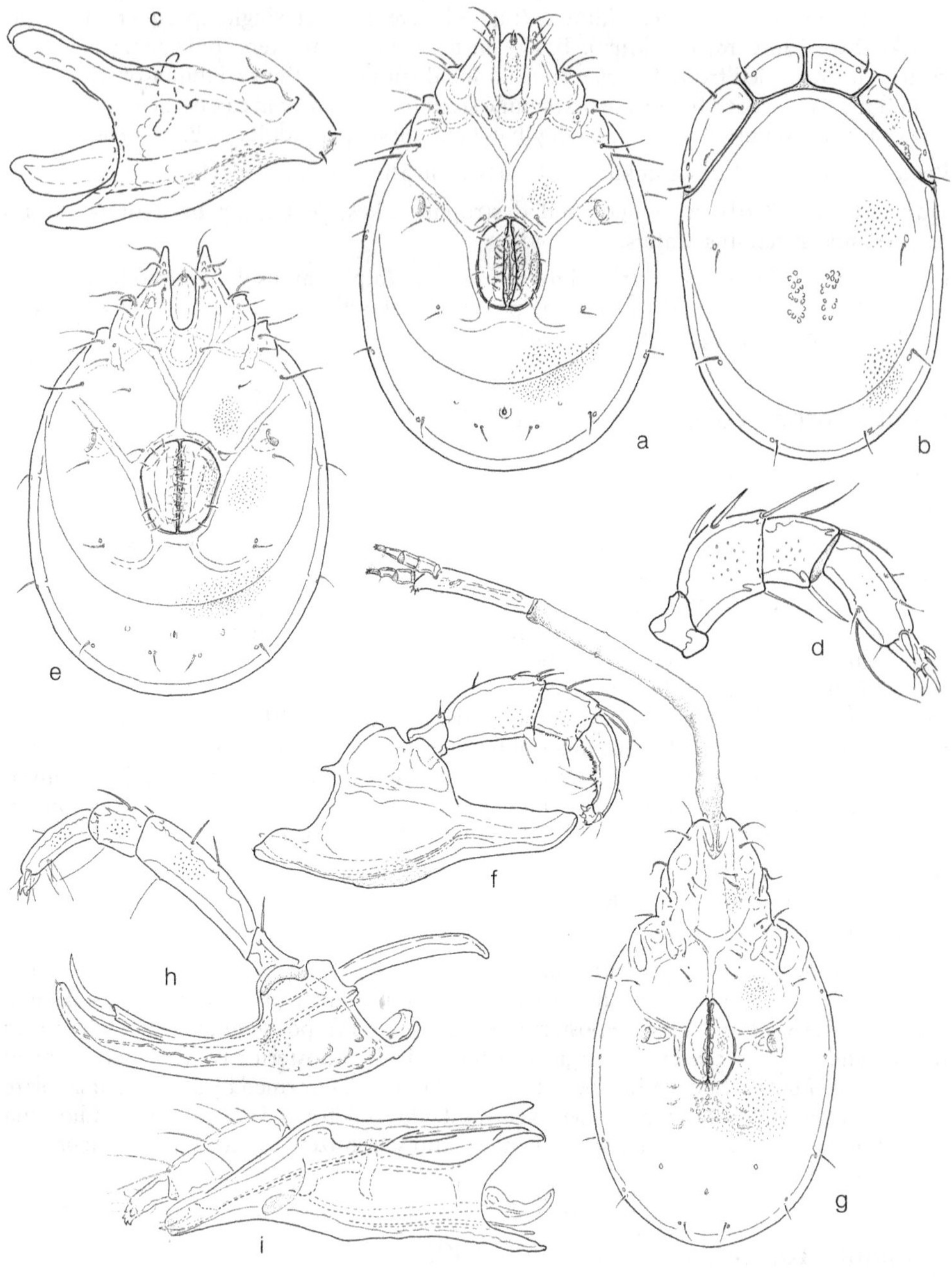

8-64: a-e, *Monatractides madritensis;* a, male ventral shield; b, male dorsal shield; c, gnathosoma; d, palp; e, female ventral shield; f, *Torrenticola ungeri* male, gnathosoma and palp; g, *Pseudotorrenticola rhynchota*, male ventral view; h, *Torrenticola (Megapalpis) remyi*, gnathosoma and palp (Gerecke & Di Sabatino 1996b); i, *Monatractides (Pinquicola) adoratus*, male gnathosoma and palp (Di Sabatino & Gerecke 1996).

Discussion: The subfamily includes four genera. In Europe only members of *Monatractides* K. VIETS, 1926l, *Pseudotorrenticola* WALTER, 1906b and *Torrenticola* PIERSIG, 1902 are present. Members of *Neoatractides* LUNDBLAD, 1941c are recorded from America, Asia and Africa. The subgeneric division of Torrenticolinae follows WILES (1997b), GERECKE & DI SABATINO (1996b) and DI SABATINO & GERECKE (1996).

Preadult stages: Larva: PRASAD & COOK (1972), WAINSTEIN (1980) I. M. SMITH (1982). Deutonymph: WALTER (1907b), MAGLIO (1909), SZALAY (1947), WALTER (1947), LUNDBLAD (1956a), TUZOVSKIJ (1981), DI SABATINO & CICOLANI (1993).

Key to genera and subgenera

1 Gnathosoma reduced in size, attached to a long membranous extrusible tube; apex of Cx-I fused forming a camerostome to which the tube is attached (*8-64 g*).
. . . . Genus *Pseudotorrenticola* (page 184)

\- Gnathosoma normal in size, not attached to a membranous tube; Cx-I forming a well developed gnathosomal bay (*8-64 a*). 2

2 (1) Gnathosoma with long posterior apodemes; rostrum very short or very long and not set off from the rest of gnathosoma; palp minute and compact, P-2 and P-3 lacking tubercles on ventral side (*8-64 d, i*); ventral seta on P-2 located distally or absent. Genus *Monatractides*
. . . . 3

\- Gnathosoma with short posterior apodemes; rostrum usually well developed, clearly set off from the base; palp slender, P-2 and P-3 usually with ventral tubercles or hyaline expansions (*8-64 f*) – if tubercles absent, ventral seta on P-2 located less distally (*8-64 h*). Genus *Torrenticola* 4

3 (2) Gnathosoma laterally compressed, not elongated posterior to palp insertion, rostrum short. P-1 and P-2 subquadratic; P-2 bearing a ventral distal seta (*8-64 d*).
. . . . Subgenus *Monatractides s.str.* (page 180)

\- Gnathosoma cylindrical, extremely elongated posterior to palp insertion; rostrum long and bent ventrally. P-1 shortened, P-2 subtriangular and without ventral seta (*8-64 i*).
. . . . Subgenus *Pinquicola* DI SABATINO & GERECKE, 1996

(Monotypic, not in the study area: *Monatractides adoratus* DI SABATINO & GERECKE, 1996 known from Italy, Greece and Turkey – BOYACI 2006).

4 (2) Rostrum various in shape but its tip never curved dorsally; chelicerae not extremely elongated (basal segment L/H-ratio < 15.0), with relatively stout, strongly curved claws; P-2 generally shorter than, in exceptional cases a little longer than P-4 (P-2/P4 L ratio < 1.3); P-2 and –3 with ventrodistal extensions bearing setae (*8-64 f*).
. . . . Subgenus *Torrenticola s. str.* (page 187)

\- Rostrum extremely elongated, its tip curved dorsally; chelicerae long and thin (basal segment L/H ratio > 15.0), with slender, stylet-like claws; P-2 distinctly longer than P-4 (P-2/P-4 L ratio > 1.3); P-2 and P-3 without ventrodistal extensions bearing setae (*8-64 h*).
. . . . Subgenus *Megapalpis* (page 185)

Genus *Monatractides* K. VIETS, 1926

1897c *Rusetria* THOR (in part), Arch. Math. Naturvid. 20: 20.

1926l *Monatractides* K. VIETS, Zool. Anz. 69: 192.

1931c *Rusetriella* K. VIETS, Zool. Anz. 93: 220.

1988 *Synaptia* BADER, Rev. Suisse Zool. 1: 91.

Type species: *Torrenticola (Monatractides) uniscutata* K. VIETS, 1925h, by original designation.

Diagnosis: Idiosoma in many species relatively large. Palaearctic species with two pairs of free anterior platelets. Without glands on Cx-III. Gnathosoma with long posterior apodemes, but little developed rostrum, in ventral view strongly narrowed (*8-65 d*). Palp stout, P-2 and P-3 usually with ventral setae, but without tubercles.

Remarks: Until recently, *Monatractides* was treated as a subgenus of *Torrenticola*. A phylogenetic analysis (WILES 1997b) however demonstrated it to be an independent monophyletic genus. *Monatractides* species are recorded from all continents except Antarctica.

Preadult stages: Larva: TUZOVSKIJ (2000a). Deutonymph: WALTER (1947). Information on larval morphology and life cycle is not available for any of the species recorded from the study area.

Subgenus *Monatractides* K. VIETS, 1926

Diagnosis: Gnathosoma laterally compressed, not elongated posterior to palp insertion, rostrum short. P-1 and P-2 subquadratic; P-2 bearing a ventral seta.

Remark: This taxon is possibly in a paraphyletic relation to the only other known subgenus, *Pinquicola*. Distribution as given for the genus.

Key to species

1 Genital field slender, anterior margins of genital flaps forming a more acute angle (*8-64 a*); ventral seta on P-4 long (ca. 50 µm, > 90 % length P-4) (*8-64 d*).
. . . . *Monatractides madritensis* (page 182)

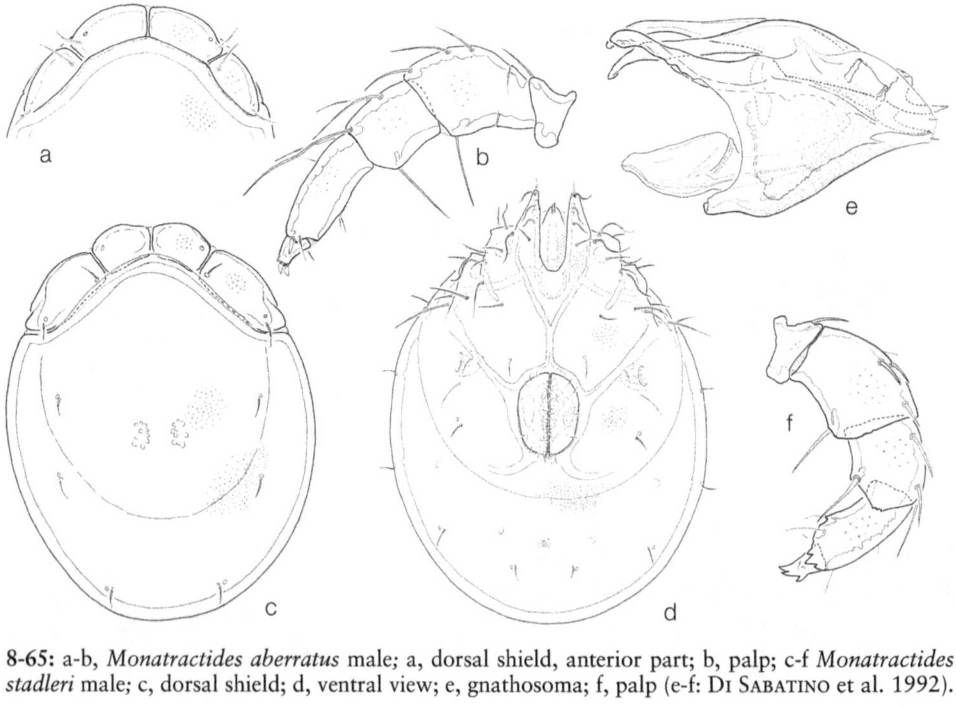

8-65: a-b, *Monatractides aberratus* male; a, dorsal shield, anterior part; b, palp; c-f *Monatractides stadleri* male; c, dorsal shield; d, ventral view; e, gnathosoma; f, palp (e-f: DI SABATINO et al. 1992).

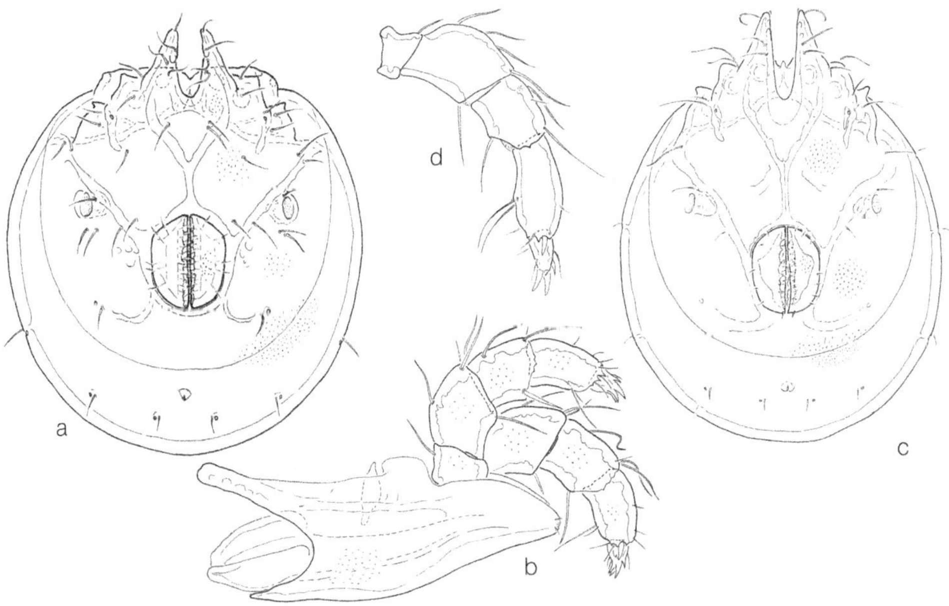

8-66: a-b, *Monatractides latissimus;* a, male ventral shield; b, female gnathosoma and palps (Di Sabatino et al. 2003); c-d, *Monatractides stenostomus* male; c, ventral shield; d, palp.

-	Genital field wider, anterior margins of genital flaps almost at right angles to longitudinal axis or forming a very obtuse angle (*8-65 d; 8-66 a, c*); ventral seta on P-4 shorter (< 90 % length P-4) (*8-65 b, f*).	2
2 (1)	Distal margin of P-3 and P-4 medially and laterally with 2-3 pointed extensions (*8-65 f; 8-66 b*); shoulder platelets distinctly longer than frontal platelets (ratio > 1.2 – *8-65 c*). .	3
-	Distal margin of P-3 and P-4 straight (*8-65 b*); frontal platelets equal in length to shoulder platelets or shoulder platelets a little shorter (ratio 0.8-1.0 – *8-65 a*). . . . . *Monatractides aberratus* (page 181)	
3 (2)	Palp larger and robust; P-2 (length > 70 µm) distinctly longer than P-4; ventral seta on P-4 short (< 30 µm) (*8-65 f*). . . . . *Monatractides stadleri* (page 183)	
-	Palp smaller; P-2 (length < 60 µm) as long as, or slightly longer than, P-4; ventral seta on P-4 longer (> 40 µm) (*8-66 b*).	4
4 (3)	Large in size and broader (idiosoma length 900-1200 µm, ratio length/width < 1.2); gnathosomal bay wide (length/width < 3.0), rounded (*8-66 a*). . . . . *Monatractides latissimus* (page 182)	
-	Smaller in size and more slender (idiosoma length 700-800 µm, ratio length/width > 1.2); gnathosomal bay slender (length/width > 3.0), pointed (*8-66 c*). . . . . *Monatractides stenostomus* (page 184)	

1. *Monatractides (Monatractides) aberratus* (Lundblad, 1941) (*8-65 a-b*)

1941c *Atractides aberratus* Lundblad, Ent. Tidskr. 62: 100.

1956a *Torrenticola (Monatractides) aberrata* Lundblad, Ark. Zool. 10: 174.

Description: Colour greenish-brownish. Shoulder platelets very short, almost of the same dimensions as frontal platelets. Suture line of Cx-IV medially directed to posterior, laterally equally curved. Excretory pore posterior to the line of primary sclerotization, Vgl-2 posterior to excretory pore. P-2 and P-3 distal margins without pointed tips. Ventral seta on P-4 slender and short. P-2 and P-4 almost equal in length. Males: Idiosoma length/width 800-900/600-700 µm; medial suture Cx-II+III short (50-60 µm); genital field large, almost roundish, length/width 170-180/150-160; length P-2, 49-53; P-4, 46-54 µm. Females: Idiosoma, length/width 850-950/670-750 µm; medial suture Cx-II+III very short (10-15 µm); genital field length/width 196/186; length P-2, 53-58; P-4, 52-54 µm. Postgenital area more extended than in male.

Habitat: Middle and high order streams, in riffles as well as in marginal pools with sandy substratum.

Distribution: Mediterranean (in the French Pyrenees reaching the southern border of the area treated); Turkey, Iran.

2. *Monatractides (Monatractides) latissimus* (K. VIETS, 1936) (*8-66 a-b*)

1936b *Atractides latissimus* K. VIETS, Arch. Hydrobiol. 29: 369.

1955a *Torrenticola (Monatractides) latissima erlangensis* K. O. VIETS, Abh. Bayerische Akad. Wiss. 73: 25 – DI SABATINO et al. (2003)

1956b *Torrenticola (Torrenticola) latissima latissima* K. VIETS, Die Milben des Süsswassers und des Meeres: 244.

Description: Colour brownish-yellowish. Idiosoma enlarged (length/width dorsal shield ca. 1.2), shoulder platelet elongated (length ratio shoulder/frontal platelet 1.4-1.5). Line of primary sclerotization only slightly away from posterior margin of genital field. Suture line of Cx-IV partially evident and curved. Vgl-2 posterior to excretory pore. Palp small, but ventral setae on P-2, P-3 and P-4 long; P-2 slender, longer than P-4; P-4 stout (length/height 1.8). Males: Idiosoma length/width 890/740; length shoulder platelet 202; frontal platelet 145; medial suture Cx-II+III 112, genital field length/width 210/175 µm, anterior margin forming an obtuse angle; length P-2, 50; P-4, 45 µm. Females: Idiosoma length/width 960/910; length shoulder platelet 260, frontal platelet 170; medial suture Cx-II 67; length/width genital field 243/225; length postgenital area 370; gnathosoma ventrally 216, dorsally 252; P-2, 54; P-4, 52 µm.

Habitat: Middle-order streams.

Distribution: Croatia, Bosnia, SE Germany (rare, only known from four sites).

3. *Monatractides (Monatractides) madritensis* (K. VIETS, 1930) (*8-64 a-e*)

1930d *Atractides madritensis* K. VIETS, Arch. Hydrobiol. 21 (3): 371.

1944 *Atractides parvipalpis* HALBERT, Proc. R. Irish Acad. 50: 75 – DI SABATINO et al. (2009).

1944 *Atractides robustus* HALBERT, Proc. R. Irish Acad. 50: 74 – DI SABATINO et al. (2009).

1953a *Atractides bicinctus* LÁSKA, Acta Acad. scient. natur. Moravo-Silesiacae 25: 290 – DI SABATINO et al. (2009).

1984d *Torrenticola (Monatractides) hibernica* CONROY, Ir. Nat. J. 21: 329 – DI SABATINO et al. (2009).

Description: Colour greenish-brown. Idiosoma slender. Postgenital area large, excretory pore well away from the line of primary sclerotization, Vgl-2 posterior to excretory pore. Mature specimens with distinct area of secondary sclerotization. Suture lines of Cx-IV starting almost from the centre of posterior margin of genital field and directed postero-laterally. Palp small, P-2 equal in length, or only slightly shorter than, P-4. Ventral seta on

P-4 relatively long and away from distal edge. Males: Idiosoma length/width 770-850/590-630; medial suture Cx-II+III length 90-115 μm; genital field anteriorly triangular, posteriorly narrowed and rounded, length/width 155-175/108-120, length postgenital area 280-305; P-2, 52-62; P-4, 54-65 μm. Females: Idiosoma length/width 728-907/515-640; medial suture Cx-II+III length 50-60 μm; genital field pentagonal, length/width 190-205/168-178, length postgenital area 325-340; P-2, 56-60; P-4, 57-63 μm.

Habitat: Low order streams. Frequently in the hyporheic, occasionally also in surface waters (Cicolani & Di Sabatino 1990; Di Sabatino et al. 2003).

Preadult stages: Larva unknown. Deutonymph: Láska 1966 ("*Torrenticola bicincta* Láska, 1953a"); Walter 1947 ("*Atractides unguiculatus* Walter, 1947").

Distribution: Western Palaearctic.

4. *Monatractides (Monatractides) stadleri* (Walter, 1924) (*8-65 c-f*)

1924a *Atractides stadleri* Walter, Zool. Anz. 59 (3-4): 107.

1957a *Torrenticola lata* Schwoerbel, Mitt. Bad. Landesver. Naturkde. Naturschutz (N. F.) 7 (1): 44 – Di Sabatino et al. (2003).

1958 *Torrenticola lobata* Schwoerbel, Mitt. Bad. Landesver. Naturkde. Naturschutz (N. F.) 7 (2): 143 – Di Sabatino et al. (2003).

Description: Colour brownish to dark-yellow. Idiosoma roundish. Genital field large. Suture line of Cx-IV well evident, originating from lateral edge of genital field. Line of primary sclerotization close to posterior margin of genital field. Area of secondary sclerotization extensive. Excretory pore and Vgl-2 well separated from the line of primary sclerotization, Vgl-2 posterior to excretory pore. Palp with thick, pennate ventral setae on P-2 and P-3. Distal margin of P-3 and P-4 medially and laterally with several pointed extensions. P-2 longer than P-4. Ventral setae on P-4 short, pennate, located near distal edge. Males: Idiosoma length/width 950-1120/800-950; medial suture Cx-II+III length 80-95 μm. Genital field subrectangular, length/width 190-234/180-190; ejaculatory complex length/width 310-330/115-120; length P-2, 78-90; P-4, 65-75 μm. Females: Idiosoma length/width 1000-1300/850-1070; medial suture Cx-II+III length 54-80 μm. Genital field pentagonal, length/width 210-265/200-250 μm. Area of primary and secondary sclerotization in mature specimens even more enlarged than in males; length P-2, 85-100; P-4, 65-80 μm.

Similar species: *Monatractides algeriensis* (Lundblad, 1941c), reported from north-Africa and South-Italy, is close to *M. stadleri*. Males of *M. algeriensis* are distinguishable in having a longer medial suture of Cx-II+III (length > 100 μm), a more slender chelicera (ratio length/height 7.5-8.5 vs. 6.1-6.3) and a smaller ejaculatory complex (total length < 300 μm) (Di Sabatino et al. 1992).

Habitat: Low-, middle- and summer-warm high-order streams. Often in shadowed pool reaches with accumulation of leaf litter. Very frequent in lowland running waters of the Mediterranean region (Gerecke & Di Sabatino 1996b).

Distribution: Central, Western and Southern Europe; Turkey, Iran.

5. *Monatractides (Monatractides) stenostomus* (K. VIETS, 1930) (*8-66 c-d*)

1930d *Atractides stenostomus* K. VIETS, Arch. Hydrobiol. 21 (3): 370.

1956a *Torrenticola (Monatractides) stadleri stenostoma* (K. VIETS), LUNDBLAD, Arkiv för Zool. 10 (1): 171.

Description: Males (Females unknown): Colour dark brownish. Idiosoma length/width 730-820/ 600-670; medial suture Cx-II+III length 84-86 µm. Gnathosomal bay posteriorly narrowed, maximum width 50; genital field length/width 165-180/144-154; postgenital area length 265 µm. Excretory pore in an indentation of the line of primary sclerotization; Vgl-2 posterior to excretory pore. Distal margin of P-3 and P-4 medially and laterally with several pointed tips. P-4 ventrally with two setae, one thick and short, the other long and hair-like, reaching the tip of P-5; P-2 and P-4 equal in length (ca. 55 µm).

Habitat: Middle-order lowland streams with warm water, in reaches with slow-current and sandy-stony substrata (LUNDBLAD 1956a).

Distribution: Portugal, Spain, France.

Genus *Pseudotorrenticola* WALTER, 1906

1906b *Pseudotorrenticola* WALTER, Zool. Anz. 30: 573.

1954b *Pseudotorrenticolella* LUNDBLAD, Zool. Anz. 153: 155.

Type species: *Pseudotorrenticola rhynchota* WALTER, 1906b, by monotypy.

Diagnosis: Dorsal shield formed by a unique undivided dorsal plate or with two pairs of free anterior platelets. Cx-I fused to a funnel-like camerostome, gnathosomal bay shallow. Gnathosoma long and slender, much reduced in size and located at the end of an extrusible membranous tube (when extended longer than idiosoma length).

Preadult stages: Larva: GERECKE & MARTIN (2006). Deutonymph: LUNDBLAD (1956a), WALTER (1906b).

Pseudotorrenticola rhynchota WALTER, 1906 (*8-64 g*)

1906b *Pseudotorrenticola rhynchota* WALTER, Zool. Anz. 153: 573.

Description: Colour yellow to brownish. Anterior platelets fused with the large dorsal plate forming an undivided dorsal shield. Gnathosoma of reduced dimensions and with a small rostrum. Palp segments very short, P-2 and P-3 without ventral tubercles. Males: Idiosoma length/width 700-890/450-550; dorsal shield length/width 500-720/390-490 µm; genital field elongate, anterior edge of flaps forming an acute angle, length/width 140-170/90-115 µm. Females: Idiosoma length/width 890-950, dorsal shield length/width 690-780/540-640 µm; genital field roundish, anterior margin smoothly convex, length/width 200-210/155-185 µm.

Preadult stages: Larva: GERECKE & MARTIN (2006). Deutonymph: WALTER (1906b).

Similar species: The only other species known from the Palaearctic, *P. quinquescutata* LUNDBLAD, 1954b from Spain, is characterized by the presence of two pairs of free anterodorsal platelets.

Habitat and Biology: Surface and interstitial waters of low- to middle-order streams. Adults feed on chironomid larvae (MARTIN 2005), larvae parasitic on adult Chironomidae (GERECKE & MARTIN 2006).

Distribution: Central and Southern Europe, Turkey.

Genus *Torrenticola* PIERSIG, 1896

1837 *Atractides* KOCH (*partim*), Deutschlands Crust., Myr. und Arach. 11: 10 (genus in Hygrobatidae).

1896a *Torrenticola* PIERSIG, Zool. Anz. 19: 438.

1897c *Rusetria* THOR (*partim*), Arch. Math. Naturvid. 20: 20.

Type species: *Atractides anomalus* KOCH, 1837, by original designation.

Diagnosis: Gnathosoma normal in size, not attached to a protrusible tube, with short posterior apodemes. Segments of palps slender, P-2 and P-3 usually with setae-bearing ventrodistal extensions.

Remark: The subgeneric division was discussed by several authors (E. ANGELIER 1954c; LUNDBLAD 1956a, COOK 1974; BADER 1988). According to GERECKE & DI SABATINO (1996b) and WILES (1997b), a phylogenetic classification separates all members of *Torrenticola* into two subgenera, *Torrenticola s. str.* and *Megapalpis*.

Subgenus *Megapalpis* HALBERT, 1944

1944 *Megapalpis* HALBERT, Proc. R. Irish Acad. 50, B 4: 72.

Diagnosis: Gnathosomal rostrum well developed and bent dorsally. Chelicerae long and thin, with slender, slightly curved claws; P-2 and P-4 very slender, generally P-2 longer than P-4, P-2 and P-3 without distoventral extensions.

Preadult stages and Biology: Larval morphology and life cycle unknown for all species.

Distribution: Reported from Europe, Asia and South-Africa.

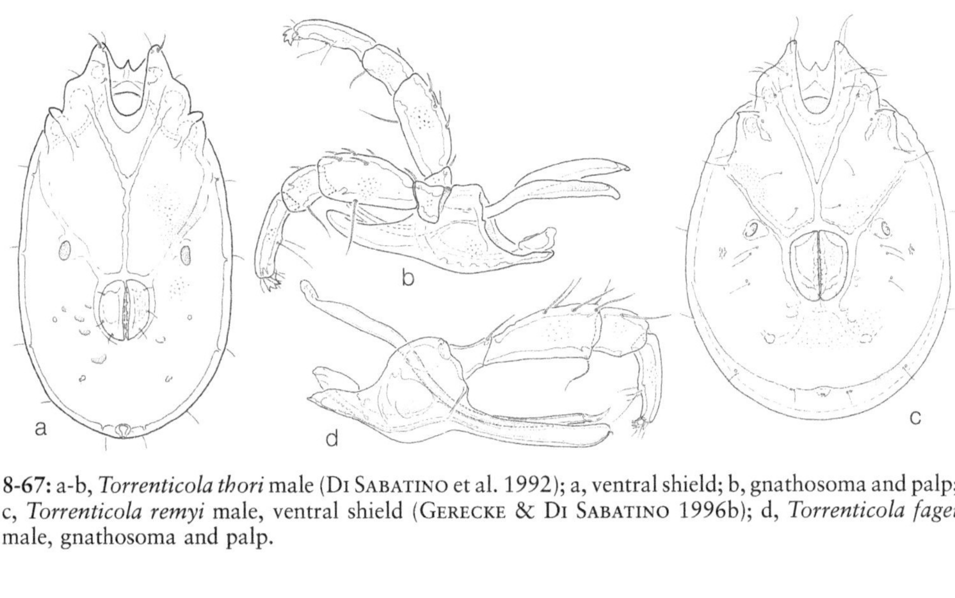

8-67: a-b, *Torrenticola thori* male (DI SABATINO et al. 1992); a, ventral shield; b, gnathosoma and palp; c, *Torrenticola remyi* male, ventral shield (GERECKE & DI SABATINO 1996b); d, *Torrenticola fagei* male, gnathosoma and palp.

Key to species

1 Idiosoma slender (*8-67 a* , dorsal shield length/width > 1.6). Cheliceral claw very long (about one half of basal segment) and evidently curved (*8-67 b*). . *Torrenticola thori* (page 187)

- Idiosoma more roundish (*8-67 c*, dorsal shield length/width < 1.5). Cheliceral claw shorter (less than one third of basal segment) and only slightly curved (*8-64 h, 8-67 d*). . . . 2

2 (1) Cheliceral claw relatively long (length ratio basal segment/cheliceral claw < 6.0; P-2 relatively long (length ratio P-2/P-4 > 1.5) (*8-64 h*). . . . . . *Torrenticola remyi* (page 186)

- Cheliceral claw relatively shorter (length ratio basal segment/claw > 7.0; P-2 relatively short (length ratio P-2/P-4 < 1.5) (*8-67 d*). . . . . *Torrenticola fagei* (page 186)

1. *Torrenticola (Megapalpis) fagei* (E. Angelier, 1949) (*8-67 d*)

1949a *Atractides fagei* E. Angelier, Bull. Mus. Hist. natur. Paris (s.2) 21 (1): 88.

Description: Colour yellowish-brown. Idiosoma roundish (dorsal shield length/width 1.3-1.5). Cheliceral claw short, ratio basal segment/claw 7.4-8.5. P-2 and P-3 without ventral tubercles or denticles, ventral seta on P-2 placed distally. Length ratio P-2/P-4, 1.3-1.4. Males: Idiosoma length/width 720-765/530-600; medial suture Cx-II+III length 98-110 µm; genital field anteriorly narrowed, length/width 180-188/138-143, postgenital area length 185-240 µm. Excretory pore and Vgl-2 located at same level on the line of primary sclerotization; length cheliceral basal segment 322, claw 38; P-2, 110-125; P-4, 85-90 µm. Females: Idiosoma enlarged posteriorly, length/width 730-870/530-610 µm; genital field pentagonal, length/width 195-215/176-193; length cheliceral basal segment 290-370, claw 35-50; P-2, 113-127; P-4, 88-95 µm.

Similar species: *Torrenticola fagei* was redescribed and compared with other species of the subgenus by Martin (2009). In the low P-2/P-4 ratio, *T. tenuirostris* K. Viets, 1936b from SE Europe is most similar, but different in the idiosoma being more slender in shape (ratio length/width dorsal shield > 1.5) and a relatively longer cheliceral claw (ratio basal segment/claw < 4.0).

Preadult stages: Larva unknown. Deutonymph: E. Angelier (1949b).

Habitat: Hyporheic of low-order streams in montane areas.

Distribution: French Pyrenees, rare.

2. *Torrenticola (Megapalpis) remyi* E. Angelier, 1954 (*8-64 h, 8-67 c*)

1954c *Torrenticola (Megapalpis) remyi* E. Angelier, Vie et Milieu 5 (1): 120.

Description: Colour greenish-yellow. Idiosoma oval (length/width dorsal shield 1.35-1.40). Excretory pore and Vgl-2 on the line of primary sclerotization. P-2 subterminal seta not supported by small tubercles or denticles. Length ratio P-2/P-4, 1.6-1.8. Males: Idiosoma length/width 717-760/490-560; medial suture Cx-II+III length 90-95 µm; genital field posteriorly narrowed, length/width 154-166/150-157, postgenital area length 234-264; length cheliceral claw 67 (ratio basal segment/claw 4.8-4.9); P-2, 150-165; P-4, 90-100 µm. Females: Idiosoma length/width 780-850/570-650; medial suture Cx-II+III length 58-85 µm; genital field subquadratic, length/width 154-166/150-157; postgenital area length 324-378 µm; posterior margin of primary sclerotization with a slight indentation near excretory pore; length cheliceral claw 90-95 (ratio basal segment/claw 5.0-5.4); P-2, 180; P-4, 105 µm.

Similar species: *Torrenticola tenuirostris* K. VIETS, 1936b from SE Europe differs in the idiosoma being more slender in shape (ratio length/width dorsal shield > 1.5), a longer cheliceral claw (ratio basal segment/claw < 4.0) and a proportionally longer P-2 (length ratio P-2/P-4, 1.4-1.5).

Habitat: Superficial and hyporheic habitats of running waters, on hard substrata.

Distribution: Southern France, Corsica, Sardinia.

3. *Torrenticola (Megapalpis) thori* (HALBERT, 1944) (*8-67 a-b*)

1944 *Atractides thori* HALBERT, Proc. R. Irish Acad. 50, B 4: 73.

1950a *Torrenticola andrei* E. ANGELIER, Vie et milieu 22 (1): 85 – GERECKE et al. (2005b).

Description: Colour brownish-yellow. Idiosoma elongated. Tip of first coxal plates blunt. Cheliceral claw very long and robust, curved. Males: Idiosoma length/width 725-810/400-470; medial suture Cx-II+III length 158-165 µm; genital field small and rounded, length/width 118/85 µm, genital flap edges bearing numerous fine setae; suture of Cx-IV laterally obliterated, postgenital area length 130 µm; excretory pore and Vgl-2 on the edge of primary sclerotization, near posterior of idiosoma; length cheliceral claw 130-140 µm (ratio basal segment/claw 1.8-2.3); palp slender, length P-2, 120-130; P-4, 105-110 µm; ratio P-2/P-4, 1.15-1.25. Females: Idiosoma length/width 930-1050/510-600; medial suture Cx-II+III short, length 30-45 µm; genital field pentagonal, posteriorly strongly converging, length/width 218-230/180-190 µm; excretory pore and Vgl-2 away from the line of primary sclerotization; length cheliceral claw 165-180 (ratio basal segment/claw 2.1-2.8); P-2, 162-170; P-4, 140-152 µm; ratio P-2/P-4, 1.02-1.25.

Habitat: Surface waters and the hyporheic of low- and middle-order streams (DI SABATINO et al. 1992).

Distribution: Central, Southern and Western Europe including British Isles; Turkey and Iran.

Subgenus *Torrenticola* PIERSIG, 1896

Diagnosis: Dorsal shield of palaearctic species composed of a large dorsal plate and two to four free anterior platelets. Gnathosoma with basal part (proximal to palp insertion) not extremely elongated, rostrum well developed and distinctly set off from gnathosomal base. Chelicerae with small claws; P-2 and P-4 not strongly differing in dimensions (length ratio P-2/P-4, 0.9-1.2).

Remarks: The subgenus *Torrenticola* represents the most diversified and species-rich taxon among the Torrenticolidae with more than 200 species described from all continents except Antarctica and Australasia. Idiosoma morphology is rather uniform in most species or species-groups; however, remarkable differences in gnathosoma and palpal morphology characterize some extraeuropean species (GOLDSCHMIDT 2007).

Key to species

All characters refer to mature specimens only.

Excluded as *species incertae* (DI SABATINO et al. 2009): *Torrenticola spinirostris* (THOR, 1897c), *T. maglioi* (KOENIKE, 1908a), *T. halberti* (THOR, 1923), *T. sandalensis* (SOKOLOW, 1926b), *T. procerivalvata* K. VIETS 1952c and *T. barsica tisialis* SZALAY, 1953a.

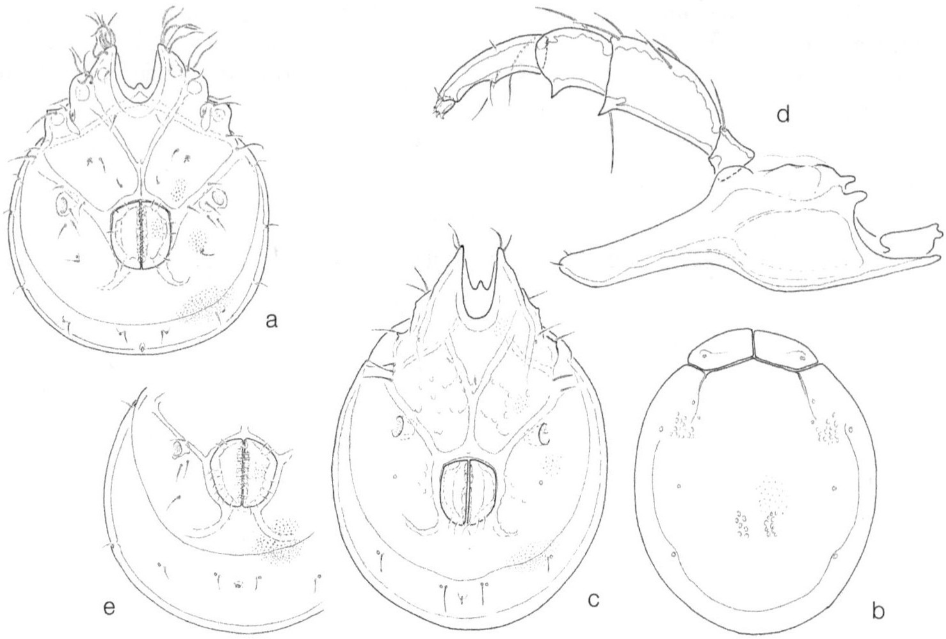

8-68: a, *Torrenticola ungeri*, male ventral shield; b-e, *Torrenticola minutivalvata;* b, male dorsal shield; c, male ventral shield; d, male gnathosoma and palp; e, female venter, detail.

1	Genital field oval or subrectangular in shape, anterior margin equally rounded or straight (*8-68 a, c*). Suture line of Cx-II+III well evident, medial length > 60 µm. Ejaculatory complex visible by transparency in the gonopore area: Males.	2
-	Genital field pentagonal in shape; anterior margin forming an obtuse angle (*8-68 e, 8-71 c*). Suture line of Cx-II+III generally shortened (medial length < 60 µm) or absent. No ejaculatory complex, but occasionally eggs visible by transparency: Females (variability of many dimensional characters still unclear – frequently, species recognition is uncertain). . .	17
2 (1)	Dorsal shield with two free anterior platelets, shoulder platelets partially fused with the large dorsal plate (*8-68 b*).	3
-	Dorsal shield with four free anterior platelets, shoulder platelets not fused with the dorsal plate (as in *8-64 b, 8-65 a*).	6
3 (2)	A group of long setae and bristles on the apex of Cx-I (*8-68 a*); ventral side of P-3 and P-4 finely serrate (*8-64 f*). . . . . *Torrenticola ungeri* (page 202)	
-	Apex of Cx-I with a few, relatively short setae; ventral side of P-3 and P-4 not serrate.	4
4 (3)	Medial suture Cx-II+III relatively long (length > 120 µm – *8-68 c*). . . . . *Torrenticola minutivalvata* (page 200)	
-	Medial suture Cx-II+III shorter (length 60-80 µm – *8-69 a, e*).	5
5 (4)	Genital field large (length > 140 µm – *8-69 a*); ejaculatory complex normal in shape (proximal and distal arms and proximal chamber well developed – *8-69 d*). . . . . *Torrenticola amplexa* (page 194)	
-	Genital field smaller (length below 130 µm – *8-69 e*); ejaculatory complex slender, proximal and distal arms very small, proximal chamber small, anterior keel reduced (*8-69 g*). . . . . *Torrenticola ischnophallus* (page 198)	

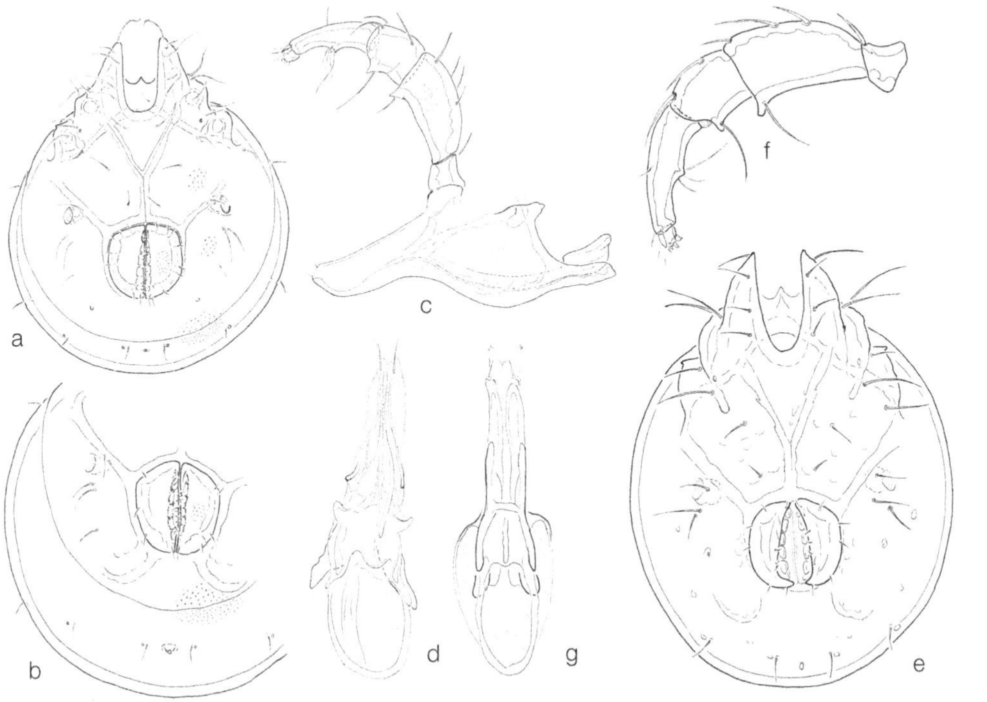

8-69: a-d, *Torrenticola amplexa;* a, male ventral shield; b, female ventral shield, detail; c, male gnathosoma and palp; d, male ejaculatory complex; e-g, *Torrenticola ischnophallus* male; e, ventral shield; f, palp; g, ejaculatory complex.

6 (2) Ventral side of P-2 and P-3 with laterally compressed, hyaline extensions (*8-70 a, c, f*) (instead of peg- or finger-like tubercles). 7

\- P-2 and P-3 ventrally with well-developed peg-like distal tubercles; without hyaline extensions (*8-70 e, 8-71 b*). 9

7 (6) Medial surface of P-4 smooth, not bearing a protuberance; ventrodistal extension of P-2 extended, distally pointed (*8-70 f*). . . . . *Torrenticola limula* (page 197)

\- Medial surface of P-4 with a finger-like protuberance bearing fine hairs; ventrodistal extension of P-2 truncate, directed ventrally (*8-70 a, c*). 8

8 (7) Anteromedial margin of P-3 smooth; P-4 stouter (*8-70 a*). . . . . *Torrenticola hyporheica* (page 198)

\- Anteromedial margin of P-3 denticulate; P-4 more slender (*8-70 c*). . . . . *Torrenticola jeanneli* (page 199)

9 (6) Gnathosomal rostrum shortened, ventrally not distinctly set off from gnathosomal base; palp robust (*8-70 e*); excretory pore away from line of primary sclerotization (*8-70 d*). . . . . *Torrenticola brevirostris* (page 196)

\- Gnathosomal rostrum longer, ventrally distinctly set off from the gnathosomal base; palp more slender (e.g., *8-71 b*); excretory pore usually not distanced from the line of primary sclerotization (*8-71 a*). 10

10 (9) Medial suture of Cx-II+III short (length 80-110, rarely exceeding 120 µm). 11

\- Medial suture of Cx-II+III longer (length generally 130-170 µm). 13

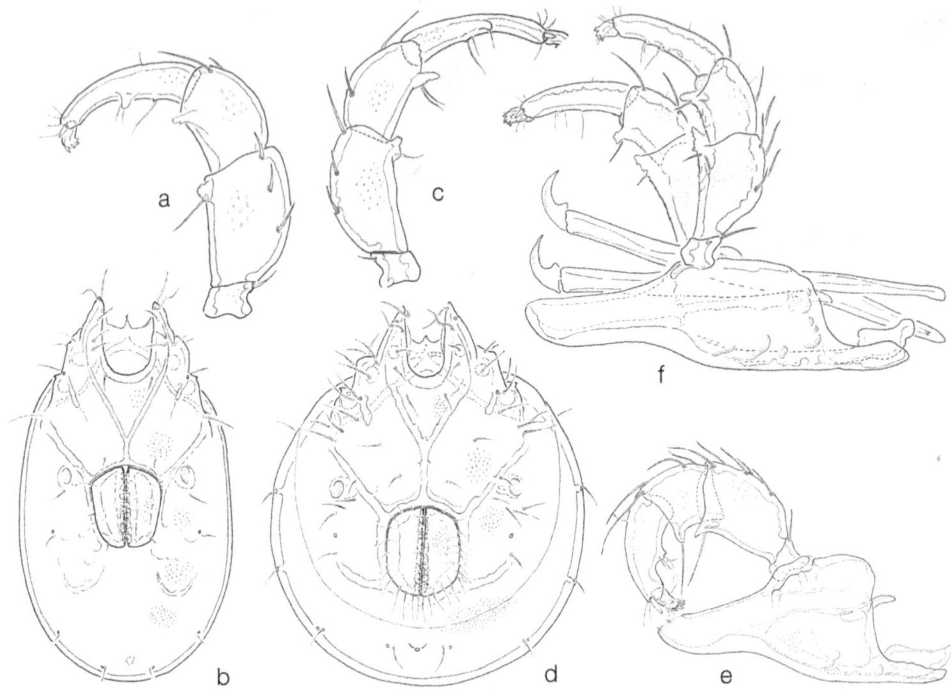

8-70: a, *Torrenticola hyporheica* male, palp; b-c, *Torrenticola jeanneli* male; b, ventral shield; c, palp; d-e, *Torrenticola brevirostris* male; d, ventral shield; e, gnathosoma and palp; f, *Torrenticola limula* male, gnathosoma with chelicerae and palps (GERECKE & DI SABATINO 1996b).

11 (10) Idiosoma roundish in shape. Postgenital area short (< 200 µm – *8-71 a*). . . . . *Torrenticola barsica* (page 195)

\- Idiosoma longish in shape. Postgenital area long (> 200 µm – *8-71 f*). 12

12 (11) Excretory pore embedded in area of primary sclerotization (*8-71 d*); ejaculatory complex with small proximal chamber (*8-71 e*). . . . . *Torrenticola similis* (page 201)

\- Excretory pore not embedded in the area of primary sclerotization (*8-71 f*); ejaculatory complex with large proximal chamber (*8-71 g*). . . . . *Torrenticola elliptica* (page 196)

13 (10) P-2 shorter than P-4; genital field small and subquadratic, posterior margins of genital flaps with long, fine setae (*8-72 a*); ejaculatory complex without proximal horns (*8-72 c*). . . . . *Torrenticola anomala* (page 194)

\- P-2 equal in length, or longer than, P-4; genital field larger and subrectangular, posterior margin of genital flaps with a few short setae only (*8-73 a; 8-74 a*); ejaculatory complex with proximal horns (*8-74 b*). 14

14 (13) Genital field very large (length/width 180-190/140-150 µm – *8-72 d*). . . . . *Torrenticola dudichi* (page 196)

\- Genital field minor in size (length/width < 180/140 µm – *8-73 a, 8-74 a*). 15

15 (14) Genital field small (length ca. 120 µm – *8-73 a*); P-2 longer than P-4 (*8-73 b*). . . . . *Torrenticola oraviensis* (page 201)

\- Genital field larger (length 140-160 µm – *8-73 d, 8-74 a*); P-2 and P-4 almost equal in length (*8-73 e*). 16

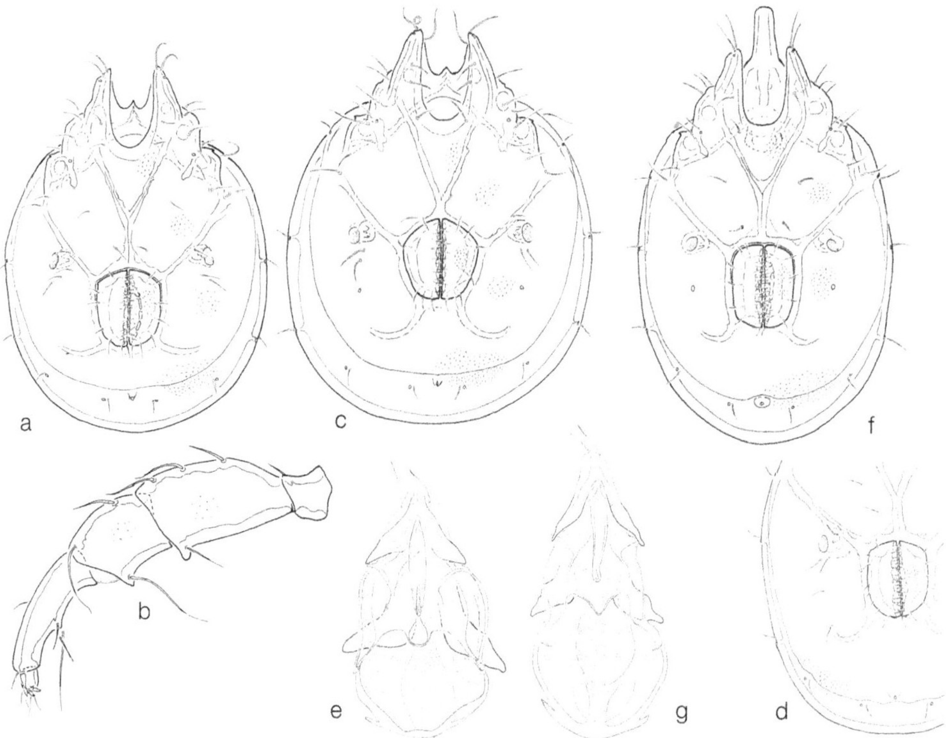

8-71: a-c, *Torrenticola barsica;* a, male ventral shield; b, male palp; c, female ventral shield; d-e, *Torrenticola similis;* d, male ventral shield, detail; e, male ejaculatory complex; f-g, *Torrenticola elliptica* male; f, ventral shield; g, ejaculatory complex.

16 (15) Ejaculatory complex small (length/width 184/78 µm), with poorly developed anterior keel (*8-73 f*). . . . . *Torrenticola microphallus* (page 200)

\- Ejaculatory complex larger (length/width 210-230/85-90 µm), with well developed anterior keel (*8-74 b*). . . . . *Torrenticola laskai* (page 199)

17 (1) Dorsal shield with one pair of free anterior platelets (shoulder platelets partially fused with dorsal plate – *8-68 b*). 18

\- Dorsal shield with two pairs of free anterior platelets (shoulder platelets not fused with dorsal plate – as in *8-64 b*). 21

18 (17) Apex of Cx-I with a group of long setae and bristles (*8-68 a*); ventral margin of P-3 and P-4 finely serrate (*8-64 f*). . . . . *Torrenticola ungeri* (page 202)

\- Apex of Cx-I with only a few or without setae; ventral side of P-3 and P-4 not serrate. 19

19 (18) Excretory pore and Vgl-2 away from line of primary sclerotization (*8-69 b*). 20

\- Excretory pore and Vgl-2 close to the line of primary sclerotization (as in *8-69 e*). . . . . *Torrenticola ischnophallus* (page 198)

20 (19) Genital field larger (length 170-180); excretory pore on the same level as Vgl-2 (*8-69 b*). . . . . *Torrenticola amplexa* (page 194)

\- Genital field generally smaller (length 165-170); excretory pore posterior to Vgl-2 (*8-68 c*). . . . . *Torrenticola minutivalvata* (page 200)

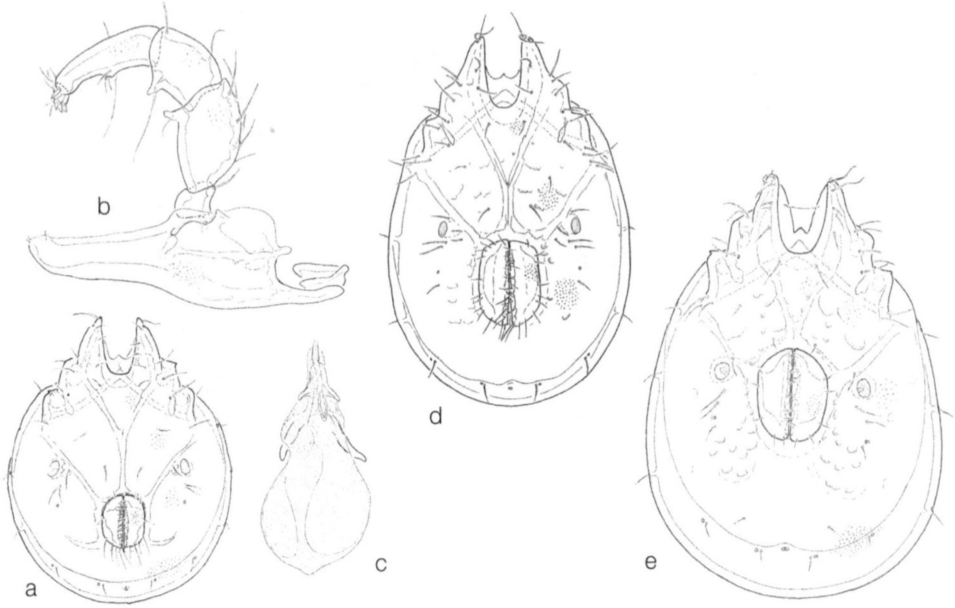

8-72: a-c, *Torrenticola anomala* male; a, ventral shield; b, gnathosoma and palp; c, ejaculatory complex; d-e, *Torrenticola dudichi;* d, male ventral shield (Di Sabatino et al. 2003); e, female ventral shield.

21 (17) P-2 and P-3 instead of the peg- or finger-like tubercles with laterally compressed, hyaline ventral extensions (*8-70 a, c, f*). 22

\- P-2 and P-3 ventrally without hyaline extensions, bearing well-developed peg-like distal tubercles (*8-70 e, 8-71 b*). 24

22 (21) Medial surface of P-4 smooth, not bearing a protuberance; ventrodistal extension of P-2 extended, distally pointed (*8-70 f*). . . . . *Torrenticola limula* (page 197)

\- Medial surface of P-4 with a finger-like protuberance bearing fine hairs; ventrodistal extension of P-2 truncate, directed ventrally. 23

23 (22) Mediodistal margin of P-3 smooth; P-4 stouter (*8-70 a*). . . . . *Torrenticola hyporheica* (page 198)

\- Mediodistal margin of P-3 denticulate; P-4 more slender (*8-70 c*). . . . . *Torrenticola jeanneli* (page 199)

24 (21) Excretory pore and Vgl-2 posterior to the line of primary sclerotization (*8-35 g*); P-2 shorter than P-4. 25

\- Excretory pore and Vgl-2 near the line of primary sclerotization (*8-72 e*); P-2 equal in length, or longer than, P-4. 26

25 (24) Gnathosomal rostrum short (*8-70 e*); palp segments robust (length/height P-4 < 2.8). . . . . *Torrenticola brevirostris* (page 196)

\- Gnathosomal rostrum long (*8-72 b*); palp segments more slender (length/height P-4 > 3.0). . . . . *Torrenticola anomala* (page 194)

26 (24) Idiosoma slender (dorsal shield length/width > 1.3) (*8-71 f*). 27

\- Idiosoma roundish (dorsal shield length/width < 1.3) (*8-72 e, 8-74 c*). 28

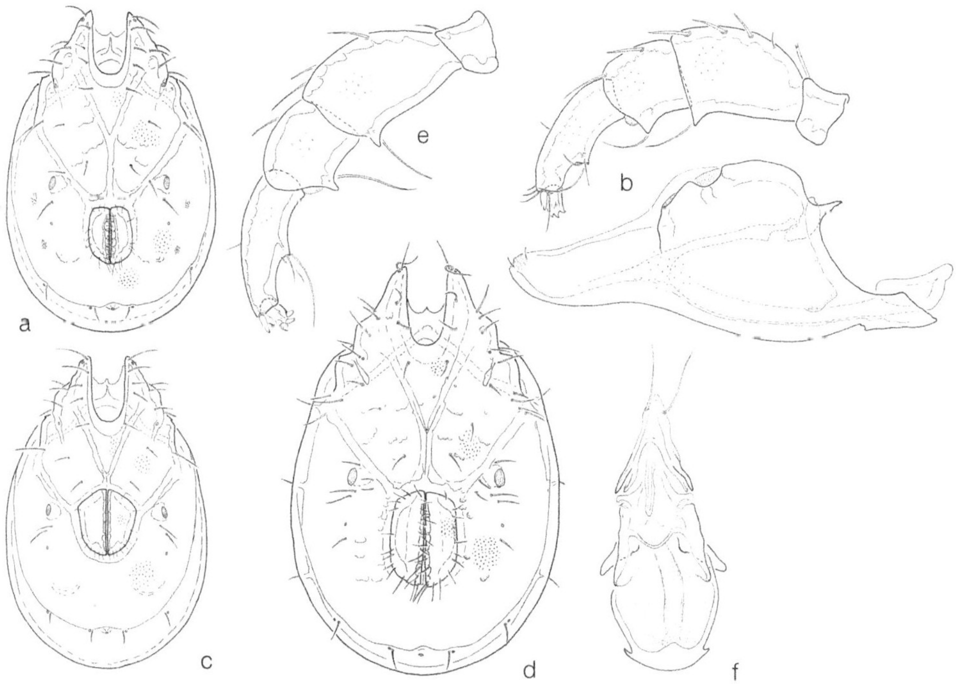

8-73: a-c, *Torrenticola oraviensis* (DI SABATINO et al. 2003); a, male ventral shield; b, male gnathosoma and palp; c, female ventral shield; d-f, *Torrenticola microphallus* male; d, ventral shield; e, palp; f, ejaculatory complex.

27 (26) Excretory pore embedded in the area of primary sclerotization (*8-71 d*). . . . . *Torrenticola similis* (page 201)

\- Excretory pore not embedded in the area of primary sclerotization (*8-71 f*). . . . . *Torrenticola elliptica* (page 196)

28 (26) P-2 and P-4 almost equal in length (*8-73 e, 8-74 d*). 29

\- P-2 generally longer than P-4 (*8-71 b, 8-73 b*). 30

29 (28) Genital field enlarged (*8-74 c*, ratio length/width < 1.1). . . . . *Torrenticola laskai* (page 199)

\- Genital field more slender (ratio length/width > 1.1). . . . . *Torrenticola microphallus* (page 200)

30 (28) Genital field enlarged, ratio length/width < 1.1; Vgl-2 in general located posterior to excretory pore (*8-71 c*). . . . . *Torrenticola barsica* (page 195)

\- Genital field more slender, ratio length/width > 1.1; Vgl-2 on the same level as excretory pore. 31

31 (30) Genital field large (length > 190 µm) (*8-72 e*). . . . . . *Torrenticola dudichi* (page 196)

\- Genital field smaller (length < 180 µm) (*8-73 c*). . . *Torrenticola oraviensis* (page 201)

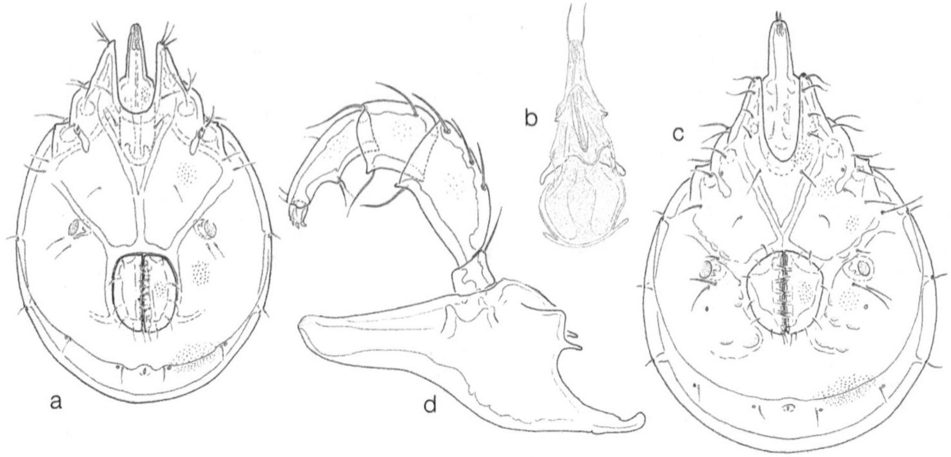

8-74: *Torrenticola laskai* (Di Sabatino et al. 2009); a, male ventral shield; b, male ejaculatory complex; c, female ventral shield; d, male gnathosoma and palp.

4. *Torrenticola (Torrenticola) amplexa* (Koenike, 1908) (*8-69 a-d*)

1908a *Atractides amplexus* Koenike, Abh. naturw. Ver. Bremen. 19 (2): 232.

Description: Colour yellowish-greenish-brown: Idiosoma almost rounded; shoulder platelets fused with dorsal plate, but in juveniles suture line between the two sclerites well visible. Excretory pore and Vgl-2 away from line of primary sclerotization. Gnathosoma with long rostrum, ventral margin curved. P-2 and P-4 equal in length or P-2 slightly longer than P-4. Males: Idiosoma length/width 640-680/500-540 µm; medial suture Cx-II+III short (length 64-94 µm); genital field subquadratic (length/width 150-152/148-150 µm). Females: Idiosoma length/width 700-800/600-700; medial suture Cx-II+III length 30; genital field length/width 170-180/155-165 µm.

Preadult stages: Larva: Tuzovskij (1981, 1982). Deutonymph: Tuzovskij (1981, 1990).

Habitat and Biology: Preferably in middle- and high order streams with sediments rich in organic material. Life cycle unknown.

Distribution: Palaearctic.

5. *Torrenticola (Torrenticola) anomala* (Koch, 1837) (*8-35 g, 8-72 a-c*)

1837 *Atractides anomalus* Koch, Deutschlands Crustaceen, Myriapoden und Arachniden 11: 10.

1933c *Torrenticola longirostris* (Szalay), Zool. Anz. 104 (7-8): 201 – Cicolani & Di Sabatino (1990).

1953a *Torrenticola piriformis* (Láska), Acta Acad. scient. natur. Moravo-Silesiacae 25 (9): 285 – Cicolani & Di Sabatino (1990).

Description: Colour yellowish-reddish-brown. Idiosoma roundish. Suture line of Cx-IV well evident, medially forming a right angle to genital field, laterally curved and directed anteriorly. Ventral margin of gnathosoma almost straight, rostrum well developed. P-2 shorter than P-4 (length ratio P-2/P-4, 0.86-0.96), Males: Idiosoma length/width 600-800/480-620; medial suture Cx-II+III length 130-160 µm; genital field almost round, small (length/width 114-133/90-117 µm), genital flaps with long setae at lateral and posterior margins; ejaculatory complex characteristic, proximal and distal arms short, proximal chamber large, proximal horns reduced; postgenital area short (length 80-130 µm); excretory pore and Vgl-2 on limit of primary sclerotization, near posterior margin of idiosoma; length P-2, 84-112; P-4, 92-116 µm. Females: Idiosoma length/width 720-930/530-730; medial suture Cx-II+III length 50-70 µm; genital field subrectangular, length/width 145-180/133-168 µm; suture line of Cx-IV well evident, medially starting in posterior direction from genital field; postgenital area extended (length 180-210 µm); excretory pore and Vgl-2 located posterior to line of primary sclerification; length P-2, 97-121; P-4, 104-131 µm.

Similar species: Males of *T. nana* Di Sabatino & Gerecke, 2003 (Turkey, Israel, Iran) and *T. oraviensis* are similar in the elongated medial suture Cx-II+III and the genital field reduced in size. They differ from *T. anomala* in the absence of long genital setae and a normally-developed ejaculatory complex. Both sexes of *T. nana* and *T.oraviensis* also differ in a relatively longer P-2 (P-2/P-4 length ratio > 1.0), but see also discussion under *T.oraviensis*.

Preadult stages: Larva: Wainstein (1980). Deutonymph: Walter (1918, "*Atractides angustirostris*"); K. Viets (1936d).

Habitat and Biology: Middle-order streams; often in riffles, but frequently also in pools; eurythermous. Life cycle unknown.

Distribution: Holarctic.

6. *Torrenticola (Torrenticola) barsica* (Szalay, 1933) (*8-71 a-c*)

1933b *Atractides barsicus* Szalay, Zool. Anz. 103 (7-8): 171.

1936b *Atractides dudichi cognatus* K. Viets, Arch. Hydrobiol. 29: 368 – Di Sabatino et al. (2003).

1955e *Torrenticola mucronata* K. Viets, Abh. naturw. Ver. Bremen 34 (1): 8 – Cicolani & Di Sabatino (1990).

Description: Colour brownish-yellowish-reddish. Idiosoma roundish. Gnathosoma with strongly curved ventral margin, rostrum well developed. P-2 longer than P-4 (ratio P-2/P-4 ca. 1.1). Males: Idiosoma length/width 700-800/500-700; medial suture Cx-II+III length 105-110 µm; genital field quadrangular with straight anterior margin forming anterolateral right angles, length/width 150-165/133-147 µm; ejaculatory complex with small, rounded proximal chamber, anterior keel, proximal and distal arms well developed, length/width 215/86 µm; suture line of Cx-IV evident, medially starting from posterior margin of genital field in a right angle to the main idiosoma axis. Excretory pore located on the margin of primary sclerotization, Vgl-2 frequently posterior to excretory pore; postgenital area short (length 130-170 µm). Length P-2, 102-119; P-4, 92-107 µm. Females: Idiosoma length/width 820-1000/590-700; medial suture Cx-II+III length 35-55 µm; genital field subquadratic, length/width 160-180/160-176 µm; excretory pore close to the line of primary sclerotization, Vgl-2 a little posterior to excretory pore; length P-2, 107-126; P-4, 97-110 µm.

Similar species: *Torrenticola hispanica* (Lundblad, 1941c) and *T. guadarramensis* (Lundblad, 1941c), both from the Iberian peninsula and very similar to *T. barsica*, are in need for revision.

Habitat and Biology: Low- and middle-order streams, both in riffles and pools; rheocrenes. Eurythermous. Preadult stages and life cycle unknown.

Distribution: Mediterranean, Middle East. In the French Pyrenees reaching the southwestern border of the treated area.

7. *Torrenticola (Torrenticola) brevirostris* (Halbert, 1911) (*8-70 d-e*)

1911 *Atractides brevirostris* Halbert, Proc. R. Irish Acad. Dublin 31: 16.

1955 *Torrenticola histriensis* Bader, Verh. Naturforsch. Ges. Basel 66 (1): 77 – Cicolani & Di Sabatino (1990).

Description: Colour brownish-yellowish-red. Idiosoma roundish. Suture line of Cx-IV prominent, starting at right angle from genital field, laterally curved anteriorly. Postgenital area extended. Gnathosomal rostrum not evidently set off from gnathosomal base. Palp robust and compact, P-2 shorter than P-4. Males: Idiosoma length/width 750-870/600-700; medial suture Cx-II+III short (length 105-117 µm, ratio medial suture Cx-II+III/Cx-I 0.3-0.4); genital field large, roundish (length/width 184-200/150-165 µm); ejaculatory complex large (length/width 310-320/100-120 µm); with well-developed anterior keel and proximal arms; postgenital area length in mature specimens up to 235 µm; excretory pore and Vgl-2 away from line of primary sclerotization; length P-2, 85-100; P-4, 97-109 µm (ratio P-2/P-4, 0.85-0.92). Females: Idiosoma length/width 870-1000/760-820; medial suture Cx-II+III length 10-43 µm; genital field rhombic and large (length/width 180-215/150-164 µm); postgenital area length 325-356, length P-2, 90-107; P-4, 100-120 µm.

Similar species: *Torrenticola diaspora* Gerecke & Di Sabatino, 1996 is a sister species of *T. brevirostris*, endemic to Corsica. It differs in larger idiosoma dimensions (idiosoma length > 900 µm), a longer medial suture of Cx-II in males (> 130 µm) and a more robust P-4 with smaller ventral denticles.

Habitat and Biology: Higher order streams with sandy or gravel substratum at low to medium elevation. Preadult stages and life cycle unknown.

Distribution: Holarctic.

8. *Torrenticola (Torrenticola) dudichi* (Szalay, 1933) (*8-72 d-e*)

1933b *Atractides dudichi* Szalay, Zool. Anz. 103: 173.

Description: Colour yellowish-brown. Idiosoma of large dimensions, rounded. Genital field large. Gnathosoma with relatively short rostrum and weakly curved ventral margin. P-2 longer than P-4 (length ratio P-2/P-4, 1.1-1.2). Males: Idiosoma length/width 760-830/540-560; medial suture Cx-II+III length 140-150; genital field length/width 190-200/140-148 µm; ejaculatory complex as given for *T. similis* (*8-71 e*), but with proximal chamber larger, length/width 282/137 µm; posterior suture line of Cx-IV weakly curved; postgenital area short (length 120-170 µm); length P-2, 102-107; P-4, 87-100 µm. Females: Idiosoma length/width 900-1050/600-700; medial suture Cx-II+III length 40-50; genital field length/width 200-210/160-180 µm; excretory pore only slightly away from line of primary sclerotization; postgenital area length 180-200 µm.

Habitat and Biology: Low order streams at middle elevations. Preadult stages and life cycle unknown.

Distribution: Central and Southeastern Europe.

9. *Torrenticola (Torrenticola) elliptica* Maglio, 1909 (*8-71 f-g*)

1909 *Torrenticola elliptica* Maglio, Atti Soc. ital. Sci. natur., Pavia 48: 291.

Description: Colour greenish-yellowish-brown. Idiosoma elongated, with subparallel lateral margins, postgenital area length > 200 µm. Suture of Cx-IV hardly evident. Excretory pore and Vgl-2 on the posterior margin of primary sclerotization near posterior end of idiosoma. Gnathosoma ventral margin strongly curved. P-2 and P-4 almost of the same length. Males: Idiosoma length/width 780-880/560-640; medial suture Cx-II+III length 100-117 µm; genital field large, subrectangular in shape, length/width 172-195/133-145 µm; ejaculatory complex with large proximal chamber, length/width 280/125; length postgenital area 200-250; P-2, 95-112; P-4, 95-114 µm. Females: Idiosoma length/width 810-990/570-670; medial suture Cx-II+III length 35-51; genital field length/width 176-207/150-180; postgenital area length 250-360 µm.

Similar species: *Torrenticola meridionalis* Di Sabatino & Cicolani, 1990 (Italy, Croatia, Montenegro), similar in the extended postgenital area of males, was separated from *T. elliptica* due to a less slender idiosoma, and, in males, the relatively shorter medial suture of Cx-III and stouter genital field. As for the first two characters, in *T. elliptica* a larger variability range could be stated (Di Sabatino et al. 2009), *T. meridionalis* remains distinguishable in the male sex only, exclusively based on the stouter genital field (length/width ratio 1.2, in *T. elliptica* > 1.25).

Preadult stages: Larva unknown. Deutonymph: Maglio (1909); Sokolov (1940a).

Habitat and Biology: Low order streams, preferably in riffle areas, occasionally in springs. Nymphal stages frequently in the hyporheic (Gerecke et al. 2005b). Life cycle unknown.

Distribution: Palaearctic. In the Mediterranean area confined to refugial habitats at high elevation (Di Sabatino et al. 1992).

10. *Torrenticola (Torrenticola) limula* Gerecke & Di Sabatino, 1996 (*8-70 f*)

1996 *Torrenticola (Torrenticola) limula* Gerecke & Di Sabatino, Arch. Hydrobiol. Suppl. 107: 310.

Description: Colour yellowish-brown with reddish areas. Idiosoma rounded. Suture lines of Cx-IV extending back beyond posterior margin of genital field, laterally curved. Excretory pore away from primary sclerotization, Vgl-2 on the level of excretory pore. Gnathosoma ventral margin only slightly curved, rostrum well developed and slender. P-2 nearly equal in length as P-4 (P-2/P-4 ratio 0.9-1.0), ventral margin with a fine denticulation both in the proximal half of the segment, distally with a laterally compressed, anteriorly directed hyaline projection not clearly set off from the proximal margin. P-3 with a subrectangular, apically serrated ventrodistal projection; medial margin of P-4 without a protuberance. Males: Idiosoma length/width 610-740/450-540; medial suture Cx-II+III length 100-125 µm; genital field anteriorly nearly straight, length/width 126-153/112 126; length postgenital area 120-150; P-2, 80-92; P-4, 88-100 µm. Females: Idiosoma length/width 670-780/480-560; medial suture Cx-II+III length 58-85 µm; genital field pentagonal, length/width 144-164/144-157; length postgenital area 184-243 µm; P-2, 80-95; P-4,83-100.

Similar species: Two further species of the *serratipalpis*-group (characterized by the longish, serrate ventrodistal extension on P-2) are known from Europe, *T. gomerae* Lundblad, 1969 from the Canary islands and *T. xylurgella* Gerecke & Di Sabatino, 1996 from Sardinia and Corsica. They differ both in the ventral denticulation of P-2 not extending to the proximal half of the segment, and extension on P-2 set off from the ventral segment margin, excretory pore included into primary sclerotization and male ejaculatory complex with shorter proximal chamber and stronger arms.

Habitat and Biology: Weakly seeping springs and low order streams, also in the hyporheic interstitial. Preadult stages and life cycle unknown.

Distribution: Western Mediterranean (Corsica, Sardinia, Spain, Di Sabatino et al. 2009). In the Spanish Pyrenees reaching the southwestern border of the treated area.

11. *Torrenticola (Torrenticola) hyporheica* Di Sabatino & Cicolani, 1993 (*8-70 a*)

1950a *Atractides ramiger* Angelier, E., Bull. Mus. Paris, 22: 88 *nec* Szalay, 1947 – Gerecke & Di Sabatino (1996b).

1993 *Torrenticola (Torrenticola) hyporheica* Di Sabatino & Cicolani, Annls Limnol. 29: 31-39.

Description: Colour yellowish-brownish. Idiosoma elongated, ocular pigment reduced or absent. Suture lines of Cx-IV extending far beyond posterior margin of genital field, laterally only slightly curved. Excretory pore and Vgl-2 on the same level near the terminal end of idiosoma, just posterior to posterior margin of primary sclerotization. Gnathosoma ventral margin moderately curved, rostrum well developed. P-2 shorter than, or equal in length to, P-4, with a ventrodistal conical hyaline projection bearing a short seta; anteromedial margin of P-3 smooth; P-4 robust, medial margin with a protuberance bearing fine hair-like setae directed distally or ventrally and located distal to the ventral tubercles. Males: Idiosoma length/width 780/460; medial suture Cx-II+III length 65 µm; genital field small, rectangular, length/width 127/102; length postgenital area 255; P-2, 137; P-4, 142 µm. Females: Idiosoma length/width 735-785/420-430; medial suture Cx-II+III length 45-63 µm; genital field pentagonal, length/width 155/130; length postgenital area 250 µm; P-2, 102; P-4, 115.

Similar species: *Torrenticola ramigera* Szalay, 1947, described from several sites in Hungary, is still poorly known. Following the original description, it differs in having less developed distoventral extensions on P-2 and P-3. See also discussion under *T. jeanneli*.

Preadult stages: Larva unknown. Deutonymph: Di Sabatino & Cicolani (1993).

Habitat and Biology: Low- and middle order streams; only in the hyporheic. Life cycle unknown.

Distribution: Sicily, French Pyrenees, Corsica (? – Gerecke & Di Sabatino 1996b). In the French Pyrenees reaching the southwestern border of the treated area.

12. *Torrenticola (Torrenticola) ischnophallus* Lundblad, 1956 (*8-69 e-g*)

1956a *Torrenticola (Torrenticola) ischnophallus* Lundblad, Arkiv Zool. 10 (1): 167.

Torrenticola amplexa auct., nec Koenike, Abh. naturw. Ver. Bremen 19 (2): 232 – Di Sabatino et al. (2009).

Description: Colour yellowish-brown. Shoulder platelets fused with dorsal shield. Suture line of Cx-IV faint or obliterated. Excretory pore and Vgl-2 only slightly away from the line of primary sclerotization. Ventral margin of gnathosoma only slightly curved. Males: Idiosoma length/width 530/380; medial suture Cx-II+III length 64 µm; genital field small and subquadratic (length/width 106/90 µm); ejaculatory complex with short proximal and distal arms, proximal chamber small, proximal horns and anterior keel completely reduced. P-2 and P-4 almost equal in length (81-83 µm). Females: Idiosoma length/width 710/550; medial suture Cx-II+III length 13 µm; genital field subquadratic (length/width 169/154 µm); P-2 slightly longer than P-4 (105/98 µm).

Habitat and Biology: Higher order streams with fast current on sandy or stony substrata. Preadult stages and life cycle unknown.

Distribution: French Alps, Germany; frequently confused with *T. amplexa* and only recently detected outside the type locality, distribution unclear (Di Sabatino et al. 2009).

13. *Torrenticola (Torrenticola) jeanneli* (MOTAŞ & TANASACHI, 1947) (*8-70 b-c*)

1947b *Atractides jeanneli* MOTAŞ & TANASACHI, Bull. sect. scient. Acad. Roumaine 29 (8): 506.

Description: Colour yellowish-brown. Idiosoma elongated, ocular pigment reduced. Postgenital area very large. Excretory pore on the limit of primary sclerotization near posterior idiosoma margin. Gnathosoma with rostrum fairly bent dorsally. P-2 and P-3 both with hyaline conical ventrodistal expansions; distomedial? anteromedial margin of P-3 with denticles; P-4 slender, medially with a distally-directed finger-like protuberance located distal to the ventral tubercle. Males: Idiosoma length/width 718-783/413-473; medial suture Cx-II+III length 105-110 µm; genital field small and rectangular, length/width 120/106 µm; posterior suture of Cx-IV in its medial part perpendicular to the longitudinal body axis, laterally distinctly curving anteriorly; length P-2, 107-119; P-4, 116-135 µm. Females: Idiosoma length/width 780-795/480-502; medial suture Cx-II+III length 25-30; genital field length/width 140-156/120-130 µm; posterior suture of Cx-IV medially directed posteriorly, laterally curved.

Similar species: *Torrenticola hyporheica* DI SABATINO & CICOLANI, 1993 is similar in the presence of hyaline expansions on P-2 and P-3 and a medial protuberance on P-4, but differs in stouter palp segments and P-3 with a smooth distal mediodistal margin.

Preadult stages: Larva unknown. Deutonymph: VAN RENSBURG (1971).

Habitat and Biology: Low- and middle- order streams, only known from the hyporheic. Life cycle unknown.

Distribution: Central and Southeastern Europe. Only known from a few sites.

14. *Torrenticola (Torrenticola) laskai* DI SABATINO, 2009 (*8-74 a-d*)

1954c *Torrenticola lativalvata* ANGELIER, E., Vie et Milieu, Suppl. 5: 110, *nec* K. VIETS, 1952c.

1955d *Torrenticola lativalvata* LÁSKA, Publ. Fac. Sci. Univ. Masaryk 364: 223, *nec* K. VIETS, 1952c.

1996b *Torrenticola lativalvata* "*sensu* ANGELIER", GERECKE & DI SABATINO, Arch. Hydrobiol. Suppl. 107: 307.

2003 *Torrenticola lativalvata* "*sensu* ANGELIER", DI SABATINO et al., Arch. Hydrobiol. Suppl. 139: 409.

2009 *Torrenticola laskai* DI SABATINO, in DI SABATINO et al., Zootaxa 2266: 27.

Description: Idiosoma roundish; dorsal shield with shoulder plates separate from dorsal plate; palp segments P-2 and P-4 nearly equal in length; extensions on P-2 and P-3 narrow, not laterally flattened. Males: Idiosoma length/width 680-830/500-590; dorsal shield 560-675/ 420-475 µm; medial suture Cx-II+III length 112-150 µm, as long as, or only slightly shorter than, medial suture of Cx-I (length ratio Cx- I/Cx-II+III 1.0-1.2, relation idiosoma length/medial suture length Cx-II+III 5.0-6.5); genital field subrectangular, anteriorly slightly convex or straight, length/width 140-160/120-135, maximum width in the anterior third; ejaculatory complex normal in shape, length ca. 220; P-2, 92-112; P-4, 92-110 µm. Females: Idiosoma length/width 840-970/600-730, median suture Cx-II+III length 35-50; length/width genital field 185-200/175-190; length P-2, 109-130; P-4, 105-120 µm.

Similar species: Males of *T. barsica*, a species often confused with *T. laskai*, differ in a shorter medial suture line Cx-II+III and the genital field with maximum width at the anterior margin. In both sexes, P-2 is longer than P-4, but in questionable cases distinction of females of the two species may be impossible.

Habitat and Biology: Low and middle order streams. Preadult stages and life cycle unknown.

Distribution: Central Mediterranean – SE Europe. In the area treated recorded so far only once, from the Czech Republic, but possibly wider distributed and frequently confused with other, similar species.

15. *Torrenticola (Torrenticola) microphallus* Lundblad, 1956 (*8-73 d-f*)

1956a *Torrenticola microphallus* Lundblad, Arkiv Zool. 10 (1): 151.

1967 *Torrenticola elliptica microphallus* K. O. Viets, in ILLIES, J. (ed.) "Limnofauna Europea": 134.

Description: Colour greenish-yellowish-brown. Suture of Cx-IV not evident. Gnathosoma ventral margin moderately curved. P-2 and P-4 almost equal in length. Males: Idiosoma length/width 760/490; medial suture Cx-II+III length 133-153 µm; genital field rectangular, length/width 157/114; postgenital area length 156 µm; ejaculatory complex length/width 184/78 µm, with poorly developed anterior keel; excretory pore and Vgl-2 on the margin of primary sclerotization; length P-2, 95; P-4, 93 µm. Females: Idiosoma length/width 874/617; medial suture Cx-II+III length 40 µm; genital field large, pentagonal, length/width 190/165; postgenital area length 315 µm; posterior suture line of Cx-IV well evident; excretory pore and Vgl-2 slightly distanced from the margin of primary sclerotization; length P-2, 109; P-4, 105 µm.

Habitat and Biology: Low- to high-order streams from 500 to 2300 m asl; eurythermous. Preadult stages and life cycle unknown.

Distribution: Iberian peninsula and Southwestern France – a record from Germany (K. O. Viets 1957) rejected by Di Sabatino et al. (2003).

16. *Torrenticola (Torrenticola) minutivalvata* Lundblad, 1956 (*8-68 b-e*)

1956a *Torrenticola (Torrenticola) amplexa minutivalvata* Lundblad, Arkiv Zool. 10 (1): 166.

2009 *Torrenticola (Torrenticola) minutivalvata* Di Sabatino et al., Zootaxa 2266: 28.

Description: Colour yellowish-brown. Idiosoma large and slender. Shoulder platelets fused with dorsal plate. Excretory pore and Vgl-2 clearly posterior to the line of primary sclerotization, excretory pore posterior to Vgl-2. Suture of Cx-IV curved. Gnathosoma rostrum long, ventral margin in lateral view curved. P-2 longer than P-4. Males: Idiosoma length/width 802/600; medial suture Cx-II+III length 135 µm; genital field small and subrectangular, length/width 135/120; length postgenital area 180; P-2, 115; P-4, 100 µm. Females: Idiosoma length/width 840-860/640-668 µm; genital field small and pentagonal in shape, length/width 165-170/154-163; length postgenital area 319; P-2, 115-120; P-4, 102-108 µm.

Habitat and Biology: Low- and middle-order streams. Preadult stages and life cycle unknown.

Distribution: French Pyrenees. Only known from a few sites.

17. *Torrenticola (Torrenticola) oraviensis* LÁSKA, 1953 (*8-73 a-c*)

1953a *Torrenticola dudichi oraviensis* LASKA, Acta Acad. Sc. Nat. Moravo-Silesiacae 25 (9): 287.

Description: Colour yellowish-brown. Idiosoma large and oval. Ventral margin of gnathosoma almost straight. Palp minute and compact, segments robust. P-2 longer than P-4. Males: Idiosoma length/width 680-720/450-500; medial suture Cx-II+III length 126-148 µm; genital field small, subrectangular, length/width 125-130/95-108; postgenital area length 130-150 µm; ejaculatory complex length/width 168-180/90-100 µm; excretory pore and Vgl-2 on the line of primary sclerotization; suture of Cx-IV not evident; length P-2, 85-95; P-4, 85-87). Females: Idiosoma length/width 795-829/580-605; medial suture Cx-II+III length 20-30 µm; genital field pentagonal, length/width 175-180/155-165; postgenital area length 280-300 µm; posterior suture line of Cx-IV well evident; excretory pore and Vgl-2 slightly away from the line of primary sclerotization; length P-2, 105-109; P-4, 90-96 µm.

Similar species: *Torrenticola nana* DI SABATINO & GERECKE, 2003 (Eastern Mediterranean) is similar in the elongated medial suture of Cx-II+III and reduced size of genital field in males, but differs in having a smaller idiosoma, a more slender P-2 (length/height ratio 1.8-2.0 vs 1.5-1.6), a smaller genital field (length < 110 µm vs > 120 µm with more rounded lateral edges. Furthermore, females of *T. nana* are characterized by the complete reduction of the medial suture Cx-II+III.

Habitat and Biology: Low- and middle-order streams. On sandy substrata. Preadult stages and life cycle unknown.

Distribution: Southern, Central and Southeastern Europe.

18. *Torrenticola (Torrenticola) similis* (K. VIETS, 1939) (*8-71 d-e*)

1939a *Atractides ellipticus similis*, K. VIETS, Zool. Anz. 125 (1-2): 9.

Description: Colour greenish or yellowish brown. Idiosoma elongated (length/width dorsal shield 1.54-1.66), length of postgenital area > 220 µm. Excretory pore in a medioposterior indentation of primary sclerotization, Vgl-2 posterior to excretory pore. Gnathosoma ventral margin strongly curved. P-2 and P-4 almost equal in length. Males: Idiosoma length/width 930/600; medial suture Cx-II+III length 110-117 µm; genital field large and rectangular in shape, length/width 190/140; ejaculatory complex with small proximal chamber and robust proximal arms, length/width 282/130; postgenital area length 276 µm; suture of Cx-IV in the lateral part curved and partially obliterated; length P-2, 95-100; P-4, 95-100 µm. Females: Idiosoma length/width 980-990/630-670; medial suture Cx-II+III length 51 µm; genital field large, length/width 203/188; postgenital area length 300 µm; posterior suture line of Cx-IV not evident; length P-2, 120; P-4, 116 µm.

Similar species: With an elongated idiosoma and large postgenital area *T. similis* is very close to *T. elliptica*. However, it can be distinguished by the excretory pore embedded in the area of primary sclerotization and by the smaller proximal chamber of the male ejaculatory complex.

Preadult stages: Larva unknown. Deutonymph: K.VIETS (1939a); WALTER (1944).

Habitat and Biology: Low-order streams. Preferably in riffles, cold-stenothermous. Life cycle unknown.

Distribution: Pyrenees, Alps, Apennines, Central, Eastern and Southeastern Europe.

19. *Torrenticola (Torrenticola) ungeri* (SZALAY, 1927) (*8-64 f, 8-68 a*)

1927b *Atractides ungeri* SZALAY, Allat. Krzlem. 47 (1-2): 72.

Description: Colour reddish- or yellowish-brown. Shoulder platelets fused to the large dorsal plate. Tip of Cx-I with numerous long, stout setae. Suture of Cx-IV in the medial part directed posteriorly, laterally curved. Excretory pore and Vgl-2 away from the line of primary sclerotization. Gnathosoma ventral margin curved, rostrum well developed. P-2 longer than P-4, slender and curved. Ventral margin of P-3 and P-4 with fine serration, P-4 in addition with 6-7 small ventral tubercles bearing fine setae. Males: Idiosoma length/width 700-750/540-560; medial suture Cx-II+III length 70-80 µm; genital field roundish, length/width 140-150/120-130 µm; ejaculatory complex with small proximal chamber and well-developed arms and proximal horns, length/width 203/63; length postgenital area 200-210; P-2, 110-118; P-4, 95-100 µm. Females: Idiosoma length/width 750-935/600-680; medial suture Cx-II+III length 30-40 µm; genital field pentagonal and slender, length/width 180/170 µm; posterior suture of Cx-IV more accentuated than in males; length P-2, 120-130; P-4, 105-110 µm.

Habitat and Biology: Surface and interstitial waters. In the Mediterranean area with a preference for superficial running waters at low to medium elevations. Life cycle unknown.

Distribution: Western Palaearctic except for Fennoscandia, Iran, Turkey.

Following pages:

Hydrachnidia, general morphology and terminology

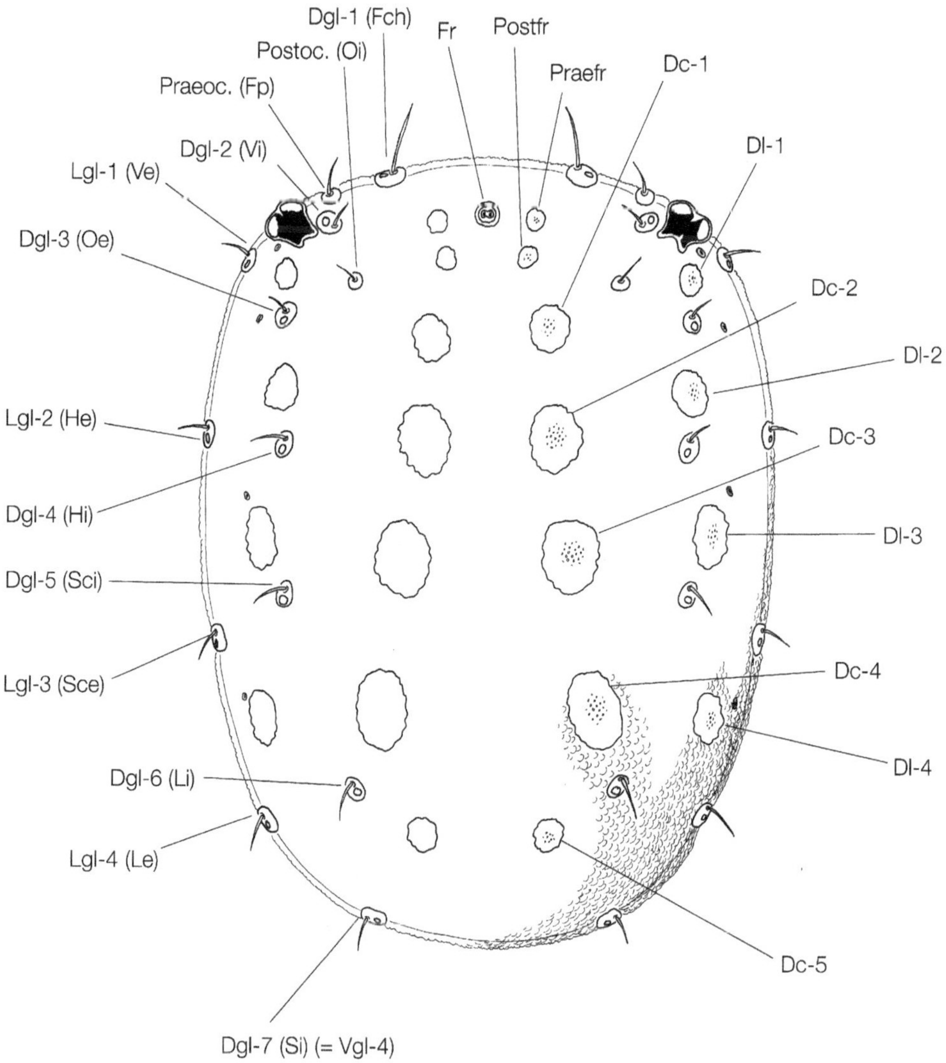

8-75: Hydrachnidia, plesiotypical idiosoma organization, dorsal view (from DAVIDS et al. 2007). Terminology following LUNDBLAD 1927 (in parentheses: following ZAHVATKIN 1952, after TUZOVSKIJ 1987). Explanation of abbreviations:

ZAHVATKIN 1952	LUNDBLAD 1927
-	Fr: Frontale
Fch: frontalis chelicerarum	Dgl-1: Dorsoglandulare 1 (Antenniforme)
Vi: verticalis interna	Dgl-2: Dorsoglandulare 2
Oe: occipitalis externa	Dgl-3: Dorsoglandulare 3
Hi: humeralis interna	Dgl-4: Dorsoglandulare 4
Sci: scapularis interna	Dgl-5: Dorsoglandulare 5
Li: lumbalis interna	Dgl-6: Dorsoglandulare 6
Si: sacralis interna	Dgl-7: Dorsoglandulare 7 (= Vgl-4, Ventroglandulare 4)
Ve: verticalis externa	Lgl-1: Lateroglandulare 1
He: humeralis externa	Lgl-2: Lateroglandulare 2
Sce: scapularis externa	Lgl-3: Lateroglandulare 3
Le: lumbalis externa	Lgl-4: Lateroglandulare 4
Oi: occipitalis interna	Postoc: Postoculare
Fp: frontalis pedipalporum	Preoc.: Preoculare
-	Dc-1-4: Dorsocentralia 1-4
-	Dl-1-4: Dorsolateralia 1-4

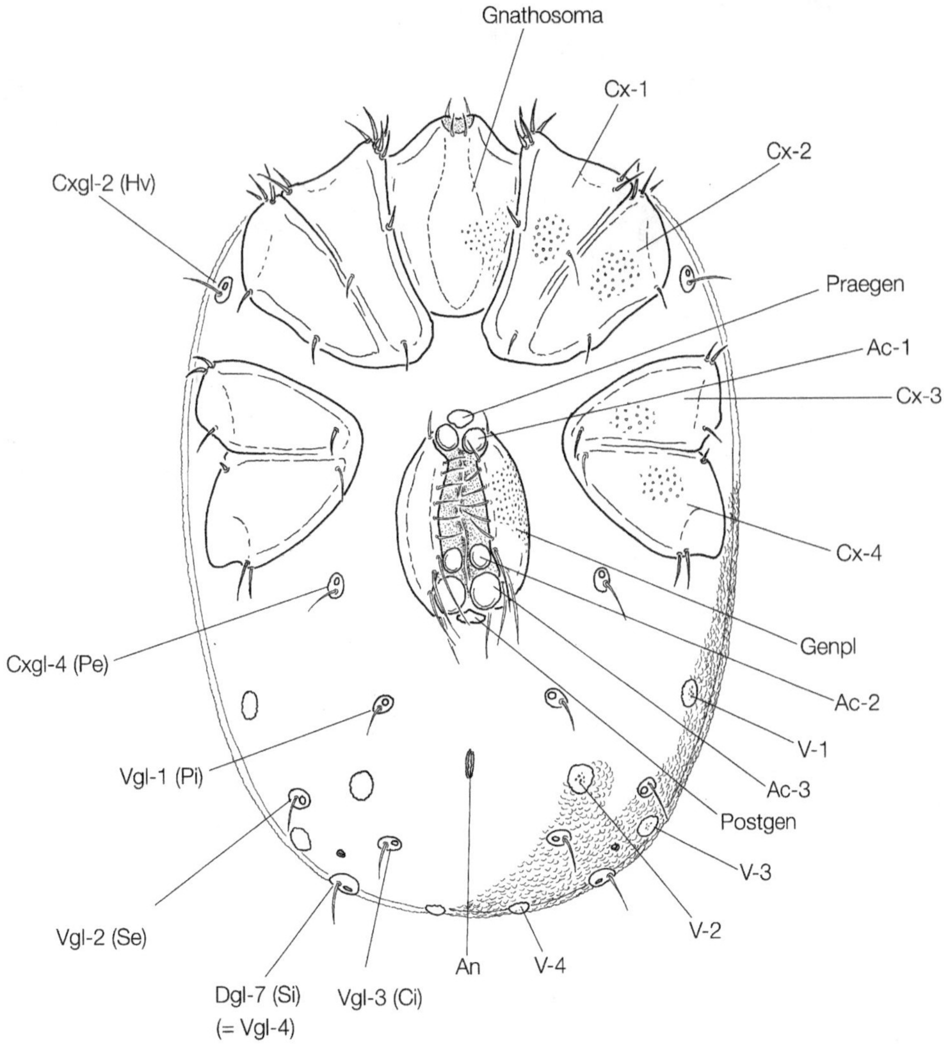

8-76: Hydrachnidia, plesiotypical idiosoma organization, ventral view (from Davids et al. 2007). Terminology following Lundblad 1927 (in parentheses, following Zahvatkin 1952, after Tuzovskij 1987).

Explanation of abbreviations:
Ac-1-3: Acetabulum 1-3
An: Anus
Genpl: Genital plate
Pregen: Pregenital plate
Postgen: Postgenital plate

Zahvatkin 1952	Lundblad 1927
-	Cx-1-4: first to fourth coxae
Hv: humeralis ventralis	Cxgl-2: Coxoglandulare 2 (Cxgl-1 and -3 not existing)
Sce: scapularis externa	Cxgl-4: Coxoglandulare 4
Si: sacralis interna	Dgl-7: Dorsoglandulare 7 (= Vgl-4: Ventroglandulare 4)
Pi: praeanalis interna	Vgl-1: Ventroglandulare 1
Se: sacralis externa	Vgl-2: Ventroglandulare 2
Ci: caudalis interna	Vgl-3: Ventroglandulare 3
-	V-1-4: Ventralia 1-4

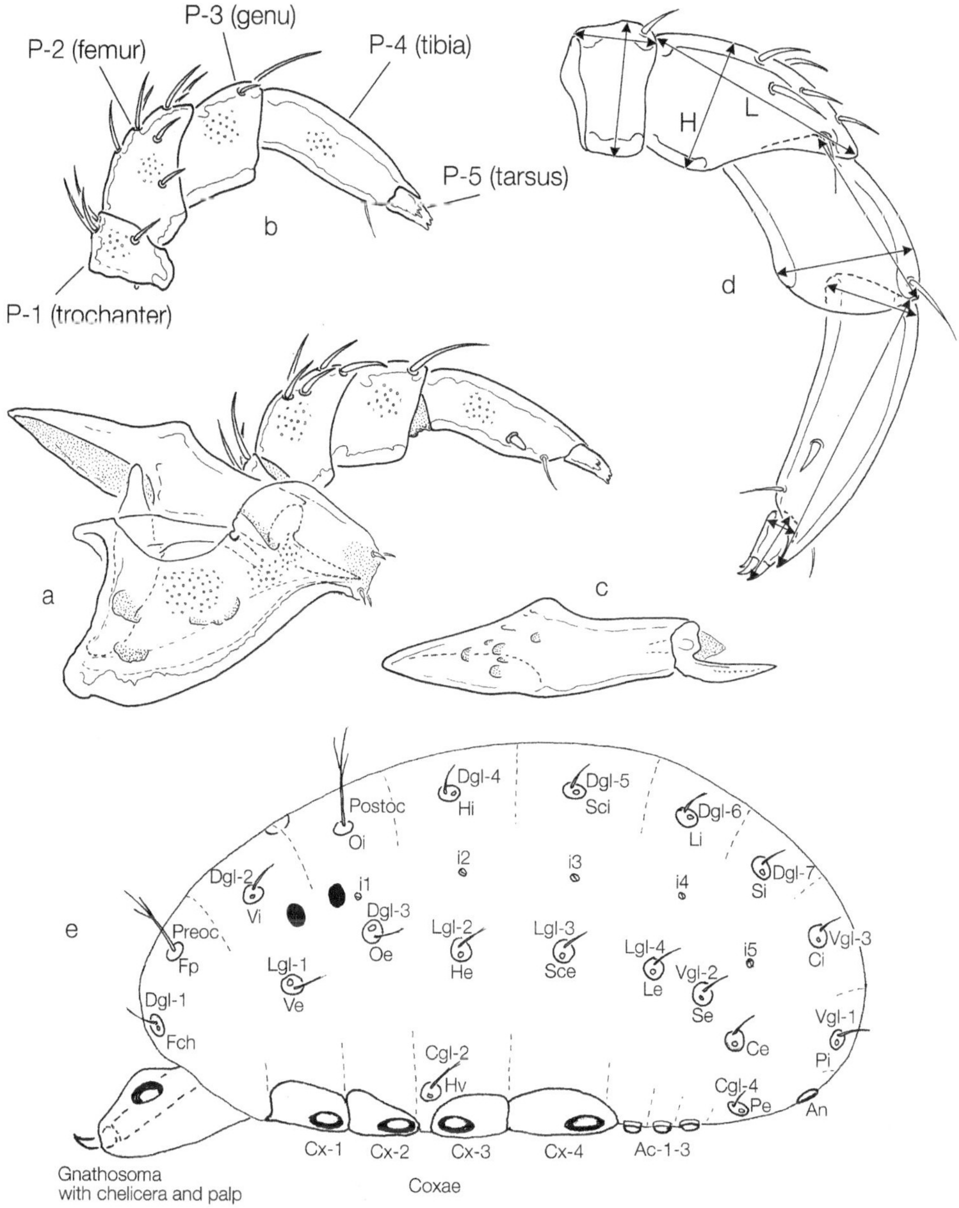

8-77: a-d, Hydrachnidia, gnathosoma and its appendages in lateral view, example of *Thyopsis cancellata* (Protz, 1896) (from Davids et al. 2007); a, gnathosoma and left palp; b, right palp laterally (indicating numbering of segments and traditional segment terminology); c, chelicera; d, palp indicating important measurement distances (L = length, H = height); e, Hydrachnidia, schematical lateral view; dashed lines indicate hypothetical segment borders (for abbreviations see Figs 8-75-76, frontal eye posterodorsally, paired lateral eyes posterolaterally from Dgl-2; setae *Ce* reduced in most taxa, developed only in postlarval instars of *Limnochares* and Pontarachnidae; after Tuzovskij 1987).

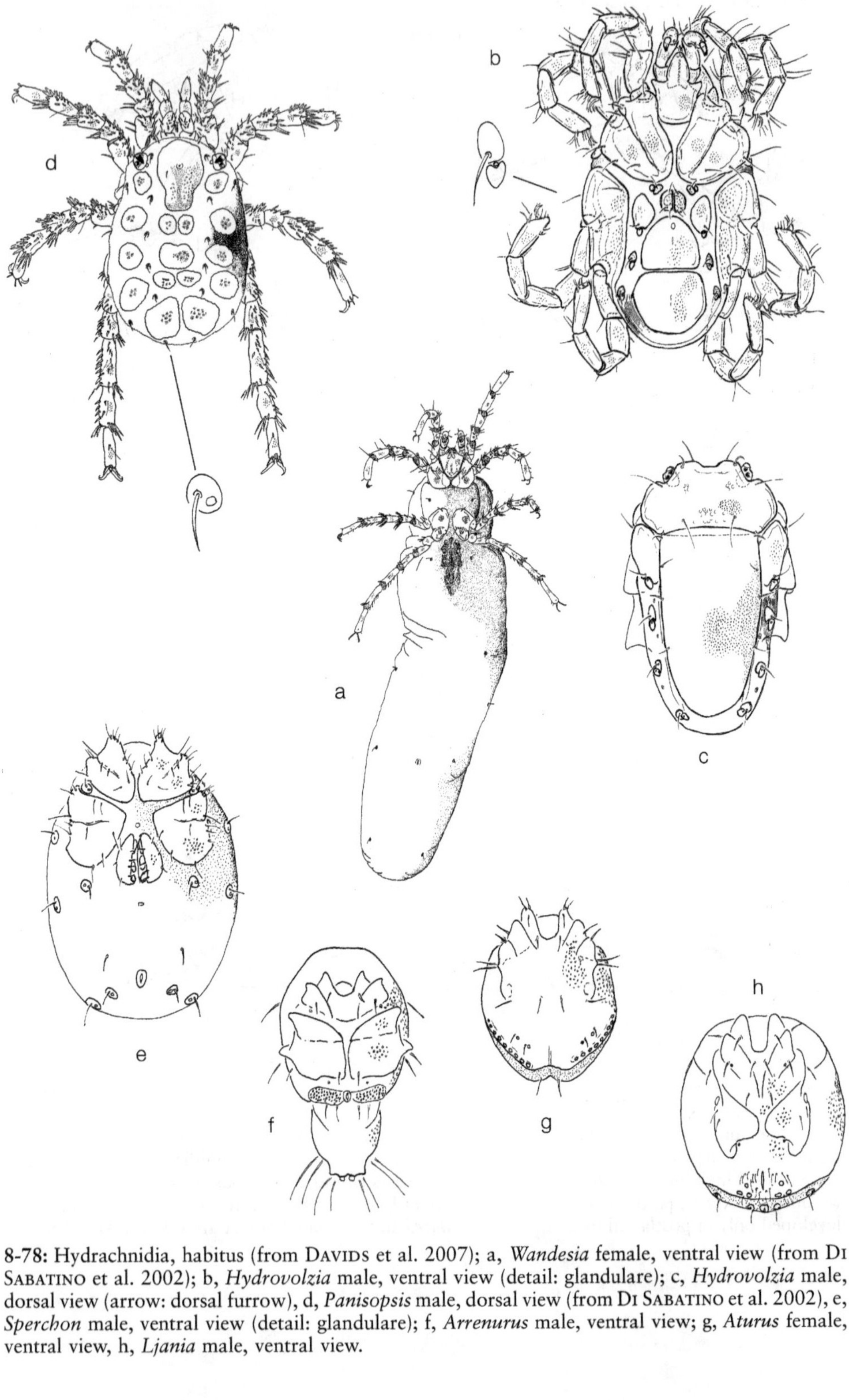

8-78: Hydrachnidia, habitus (from DAVIDS et al. 2007); a, *Wandesia* female, ventral view (from DI SABATINO et al. 2002); b, *Hydrovolzia* male, ventral view (detail: glandulare); c, *Hydrovolzia* male, dorsal view (arrow: dorsal furrow), d, *Panisopsis* male, dorsal view (from DI SABATINO et al. 2002), e, *Sperchon* male, ventral view (detail: glandulare); f, *Arrenurus* male, ventral view; g, *Aturus* female, ventral view, h, *Ljania* male, ventral view.

Hydrachnidia, key to superfamilies

1 Legs with three claw-like, sickle-shaped appendages: paired, ventrally pectinate, true claws, and an empodium similar in shape, but without pectination; idiosoma vermiform, without muscle attachment sclerites; rod-shaped frontal sclerite (*8-79 a*) with one unpaired, and three paired setae; eyes strongly reduced or completely absent; legs covered by rather uniform hairs over the segment surface ("hypertrichous"); palp compact (*8-80 a*), P-1-3 fused, P-4+5 forming a functional unit; P-5 without claws, bearing one stiff terminal seta; pale reddish species of interstitial waters. . . . . **Stygothrombioidea** (Vol. I, page 289)

\- Legs with two (or occasionally less) claw-like appendages (*8-79 h-k*): empodium not claw like; claws rarely with ventral pectination; idiosoma rarely vermiform (*8-78 a*), with or without muscle attachment sclerites; if a rod-shaped frontal sclerite is present (e.g., *8-79 b*), never bearing an unpaired anterior seta; eyes various, in most cases with paired lateral eyes, sometimes also with an unpaired frontal eye (*8-79 e, g*); legs rarely hypertrichous, in general with distal margin setae stronger than setae on remaining segment surface (*8-79 h-k*); palp various (*8-80 b, e-f, h-i, l-o*), P-5 generally with claws, rarely bearing only one stiff terminal seta. 2

2 (1) Palp chelate (P-4 forming a chela with P-5) <u>and</u> P-1 longer than P-2, P-3 longer than P-4, P-4 reduced in size, maximum height of palps in P-1, the following segments gradually tapering (*8-80 b*); gnathosoma in ventral view characteristically narrowed at the base of the slender, pointed rostrum (*8-80 b, d*), chelicerae needle-like (basal segment + claw fused to a single unit, *8-80 c*); genital field (*8-81 c*) with maximum width in the anterior part, bearing numerous acetabula located anteriorly from, or on the level of, the anterior margin of the gonopore; rather large, globular red species of standing waters. . . . . **Hydrachnoidea** (Vol. I, page 333)

\- If palp chelate, neither P-1 longer than P-2, nor P-4 shorter than P-3, maximum height of palps in general in segments P-2-4 (*8-80 e-f, h-i, l-o*); gnathosomal rostrum, if present, generally not pointed (*8-80 i*), chelicerae two-segmented, not needle-like (*8-80 k, p*); genital field various in shape, if numerous acetabula present, they are arranged in a different manner or the genital field has not its maximum width anteriorly from the gonopore (*8-81 a-b, e, h, m-o*); larger or smaller species, various in colour and shape. 3

3 (2) Lateral eyes, if present, incorporated into a frontal sclerite (*8-79 b-d, f*); palps not chelate (*8-80 e-f, h*), in some groups transformed by fusion of segments (e.g., *8-80 f*), claws of P-5 straight (*8-80 h*) or reduced to fine tips (*8-80 f*); mouth opening surrounded by a circular, membranous fringe (*8-80 g*) or (in Apheviderulicidae, so far not recorded from the study area, *8-80 f*) reduced to a fine slit in the centre of a conspiciously enlarged gnathosoma; acetabula numerous, or on roundish sclerites flanking the gonopore (*8-81 a*), or in the membranous perigenital area (*8-81 b*), or extremely reduced in size scattered all over the membranous idiosoma surface (*8-81 d*); glandular openings occasionally with semilunar, fan-shaped openings (*8-79 f, 8-81 a*); large, red species of temporary and permanent standing waters (outside the study area also pale inhabitants of springs and interstitial waters). . . . . **Eylaoidea** (Vol. I, page 301)

\- Lateral eyes, if present, not incorporated into a frontal sclerite (but occasionally very close to such a sclerite, *8-79 g*); palps various in shape, P-5 in general with hook-like distal claw setae (*8-80 i, l-o*); mouth opening various, but neither surrounded by a circular fringe, nor in the centre of an extremely enlarged idiosoma; acetabula in three pairs or numerous (*8-81 e-o*), in rare cases invisible by light microscope; glandular openings never fan-shaped (*8-79 e, g*). 4

4 (3) I/II-L directed anteriorly, III/IV-L directed posteriorly, the two leg groups separated by a large interspace (*8-78 b*); genital field minute, in the centre of the ventral idiosoma, devoid of acetabula (these numerous, mostly invisible in light microscope and inserted as minute cups on the medial coxal surface); in the postgenital area two large unpaired plates, the anterior one with the anal pore, or all perigenital plates and platelets fused to a shield embracing the genital field; dorsum with two shields in "T" position (*8-78 c*); glandular openings and setae on paired paletelets (*8-78 b*, arrow); flattened inhabitants of springs, cascades and interstital waters. . . . . **Hydrovolzioidea** (Vol. I, page 294)

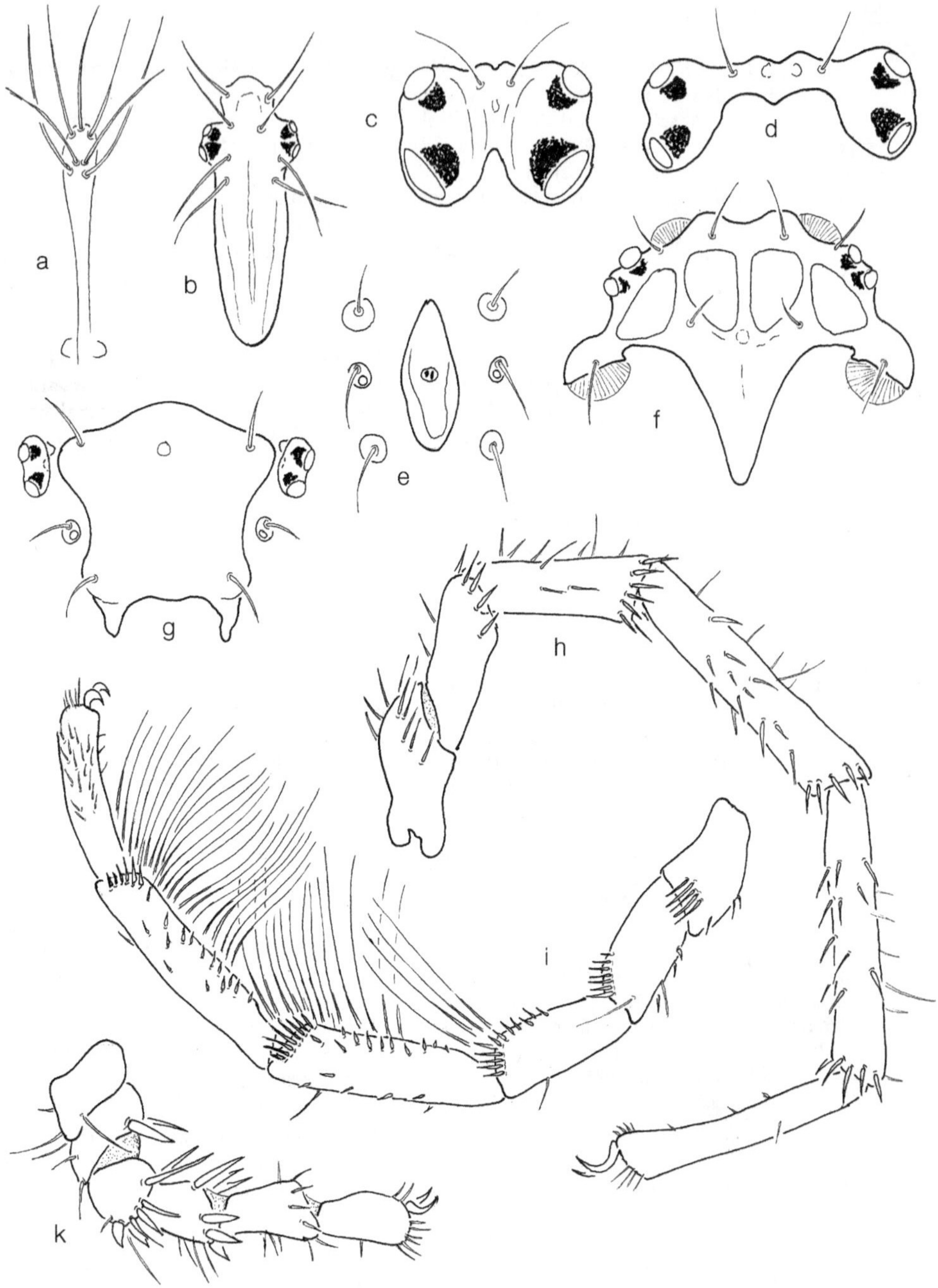

8-79: Hydrachnidia (from DAVIDS et al. 2007); a-g, frontal sclerites, h-k, legs; a, *Stygothrombium*; b, *Limnochares*; c-d, *Eylais*; e, *Euthyas*; f, *Piersigia*; g, *Hydryphantes*; h, *Parathyas* IV-L; i, *Hydryphantes* (*Polyhydryphantes*), IV-L; k, *Trichothyas* I-L.

- If legs arranged in two groups, they are not orientated into two opposite directions, and the interspace is less extended (*8-78 a, d*); genital field various, (with the exception of the marine Pontarachnidae) always with groups of acetabula arranged in various ways (*8-81 e-o*); if dorsal shields present, not arranged in a "T" position; glandularia and their associated setae together on one single platelet (*8-79 e, g*, arrow). 5

5 (4) Palps chelate (*8-77 a-b, d, 8-80 i*, arrow: P-4 with a dorsal extension or a heavy dorsal seta dorsally flanking P-5); idiosoma smooth or with regularly arranged muscle insertion sclerites (*8-78 d*), only in exceptional cases with extended dorsal and ventral shields; acetabula in various arrangements in, or around, the genital field (*8-81 e, g-h, l*), but never embedded in a sclerotized ventral shield; diversified morphological adaptations, but mostly larger species, all kinds of inland waters. . . . . **Hydryphantoidea** (this Vol., page 1)

- Palps not chelate (*8-80 l-o*); idiosoma various; if a sclerotized ventral shield is present, it may include also some, or all acetabula (*8-78 f-h, 8-81 m-o*). 6

6 (5) Palps uncate: P-4 with a ventrodistal extension opposing P-5 (*8-80 o*) or at least P-4 ventrally bearing a projecting setal tubercle (*8-80 m*); idiosoma in general heavily sclerotized (*8-78 f*); genital field without movable genital flaps; Ac numerous, or on fixed plates flanking the gonopore (often fused to the ventral shield as given in *8-81 n*), or lined up in the membranous area flanking the gonopore, or both (*8-81 m*); various in colour and dimensions, mostly in standing waters, but also in spring and interstitial habitats. . . . . **Arrenuroidea** (Vol. III)

- Palps not uncate (*8-80 l, n*: P-4 without extensions flanking P-5); movable flaps may be present in the genital field (*8-81 f*), Ac in three pairs or numerous - or located between the flaps (*8-81 f*), or on immovable genital plates (*8-81 i-k*) - if these plates are shifted to the posterior margin and variously fused with surrounding sclerites, in ventral view acetabula may be visible tangentially (*8-78 g, 8-81 o*); various in colour and dimensions, all kinds of habitats. 7

7 (6) Acetabula located in two rows between the gonopore and movable genital flaps (*8-81 f, 2-4 f*) (if, in rare exceptions of species so far not recorded in the study area, they are fused to the flaps, a suture indicates the former line of separation between Ac and flaps); if a ventral shield is present, genital flaps remain always separate and movable. . . . . **Lebertioidea** (this Vol., page 76)

- Acetabula located on immovable plates flanking the gonopore, they are fully integrated, no suture lines are recognizable between acetabula and plate surface (*8-78 g-h; 8-81 i, k*); Ac-bearing plates may be fully integrated into the ventral shield; in this case, Ac are in three to four pairs and directed ventrally (*8-78 h*), <u>or</u> multiplied and directed caudally (*8-78 g, 8-81 o*). . . . . **Hygrobatoidea** (Vol. III)

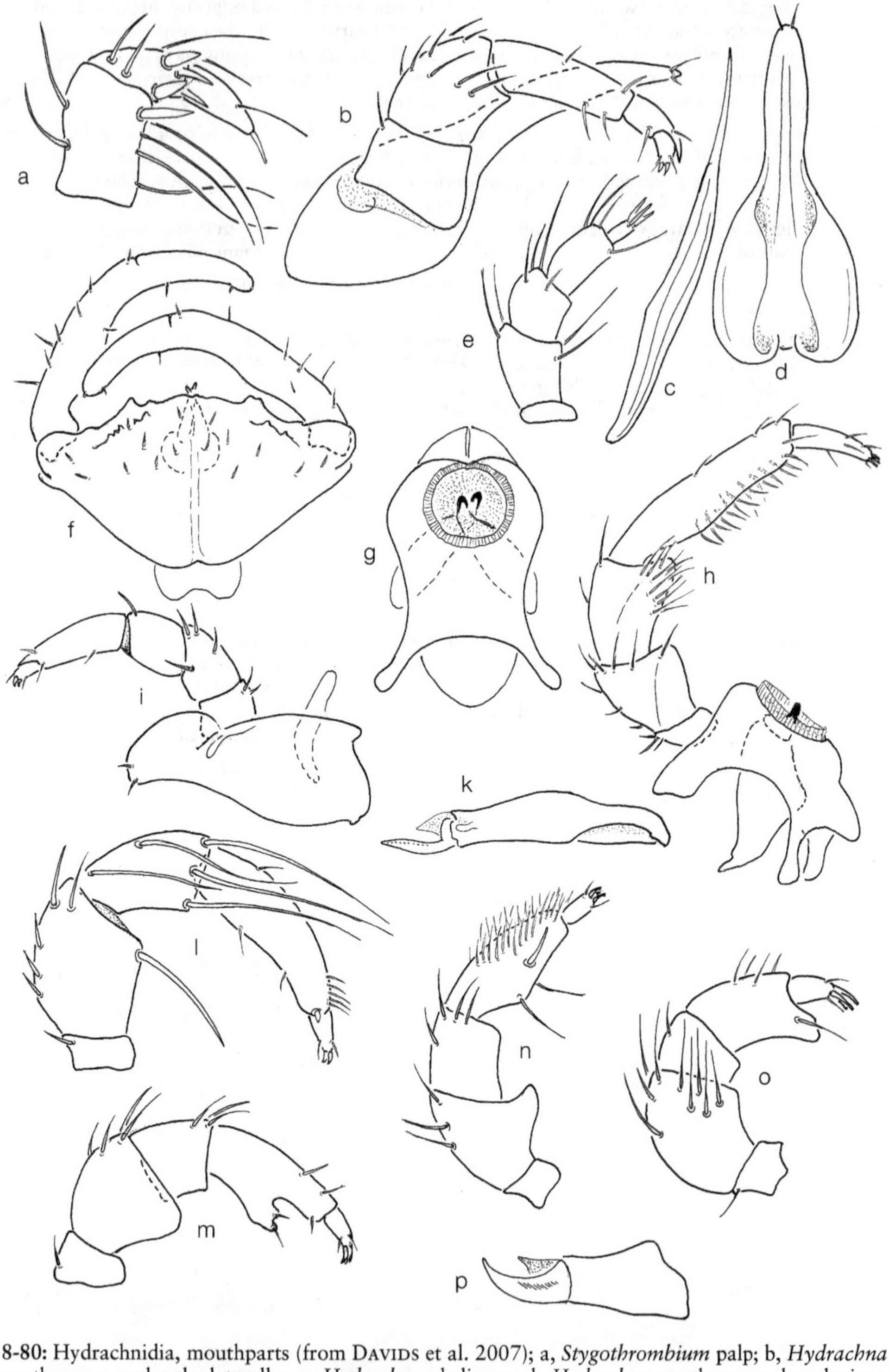

8-80: Hydrachnidia, mouthparts (from Davids et al. 2007); a, *Stygothrombium* palp; b, *Hydrachna* gnathosoma and palp laterally; c, *Hydrachna* chelicera; d, *Hydrachna* gnathosoma dorsal view; e, *Limnochares* palp; f, *Apheviderulix* gnathosoma and palps ventral view; g, *Eylais* gnathosoma ventral view; h, *Eylais* gnathosoma and palp lateral view; i, *Hydryphantes* gnathosoma and palp lateral view; k, *Hydryphantes* chelicera; l, *Lebertia* palp; m, *Mideopsis* palp; n, *Atractides* palp; o, *Arrenurus* palp; p, *Arrenurus* chelicera.

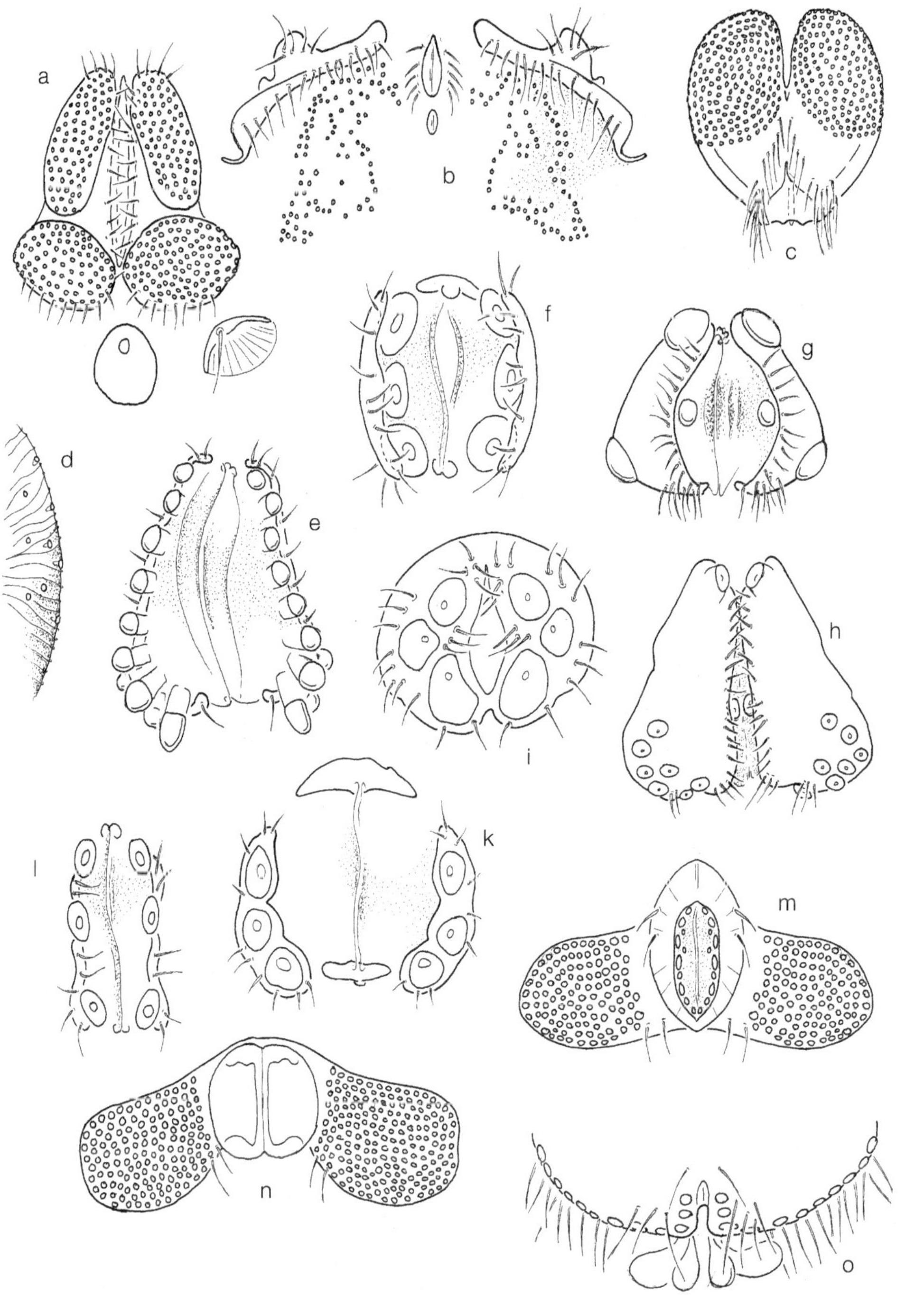

8-81: Hydrachnidia, genital fields (from Davids et al. 2007); a, *Piersigia* female; b, *Limnochares* female; c, *Hydrachna* male; d, *Eylais* male (acetabula in the integument); e, *Protzia* female; f, *Sperchon* female; g, *Hydryphantes* s.str. male; h, *Hydryphantes* (*Polyhydryphantes*) female; i, *Hygrobates* male; k, *Hygrobates* female; l, *Wandesia* (*Pseudowandesia*) female; m, *Balcanohydracarus* female; n, *Arrenurus* female; o, *Aturus* male.

Acknowledgements

The following colleagues helped with loan of collection material: A. HÄNGGI & U. WÜEST (Naturhistorisches Museum Basel); J. DUNLOP & A. FRIEDERICHS (Museum für Naturkunde der Humboldt-Universität Berlin); P. TUZOVSKIJ (Hydrobiological Institute, Borok); A.KOLIBAČ & I. MALINOVSKÝ (Moravian Museum Brno); M. HOLMES (National Museum of Ireland, Dublin); J. ALTMANN & P. JÄGER (Senckenberg-Museum Frankfurt); M. JUDSON (Musée d'Histoire Naturelle Paris); V. PEŠIĆ (University of Montenegro, Podgorica), N. JÖNSSON & T. KRONESTEDT (Naturiska Rijksmuseet, Stockholm).

ANDREJ ZAWAL (Szeczin) and R. STRYJECKI (Lublin) supported our attempts to get in loan specimens from the Polish fauna.

PETER MARTIN (Kiel) checked a former manuscript version and gave useful suggestions for improving the contents, mainly with regard to larval morphology and life cycles. Furthermore, we are thankful to D.R. COOK (Paradise Valley, U.S.A.), I. MALINOSKÝ (Brno, Czech Republic), V. PEŠIĆ (Podgorica, Montenegro), I.M. SMITH (Ottawa, Canada) and P. TUZOVSKIJ (Borok, Russia) for helpful comments and useful advice.

ANTONIO DI SABATINO thanks B. CICOLANI (L'Aquila) for continuous assistance and scientific support and the Italian MIUR (ex 60 %) to have partially funded his research.

Basic taxonomic revisions of R. GERECKE were supported by grants of the Deutsche Forschungsgesellschaft (Revisions of Hydryphantidae and Lebertiidae), the Musèe d'Histoire Naturelle Luxembourg and the EU (Highlat project SMNH Stockholm, Synthesys projects MNHN Paris).

TERRY GLEDHILL thanks the Council of the Freshwater Biological Association for continued support. The foundation Pro Acarologia Basiliensis (Basel) gave a financial contribution to the realization of some of the figure tables by C.ALLGAIER (Tübingen). The publishing house E. Schweizerbart (Stuttgart) gave permission to reproduce numerous figures from "Archiv für Hydrobiologie" and "Abhandlungen der Senckenbergischen Naturforschenden Gesellschaft".

Glossary

acetabular plate	see "genital field"
anchoral process	proximal extension of the gnathosoma serving as an attachment site of muscles which move mouth parts
antagonistic bristle	pointed dorsal projection of P-4 flanking P-5.
anterior keel	see "ejaculatory complex"
anterior ramus	see "ejaculatory complex"
apical setae	see "ejaculatory complex"
body pores	fine pores in sclerotized plates of Halacaridae and Hydrachnidia, at least in some Hydrachnidia species facilitating gas exchange processes.
cauda	extension of the posterior part of the idiosoma, more or less distinctly set off from the remainder of the body; frequently found only in males of species with sexual dimorphism in idiosoma shape; posteriolateral extensions of the cauda (found mainly in male representatives of Arrenuridae) are called "pygal lobes"
chelate	morphological state of (→) chelicera or (→) palp with two segments acting together like the branches of scissors.
coxoglandularia	see "glandularia"
distal arm	see "ejaculatory complex"
dorsal furrow	see "dorsal shield"
dorsal shield	large sclerotized shield covering the whole dorsal idiosoma, originating from the fusion of the (→) "dorsalia", in general not including (→) "lateral eyes" and (→) "lateroglandularia"; in most cases, this shield separated from the (→) ventral shield by a ring-shaped membranous area called "dorsal furrow"
dorsalia	dorsal muscle attachment sclerites; plesiotypically paired, in a total number of eleven on each side, in a regular pattern following the original segmentation: pre- and post frontalia (prefr, postfr) flanking the area of the (→) median eye , sometimes fused with the (→) frontale; five dorsocentralia (dc) paralleling the median line; four dorsolateralia (dl) arranged more laterally; sclerites may be fused in various ways; fusion of all sclerites produces the (→) dorsal shield .
dorsocentralia	see "dorsalia"
dorsolateralia	see "dorsalia"
dorsoglandularia	see "glandularia"
ejaculatory complex	distal part of the male internal genital system, responsible for the composition and deposition of "spermatophores", for morphology consult *7-4* in Vol. I; plesiotypically a complex organ made up of unpaired keels ("an-

	terior, posterior keel"; lat. *carina anterior, posterior*), paired sclerite arms ("distal, proximal arm", lat. *brachium distale, proximale*; "proximal processus", lat. *processus proximalis*), a proximal vesicular structure ("proximal chamber", lat. *cella proximalis*), and an apical sclerite bridge ("anterior ramus", lat. *ramus anterior*) bearing paired setae ("apical setae", lat. *setae apicales*), in many clades modified for indirect or direct spermatophore transmission. (= "genital skeleton"; [German] "Penisgerüst")
genital field	area surrounding the opening of the reproductive system ("gonopore"), consisting of the following elements: paired sclerotized plates called "acetabular plates" or "genital plates" ([German] "Genitalplatten") when acetabula are located on their surface, but "genital flaps" or "genital valves" ([German] "Genitalklappen") when acetabula are placed beneath them and can be covered by the door-like closing flaps; small unpaired anterior ("pregenital") and posterior ("postgenital") sclerites delimiting the slit-shaped gonopore; occasionally, a depressed area surrounding the male gonopore is called "genital pit"
genital flap, pit, plate, valve	see "genital field"
genital skeleton	see "ejaculatory complex"
glandulae limnesiae	see "glandularium"
glandularium	defence organ typical for, and a synapomorphy of, Hydrachnidia; consisting of a round plate bearing a sensory seta and a glandular opening; plesiotypically, glandularia are distributed over the idiosoma in a characteristic ground pattern reflecting the original segmentation of the arthropod body: six pairs of "dorsoglandularia" flanking the median line, four pairs of "lateroglandularia" in row along the lateral line, four pairs of "coxoglandularia" placed on or near the coxae, and four pairs of "ventroglandularia" surrounding the anal area; a particular pair of glandularia found in species of the family Limnesiidae on coxae 4, occasionally called "glandulae limnesiae"; for further terminological details, see *8-78-79*.
hypostomal setae	pair of setae on the ventral surface of the gnathosomal rostrum in water mite larvae
lateroglandularia	see "glandularium"
petiole	small projecting sclerite at the posterior end of the idiosoma, mainly found in males of *Arrenurus*-species, often combined with the formation of a cauda (see there)
posterior keel	see "ejaculatory complex"
postfrontalia	see "dorsalia"
postgenital sclerite	see "genital field"

postocularia	paired, knob-shaped, seta-bearing sclerites in the ground plan of Hydrachnidia located posteriorly from the lateral eyes
prefrontalia	see "dorsalia"
pregenital sclerite	see "genital field"
preocularia	paired seta-bearing sclerites in the ground plan of Hydrachnidia located anteriorly from the lateral eyes
proximal arm	see "ejaculatory complex"
proximal chamber	see "ejaculatory complex"
proximal processus	see "ejaculatory complex"
pygal lobes	see "cauda"
subchelate	morphological state of palp with the terminal segment P-5 acting against P-4 like a clasp-knife (occasionally called "uncate")
swimming setae	elongated setae arranged mainly on III-L-4/5 and IV-L-4/5, more rarely also on the anterior legs and/or on more proximal segments; in general in rows or as single elements distally on the anterior segment surface, reaching in length at least 50 % of the following segment. (occasionally called "swimming hairs")
uncate	see "subchelate"
ventralia	paired ventral muscle attachment sclerites; plesiotypically found in water mites in the circumanal area in a total number of four on each side ("ventrale 1-4"); sclerites may be fused in various ways; fusion of all sclerites and the coxae produces the ventral shield.

References

ANDRÉ, M. & NAUDO, M.H. (1962): Un nouveau sous-genre de *Wandesia* (Hydrachnellae: Protziidae) *Euwandesia sensitiva* nov. Acarologia 4 (4): 595-604.

ANGELIER, E. (1947): Note sur deux Hydrachnelles (Acariens) hypogés des sables littoraux. Bull. Mus. Hist. natur., Paris (s. 2) 19: 446-452.

ANGELIER, E. (1949b): Hydracariens phréaticoles des Pyrénées orientales. Mem. Mus. Hist. natur., Paris (n. s.), 29 (2): 227-258.

ANGELIER, E. (1951b): Diagnoses sommaires d'Hydracariens psammiques nouveaux de Corse. Bull. Mus., Paris (s. 2) 23 (5) : 508-510.

ANGELIER, E. (1954c): Acariens (Hydrachnellae et Porohalacaridae) des eaux superficielles. In: ANGELIER, E. (ed.): Contribution à l'étude de la faune d'eau douce de Corse. Vie et Milieu 5 (1): 74-148.

ANGELIER, E. (1959): Les eaux douces de la Corse. Vie et Milieu, Suppl. 8 : 1-56.

ANGELIER, E., ANGELIER, M.-L. & LAUGA, J. (1985): Recherches sur l'écologie des Hydracariens (Hydrachnellae, Acari) dans les eaux courantes. Annls. Limnol. 21 (1): 25-64.

BADER, C. (1955): Hydracarinen-Diagnosen aus dem Nachlaß von C. WALTER. Verh. naturforsch. Ges. Basel 66 (1): 61-84.

BADER, C. (1974): Die Sperchoniden der *glandulosus*-Gruppe (Acari, Prostigmata). Arch. Hydrobiol. 73 (4): 470-491.

BADER, C. (1975): Die Wassermilben des schweizerischen Nationalparkes. 1. Systematisch-faunistischer Teil. Ergebn. wiss. Unters. schweiz. Nationalpark 14: 1-270.

BADER, C. (1980h): Mites of Saudi Arabia. Hydrachnellae, Fam. Nilotoniidae. Fauna of Saudi Arabia 2: 49-56.

BADER, C. (1981a): *Panisus*-Studien: I. Zur Revision der Gattung *Panisus* KOENIKE, 1896 (Acari, Actinedida, Hydrachnellae). Entomologica Basiliensia 6: 51-57.

BADER, C. (1988): Die Torrenticolidae (Acari, Hydrachnellae). Eine abklärende Studie über eine schwierige Wassermilben-Familie. Revue Suisse Zool. 95 (1): 87-98.

BARROIS, T. (1889b): Notes hydrachnologiques. I. Remarques sur le *Diplodontus scapularis* DUGÈS. - Rev. biol. Nord France 1 (10): 393-400. II. De l'identité des *Nesaea uncata* KOENIKE et *Nesaea viridis* KOCH. - Rev. biol. Nord France 1 (11): 430-432. III. Sur l'adaption des Hydrachnides aus eaux saumâtres. - Rev. biol. Nord France 1 (11): 432-433.

BERTUZZI, E., LAZZARA, M. & GERECKE, R. (2004): Prime indicazioni sugli idracari del lago di Tovel (Trentino). Poster. 3.Tagung des Naturmuseums Bozen.

BESCH, W.K. (1964): Systematik und Verbreitung der südamerikanischen rheophilen Hydrachnellen. Beitr. Neotrop. Fauna 3: 77–194.

BESSELING, A.J. (1939): [De levenswijze van *Sperchon setiger* S. THOR]. Ent. Ber., Amsterdam 10: 143-144. (in Dutch)

BESSELING, A.J. (1941): [Nederlandsche Hydrachnidae. Genus *Hydryphantes*]. Entom. Ber., Amsterdam 10 (238): 306-310. (in Dutch)

BESSELING, A.J. (1964). [De Nederlandse Watermijten (Hydrachnellae LATREILLE, 1802)]. Monogr. Nederl. Entomol. Ver. 1: 1-199. (in Dutch)

BIESIADKA, E. (1972): Sur quelques Hydracariens nouveaux recueillis en Pologne. Bull. Acad. Pol. Sci. Ser. Sci. Biol. Zool. Cl. II., 20 (4): 249-256.

BIESIADKA, E. (1973): [Hydracariens nouveaux ou peu connus pour la faune de Pologne]. Fragm. Faun. 18 (22) : 437-490. (in Polish)

BIESIADKA, E. & CICHOCKA, M. (1984): Studies on the morphology of larval stages of water mites (Hydracarina). 1. Description of three species of the genera *Protzia* PIERSIG and *Sperchon* KRAMER. Ann. Zool., Warsaw 38 (7): 159-171.

BIESIADKA, E. & CICHOCKA, M. (1990): Studies on the morphology of larval stages of water mites (Hydracarina). 2. Some species of the superfamily Hydryphantoidea. Annales Zoologici 43: 461-492.

BOEHLE, W.R. (1996): Contribution to the morphology and biology of larval *Panisellus thienemanni* (VIETS, 1920) (Acari: Parasitengonae: Hydrachnidia). Acarologia 37: 121-125.

BOTOŞĂNEANU, L. & NEGREA, S. (1961): Une oasis aquatique à faune relique dans la plaine du Danube inférieur. Hydrobiologia 18: 199-218.

BÖTTGER, K. (1972c): Wassermilben in der Mantelhöhle von Sußwasserschnecken. Zool. Anz. 189 (3-4): 233-235.

BOYACI, Y. (2006): New records of water mites (Acari, Hydrachnidia, Torrenticolidae) from Turkey. Zoology in Middle East 38: 115-117.

CICOLANI, B. & DI SABATINO, A. (1990): Recherches faunistiques et écologiques sur les Hydracariens de l'Apennin (Italie). Le genre *Torrenticola* PIERSIG (Acari, Actinedida, Torrenticolidae). Annls. Limnol. 26 (2-3): 153-176.

CICOLANI, B., DI SABATINO A., MICCOLI F.P., GIUSTINI, M. & FERRARESE, U. (2001): Parassitismo larvale di Acari acquatici (Acari: Actinedida: Hydrachnidia) su ditteri: osservazioni preliminari in una sorgente del Gran Sasso. Monitoraggio Biologico del Gran Sasso 2: 108-118.

Cîmpean, M. & Gerecke, R. (2006): Water mites (Acari: Hydrachnidia) from the Retezat National Park (Romania). pp. 63-74 in: Bănăduc, D., Sîrbu, I. & Curtean, A. (eds): The Retezat National Park. Transylvanian Review of Systematical and Ecological Research 3: 1-204.

Cook, D.R. (1966): The water mites of Liberia. Mem. Amer. Ent. Inst. 6: III + 1-418.

Cook, D.R. (1967): Water mites from India. Mem. Amer. Ent. Inst. 9: III + 1-411.

Cook, D.R. (1974): Water mite genera and subgenera. Mem. Amer. Ent. Inst. 21: VII + 1-860.

Cook, D.R. (1986): Water mites from Australia. Mem. Amer. Ent. Inst. 40: IV + 1-568.

Cook, D.R. (1992): Water mites (Hydracarina), mostly from driven wells in New Zealand: taxa other than the Notoaturinae Besch. Stygologia 7: 43-62.

Cramer, C. & Smith, I.M. (1993): A new species of the genus *Mamersellides* Lundblad, 1937 (Acari: Hydrachnida) with remarks on the family Anisitsiellidae. Can. Ent. 125: 769-783.

Crema, S., Ferrarese, U., Golo, D., Modena, P., Sambugar, B. & Gerecke R. (1996): Ricerche sulla fauna bentonica ed interstiziale di ambienti sorgentizi in area alpina e prealpina. Centro di Ecologia Alpina, Viote-Monte Bondone, Report 8: 1-104.

Crowell, R.M. (1952): New records of rheophilic water-mites from northwestern North Carolina. Journ. Elisha Mitchell Sci. Soc. 68 (2): 191-194.

Daday, E. von (1898a): Édesvizi mikroskopi állatok Ceylonból. Math. Természett. Értesitö, Budapest 16 (1): 89-100.

Davids, C. (1979): [De watermijten (Hydrachnellae) van Nederland. Levenswijze en voorkomen]. Wetensch. Meded. K.N.N.V. 132: 1-78. (in Dutch)

Di Sabatino, A. & Cicolani, B. (1990): *Torrenticola meridionalis* n.sp., a new species of water mites (Acari, Actinedida, Torrenticolidae) from the Calabrian Apennines (South Italy). Lauterbornia 5: 43-48.

Di Sabatino, A. & Cicolani, B. (1993): On the presence of the family Torrenticolidae Piersig (Acari, Hydrachnidia) in interstitial waters of Sicily (South Italy): description of a new species. Annls. Limnol. 29 (1): 31-39

Di Sabatino, A. & Gerecke, R. (1996): An interesting new water mite (Acari, Actinedida, Torrenticolidae) from springs and interstitial waters of Corsica and Sardinia. Aquatic Insects 18 (3): 185-191.

Di Sabatino, A., Gerecke, R. & Cicolani, B. (1992): The water mites of the family Torrenticolidae Piersig, 1902 (Acari, Actinedida) in springs and running waters of Sicily (South Italy). Arch. Hydrobiol., Suppl. 90 (2): 253-282.

Di Sabatino, A., Gerecke, R., Smit, H., Pesic, V. & Panesar, A. (2003): Water mites of the family Torrenticolidae (Acari, Actinedida, Hydrachnidia) from the Eastern Mediterranean. Arch. Hydrobiol., Suppl. 139: 393-431.

Di Sabatino, A., Gerecke, R., Gledhill, T. & Smit, H. (2009): On the taxonomy of water mites (Acari: Hydrachnidia) described from the Palaearctic, part 2: Hydryphantoidea and Lebertioidea. Zootaxa 2266: 1-34.

Di Sabatino, A., Gerecke, R., Gledhill, T. & Smit, H. (2010): The taxonomic status of the water mite genera *Todothyas* Cook and Parathyas Lundblad – supplement to Di Sabatino *et al.* (2009). Zootaxa 2361: 68

Dittmar, H. (1955): Ein Sauerlandbach. Arch. Hydrobiol. 50 (3-4): 305-552.

Dugès, A.-L.D. (1834b): Deuxième mémoire sur l'ordre des Acariens. Recherches sur l'ordre des Acariens. Rémarques sur la famille des Hydracnés. Annales des sciences naturelles, Paris (série 2), Zoologique 1: 144-174.

Efford, I.E. (1963): The parasitic ecology of some watermites. J. Anim. Ecol. 32: 141-156.

Gerecke, R. (1988): *Limnolegeria longiseta* Motaş (Acari, Actinedida): Zur Morphologie, systematischen Stellung und Ökologie einer wenig bekannten Wassermilbe, mit Erstbeschreibung der Nymphe. Arch. Hydrobiol. 112: 611-626.

Gerecke, R. (1991a): Taxonomische, faunistische und ökologische Untersuchungen an Wassermilben (Acari, Actinedida) aus Sizilien unter Berücksichtigung anderer aquatischer Invertebraten. Lauterbornia 7: 1-304.

Gerecke, R. (1991b): Über drie Arten der Gattung *Bandakia* Thor 1913 aus Mittel- und Südeuropa. Senckenbergiana biol. 71 (1-3): 139-153.

Gerecke, R. (1992): A remarkable new *Nilotonia*-species from the Pollino Mountains (Southern Italy). Spixiana 15: 97-105.

Gerecke, R. (1993): The water mites of the genus *Partnunia* Piersig, 1896 (Acari, Actinedida, Hydryphantidae). Entom. Basiliensia 15: 13-44.

Gerecke, R. (1994): New records of water mites from springs and running waters in the Mediterranean region (Acari, Actinedida: Anisitsiellidae, Aturidae, Momoniidae). Annls. Limnol. 30: 267-284.

Gerecke, R. (1996a): Untersuchungen über Wassermilben der Familie Hydryphantidae (Acari, Actinedida) in der Westpalaearktis. I. Beitrag zur Kenntnis der Gattung *Protzia* Piersig, 1896 (Acari, Actinedida, Hydryphantidae). Arch. Hydrobiol., Suppl. 77: (3-4): 271-336.

Gerecke, R. (1996b): Untersuchungen über Wassermilben der Familie Hydryphantidae (Acari, Actinedida) in der Westpalaearktis. II. Die Wassermilben der Familie Hydryphantidae Piersig, 1896 in den Mittelmeerländern. Arch. Hydrobiol. Suppl. 77: (3-4): 337-513.

Gerecke, R. (1999): Further studies on hydryphantoid water mites in the W Palaearctic region (Acari, Actinedida). Arch. Hydrobiol., Suppl. 121: 119-158.

Gerecke, R. (2002): The water mites (Acari, Hydrachnidia) of a little disturbed forest stream in southwest Germany - a study on seasonality and habitat preference, with remarks on diversity patterns in different geographical areas. In: Bernini, F., Nannelli, R., Nuzzaci, G. & De Lillo, E. (eds): Acarid phylogeny and evolution. Adaptations in mites and ticks: 69-89. Kluwer Academic Publishers.

Gerecke, R. (2004a): Taxonomy and Phylogeny in African water mites of the genus *Diplodontus* Dugès, 1834 (Acari, Hydrachnidia, Hydryphantidae). Ann. Limnol. – Int. J. Lim. 40 (1): 71-85.

GERECKE, R. (2004b): The water mites of Madagascar (Acari, Hydrachnidia): a revised list completed by original material conserved at the Muséum national d'Histoire naturelle Paris. Zoosystema 26 (3): 393-418.

GERECKE, R. (2006): Wassermilben. pp. 58-61. In: Regierung von Schwaben (Hrsg.): Das Benninger Ried. Insel der Vielfalt: 1-104. Augsburg.

GERECKE, R. (2008): A new species and subgenus of *Lebertia* (Acari: Hydrachnidia: Lebertiidae) from the Brenta-Adamello Natural Park (Italian Alps). Fund. Appl. Limnol. – Arch. Hydrobiol. 170: 325-332.

GERECKE, R. (2009): Revisional studies on the European species of the water mite genus *Lebertia* Neuman, 1880 (Acari: Hydrachnidia: Lebertiidae). Abhandlungen der Senckenberg Gesellschaft für Naturforschung 566: 1-144.

GERECKE, R. & DI SABATINO, A. (1996a): Water mites (Acari, Hydrachnellae) and spring typology in Sicily. Crunoecia 5: 35-41.

GERECKE, R. & DI SABATINO, A. (1996b): The water mites of the family Torrenticolidae PIERSIG (Acari, Actinedida, Torrenticolidae) in springs and running waters of Corsica and Sardinia. Arch. Hydrobiol. Suppl. 107 (3): 287-334.

GERECKE, R., JÄNTSCH, W. & SCHREIBER, U. (2005a): Neue Beiträge zur Kenntnis der Wassermilbenfauna (Acari: Hydrachnidia et Halacaridae) Sachsens. Lauterbornia, 55: 65-77.

GERECKE, R. & MARTIN, P. (2006): Spinnentiere: Milben (Chelicerata: Acari). In: GERECKE, R. & FRANZ, H. (eds): Quellen im Nationalpark Berchtesgaden. Lebensgemeinschaften als Indikatoren des Klimawandels. Nationalpark Berchtesgaden, Forschungsbericht 51: 122-149.

GERECKE, R. & MORENO, J.L. (2008): Contribution to the knowledge of the rare water mites *Nilotonia borneri* WALTER, 1922 and *Nilotonia parva* WALTER, 1931 (Acari, Hydrachnidia, Anisitsiellidae). Lauterbornia 63: 23-30.

GERECKE, R. & SCHWOERBEL, J. (1992): Water Quality and Water Mites in the upper Danube region. 1959 - 1984. In: DUSBÁBEK, F. & BUKVA, V. (eds): Modern Acarology, Vol. 1 (8th Int. Congr. Acarology, Budejovice): 483-491; Academia (Prague) & SPB Academic (The Hague).

GERECKE, R. & SMIT, H. (1994): *Partnunia tabernaculata*, spec. nov., a new water mite from the Spanish Pyrenees. Spixiana 17: 202-206.

GERECKE, R. & SMITH, I.M. (1993): Description of larvae of *Nilotonia longipora* (WALTER) (Acari: Hydrachnida) with remarks on the relationships and classification of *Nilotonia* THOR, 1905. Internat. J. Acarol. 19 (3): 259-266.

GERECKE, R. & SMITH, I.M. (1996): On the biology of the spring-dwelling water mite *Nilotonia longipora* (WALTER, 1925) (Acari, Actinedida, Anisitsiellidae). In: KROPCZYNSKA, D., BOCZEK, J. & TOMCZYK, A. (eds), The Acari - Physiological and Ecological Aspects of Acari-Host Relationships: 377-383. Dabor, Warsaw.

GERECKE, R., STOCH, F., MEISCH, C. & SCHRANKEL, I. (2005b): Die Fauna der Quellen und des hyporheischen Interstitials in Luxemburg. Unter besonderer Berücksichtigung der Acari, Ostracoda und Copepoda Ferrantia 41: 1-134.

GERECKE, R. & TUZOVSKIJ, P. (2001): The water mite *Rutripalpus limicola* SOKOLOW, 1934: New data on morphology and biology, and considerations on the systematic position of the monotypic family Rutripalpidae (Acari, Hydrachnidia). J. Natur. Hist. 35 (6): 931-944.

GEORGE, C.F. (1901d): British freshwater mites. Sci.-Gossip (n.s.) 8 (86): 44-46.

GLEDHILL, T. & VIETS, K.O. (1976): A synonymic and bibliographic check-list of the freshwater mites (Hydrachnellae and Limnohalacaridae, Acari) recorded from Great Britain and Ireland. Freshwat. Biol. Assoc. Occasional Publ. 1: 1-59.

GLEDHILL, T., COWLEY, J. & GUNN, R.J.M. (1982): Some aspects of the host: parasite relationships between adult blackflies (Diptera: Simuliidae) and larvae of the water mite *Sperchon setiger* (Acari: Hydrachnellae) in a small chalk stream in southern England. Freshw. Biol. 12: 345-357.

GOLDSCHMIDT, T. (2001): Die Wassermilbenfauna Costa Ricas (Hydrachnidia, Acari). Systematik, Ökologie, Diversität und Zoogeographie. Thesis, Wilhelms-Universität Münster, 425 pp.

GOLDSCHMIDT, T. (2004b): Studies on Neotropical Limnesiidae THOR, 1900. Part II. Tyrelliinae KOENIKE, 1910 (Acari, Actinedida, Hydrachnidia). Arch. Hydrobiol. Suppl. 151 (1-2): 25-68.

GOLDSCHMIDT, T. (2007): Studies on Latin American water mites of the genus *Torrenticola* PIERSIG, 1896 (Torrenticolidae, Hydrachnidia, Acari). Zoological Journal of the Linnean Society 150(3): 443-678.

GOLDSCHMIDT, T. & GERECKE, R. (2003): Studies on hydryphantid water mites (Acari: Actinedida: Hydrachnidia) from Central and South America. In: SMITH, I.M. (ed.), An Acarological Tribute to David R. COOK (From Yankee Springs to Wheeny Creek): 83-150. Indira Publishing House, W. Bloomfield.

HABEEB, H. (1958): North American Hydrachnellae. XLVIII-L. XLVIII. *Thyasides sphagnorum*, new species. XLIX. On our eastern forms of *Mideopsis*. L. Notes on *Wandesia*. Leaflets Acadian Biol. 17: 1-2.

HABEEB, H. (1958a): New mites from New Brunswick. Leaflets Acadian Biol. 18: 1-4.

HABEEB, H. (1963c): *Estelloxus*, a new genus in Lebertiidae. Leaflets Acadian Biol. 32: 1-2.

HALBERT, J.N. (1911): Acarina: I-Hydracarina. Clare Island Survey, part 39, I. Proc. Irish Acad., Dublin & London 31 (Sect. 2, part 39 i): 1-44.

HALBERT, J.N. (1944): List of Irish fresh-water mites (Hydracarina). Proc. Royal Irish Acad., 50, B (4): 39-104.

HARVEY, M.S. (1987): New and little-known species of the water mite genera *Tartarothyas*, *Pseudohydryphantes* and *Cyclohydryphantes* from Australia (Chelicerata: Actinedida: Hydryphantidae). Mem. Mus. Victoria 48: 107-122.

HARVEY, M.S. (1998): The Australian water mites. A guide to Families and Genera. Monographs on Invertebrate Taxonomy 4. CSIRO Publishing, Collingwood, 150 pp.

HARVEY, M.S. (1990): A review of the water mite family Anisitsiellidae in Australia (Acarina). Invertebr. Taxon. 3: 629-646.

HEUSS, K. (1966): Beitrag zur Fauna der Werra, einem salinaren Binnengewässer. Gewässer u. Abwässer 43: 48-64.

HUBAULT, E. (1927): Contribution à l'étude des invertébrés torrenticoles. Bull. Biol. France & Belg, Paris, Suppl. 9: 1-388.

HUSIATINSCHI, A. (1937d): *Distolebertia bucovinensis* nov. subgen. und fünf andere neue Lebertien aus dem Hochmoorgebiet Mihodra in der Bukowina, Rumänien. Zool. Anz. 118 (5-6): 154-166.

IMAMURA, T. (1950a): On a new species of water mite, *Partnunia uchidai* n.sp. from Hokkaido. Journ. Fac. Sci. Hokkaido Univers., Sapporo (s. 6), Zool. 10 (1): 33-37.

IMAMURA, T. (1951b): [On a water mite *Hydryphantes* sp. parasitic on a mosquito]. Kagaku. 21 (5): 250-251. (in Japanese)

IMAMURA, T. (1953b): Some water-mites from Hiroshima Prefecture. Journ. Fac. Sci. Hokkaido Univers., Sapporo (s. 6), Zool. 11 (2): 193-260.

IMAMURA, T. (1954b): Studies on water mites from Hokkaido. Journ. Hokkaido Gakugei Univers., Sapporo, Sect. B. Suppl. 1: 1-148.

JONES, R.K.H. (1967): Descriptions of the larvae of *Aturus scaber* KRAMER, *Protzia eximia* PROTZ and *Piona uncata* KOENIKE with notes on the life-histories of the latter two. Annls. Limnol. 3 (2): 231-247.

KOCH, C.L. (1842): Übersicht des Arachnidensystems, Vol. 3. Zeh, Regensburg.

KOENIKE, F. (1885a): Einige neubenannte Hydrachniden. Abh. naturw. Ver. Bremen 9: 215-223.

KOENIKE, F. (1889a): Kurze Mittheilung über ein neues Hydrachniden-Genus. Zool. Anz. 12 (300): 103-104.

KOENIKE, F. (1891b): Nomenclatorische Correctur innerhalb der Hydrachniden-Familie. Zool. Anz. 14 (354): 19-20.

KOENIKE, F. (1892b): Zwei neue Hydrachniden-Gattungen aus dem Rhätikon. Zool. Anz. 15 (399): 320-324.

KOENIKE, F. (1893b): Noch eine neue Hydrachnide aus dem Rhätikon. Zool. Anz. 16 (414): 93-96.

KOENIKE, F. (1896b): Zwei neue Hydrachniden-Gattungen nebst sechs unbekannten Arten (Vorläufige Mittheilung). Zool. Anz. 19 (510): 356-361.

KOENIKE, F. (1898b): Über *Oxus* KRAM., *Frontipoda* KOEN., und eine neue verwandte Gattung. (Vorläufige Mittheilung). Zool. Anz. 21 (556): 262-263, (557): 265-273.

KOENIKE, F. (1908a): Beitrag zur Kenntnis der Hydrachniden. Abh. naturw. Ver. Bremen 19 (2): 217-266.

KOENIKE, F. (1909c): Acarina, Milben. In: BRAUER, A., Die Süßwasserfauna Deutschlands. Jena, G. FISCHER 12:13-184.

KOENIKE, F. (1911b): Ein Fall von Sexualdimorphismus bei einer Oribatide. Abh. naturw. Ver. Bremen 20 (2): 233-235.

KOENIKE, F. (1914f): Bewohnt *Lebertia tauinsignata* (LEB.) Sig THOR in der Tat den Vierwaldstätter- und Bodensee? Zool. Anz. 45 (1): 14-16.

KOWALIK, W. (2002): The occurrence of water mites (Hydrachnidia, Acari) in saline waters from a stone coal-mine in Bogdanka (South-Eastern Poland). In: BERNINI, F., NANNELLI, R., NUZZACI, G. & DE LILLO, E. (eds), Acarid phylogeny and evolution. Adaptations in mites and ticks: 119-124. Kluwer Academic Publishers.

KRAMER, P. (1879a): Neue Acariden. Arch. Naturg. 45, I: 1-18.

LANGE, J. (1990): Lebensgemeinschaften (Merozönosen) auf der Stromsohle des südlichen Oberrheins und eine erste Bestandsaufnahme der Wassermilbenfauna (Hydracarina) im Rhein. – Albert-Ludwigs-Universität Freiburg, Faculty of Biology, diploma thesis: 1-99.

LÁSKA, F. (1953a): Einige neue Wassermilben aus dem Flußgebiete Orava in der Slowakei. Acta Acad. scient. natur. Moravo-Silesiacae 25 (9) : 281-296.

LÁSKA, F. (1955c): Nový druh rodu Lebertia Neuman (Hydrachnellae, Acari) ze slezských potoku. Přrodoved. Sborn. Ostrav. kraje, 16 (1): 46-50.

LÁSKA, F. (1957): [Beitrag zur Kenntnis der Wassermilben (Hydrachnellae) aus der Hohen Tatra. I. Die Art *Wandesia thori* SCHECHTEL, 1912]. Biológia 12 (4): 292-297. (in Czech)

LÁSKA, F. (1966): Vodule slezské oblasti (Hydrachnellae, Acari). (Wassermilben aus dem schlesischen Verwaltungsgebiete). Acta Mus. Silesiae, (A) 15: 55-141.

LEBERT, H. (1879): Description de quelques espèces nouvelles d'Hydrachnides du Lac Léman. - Mémoire posthume. In: FOREL, Matériaux. etc. (s.6, §XLIX).

LESAGE, L. & HARRISON, A.D. (1980): The biology of *Cricotopus* (Chironomidae: Orthocladiinae) in an algal-enriched stream: Part II. Effects of parasitism. Arch. Hydrobiol., Suppl. 58 (1-2): 1-25.

LETTEVALL, U. (1962): On the Hydracarina of Greenland with a description of *Lebertia (Pseudolebertia) groenlandica* n.sp. Meddelelser om Grönland 170 (1): 1-40.

LUCAS, H. (1844): Diplodonte, Eylaides, *Eylais*. In: D'ORBIGNY, C., Dictionnaire Universel d'Histoire Naturelle, 5: 52-53.

LUNDBLAD, O. (1913c): Zur Kenntnis der Larve von *Thyas dentata*. Ent. Tidskr. 34: 46-50.

LUNDBLAD, O. (1924b): Neue Hydracarinen aus Schweden. Vorläufige Mitteilung. Ent. Tidskr. 45 (1): 67-72.

LUNDBLAD, O. (1924c): Ein unerwartetes, interessantes Verhältnis in der Epimorphose einer schwedischen *Hexalebertia*. Ent. Tidskr. 45 (2): 72-79.

LUNDBLAD, O. (1927c): Die Hydracarinen Schwedens. I. Beitrag zur Systematik, Embryologie, Ökologie und Verbreitungsgeschichte der schwedischen Arten. Zool. Bidrag. 11: 181-540.

LUNDBLAD, O. (1930a): Hydracarina. Zoology of the Faroes, Copenhagen 2 (48): 1-65.

LUNDBLAD, O. (1930c): Die Hydracarinen der Insel Bornholm. Dansk. Vidensk. Selsk. biol. Meddel., Kobenhavn 8 (7): 1-96.

LUNDBLAD, O. (1933b): Zur Kenntnis von *Lundbladia petrophila* (MICHAEL) und der verschiedenen Entwicklungsrichtungen bei Thyasinen. Zool. Bidr., Uppsala 14: 219-252.

LUNDBLAD, O. (1934a): Report on Hydracarina. Mem. Connecticut Acad. Arts Sci., New Haven 10: 85-118.

LUNDBLAD, O. (1941c): Neue Wassermilben. Vorläufige Mitteilung. Ent. Tidskr. 62 (1-2): 97-121.

LUNDBLAD, O. (1942c): Die Arthropodenfauna von Madeira nach den Ergebnissen der Reise von Prof. Dr. O. LUNDBLAD Juli-August 1935. 31. Hydrachnellae. Ark. Zool. 34, A (5): 1-122.

LUNDBLAD, O. (1942d): Afrikanische Hydracarinen. Ent. Tidskr. 63 (3-4): 155-209.

LUNDBLAD, O. (1949): Hydrachnellae. Explorat. Park National Albert. 2. Miss. DAMAS. Bruxelles, 18: 1-87.

LUNDBLAD, O. (1954b): Neue südeuropäische Wassermilben. Zool. Anz. 153 (7-8): 154-159.

LUNDBLAD, O. (1956a): Zur Kenntnis süd- und mitteleuropäischer Hydrachnellen. Ark. Zool. 10 (1): 1-306.

LUNDBLAD, O. (1962): Die Hydracarinen Schwedens. II. Ark. Zool. (2) 14 (1): 1-635.

LUNDBLAD, O. (1968): Die Hydracarinen Schwedens. III. Ark. Zool. (2) 21 (1): 1-633.

LUNDBLAD, O. (1969): Indische Wassermilben, hauptsächlich von Hinterindien. Ark. Zool. 22 (10): 289-443.

MAGLIO, C. (1906): Idracnidi nuovi o poco noti dell'Italia superiore. Zool. Anz. 30: 406-413.

MAGLIO, C. (1909): Idracarini del Trentino. (Contributo alla conoscenza dell'idracnofauna alpina.). Atti soc. ital. Sci. natur., Pavia 48: 251-296.

MARSHALL, R. (1924c): Water mites of Alaska and the Canadian Northwest. Trans. Americ. microsc. Soc. 43 (4): 236-255.

MARSHALL, R. (1929c): Canadian Hydracarina. Univ. Toronto Stud. 39: 57-93.

MARTIN, P. (1997): Zur Faunistik und Substratpräferenz der Wassermilben (Hydrachnidia, Acari) zweier durch feine mineralische Substrate geprägten Bäche in Schleswig-Holstein. Faun.-Ökol. Mitt. 7: 221-237.

MARTIN, P. (2000): Larval morphology and host-parasite associations of some stream living water mites (Hydrachnidia, Acari). Arch. Hydrob., Suppl. 121: 269-320.

MARTIN, P. (2003): Larval morphology of spring-living water mites (Hydrachnidia, Acari) from the Alps. Ann. Limnol. - Int. J. Lim. 39: 363-393.

MARTIN, P. (2005): Water mites (Hydrachnidia, Acari) as predators in lotic environments. In: WEIGMANN, P., ALBERTI, G., WOHLTMANN, A. & RAGUSA, S. (eds): Acarine Biodiversity in the Natural and Human Sphere. Proceedings of the V Symposium of the European Association of Acarologists. Phytophaga 14: 307-321.

MARTIN, P. (2006): On the morphology and classification of larval water mites (Hydrachnidia, Acari) from springs in Luxembourg. Zootaxa 1138: 1-44.

MARTIN, P. (2009): A redescription of *Torrenticola* (*Megapalpis*) *fagei* (E. ANGELIER, 1949) (Acari, Hydrachnidia). Lauterbornia 67: 1-4.

MARTIN, P. & SPETH, S. (1996): Die substratspezifische Verteilung der Hydrachnidia (Acari) zweier Bäche in jungglazialen Sandergebieten Schleswig-Holsteins (Osterau und Rodenbek). Limnologica 26 (1): 93-103.

MARTIN, P. & STUR, E. (2005): Are the attachment sites of larval water mites (Acari, Hydrachnidia) on their dipteran hosts suited for host partitioning? Arch. Hydrobiol. 164: 89-107.

MARTIN, P. & STUR, E. (2006): Parasite-host associations and life cycles of spring-living water mites (Hydrachnidia, Acari) from Luxembourg. Hydrobiologia 573: 17-37.

MARTIN, P., STUR, E. & WIEDENBRUG, S. (2010): Larval parasitism of spring-dwelling alpine water mites (Hydrachnidia, Acari): a study with particular reference to chironomid hosts. Aquatic Ecology DOI 10.1007/s10452-009-9301-4.

MEYER, E. (1985): Der Entwicklungszyklus von *Hydrodroma despiciens* (O.F. MÜLLER, 1776) (Acari: Hydrodromidae). Arch. Hydrobiol., Suppl. 66 (3): 321-453.

MEYER, E. & SCHWOERBEL, J. (1981): Untersuchungen zur Phänologie der Wassermilben (Hydracarina) des Mindelsees. Arch. Hydrobiol., Suppl. 59: 192-251.

MIGOT, A. (1926a): Diagnoses sommaires d'Hydracariens nouveaux des Pyrénées. Bull. Soc. Entomol. France, 41: 52-56.

MIGOT, A. (1926b): Sur la faune française des hydracarides. Bull. Soc. Zool. France 51 (2): 91-134.

MIGOT, A. (1926c): Le premier Halacaride d'eau douce de la faune française, *Walterella gallica* n. sp. Bull. Soc. Zool. France, 51 (5): 376-383.

MITCHELL, R. (1953): A new species of *Lundbladia* and remarks on the family Hydryphantidae (Water mites). Amer. Midland Naturalist. 49 (1): 159-170.

MITCHELL, R. (2003): The vernal pond mite fauna of Blacklick Woods Metro Park, Columbus, Ohio. Part I. In: SMITH, I. M. (ed.), An Acarological Tribute to David R. COOK (From Yankee Springs to Wheeny Creek: 183-199. Indira Publishing House, W Bloomfield).

MONTI, R. (1904a): Di un'altra nuova specie di "*Lebertia*" e di alcune idracne nuove per la fauna italiana. Rend. Ist. Lombard. Sci. Lett., (s.2) 37: 1-14.

MORENO, J.L., GERECKE, R. & TUZOVSKIJ, P. (2004): The larval stage and life cycle of *Diplodontus scapularis* DUGÈS, 1834, and *D. semiperforatus* WALTER, 1925 (Acari: Hydrachnidia), with remarks on their habitat preference and distribution in the south west Palaearctic region. Aquatic Insects 26: 199-206.

MORENO, J.L., GERECKE, R. & TUZOVSKIJ, P. (2008): Biology and taxonomic position of an ovovivaparous water mite (Acari: Hydrachnidia) from a hypersaline spring in southern Spain. Aquatic Insects 30: 307-317.

MOTAŞ, C. (1928f): Contribution à la connaissance des Hydracariens français, particulièrement du Sud-Est de la France. Trav. Labor. Piscicult. Univ. Grenoble 20: 1-373.
MOTAŞ, C. (1928g): Considérations générales sur la répartition des hydracariens dans les eaux stagnantes et courantes et sur leur distribution géographique particulierèment dans le sud-est de la France. Bulletin de la Société du Dauphiné 49: 1-65.
MOTAŞ, C. (1929): Sur le développement postembryonnaire de *Calonyx brevipalpis* (MAGLIO) et sur les affinités des Protziidae. Ann. Scientif. Univ. Jassy 16 (1-2) : 250-262.
MOTAŞ, C. (1959): [Description de trois Hydracriens nouveaux]. Acad. Republ. Pop. Rom., Omagiu lui Traian SĂVULESCU cu prilejul împlinirii a 70 de ani : 473-498. (in Romanian)
MOTAŞ, C. (1961): Hydrachnellae. The Zoology of Iceland 3 (56): 1-26. Munksgaard, Copenhagen/Reykjavik.
MOTAŞ, C. & ANGELIER, C. (1927d): Hydracariens recueillis dans le Massif Central. Trav. Labor. Piscicult. Univ. Grenoble 19: 121-137.
MOTAŞ, C. & ŞOAREC, J. (1939a): Sur deux nouveaux Hydracariens français recueillis dans les Pyrénées. Ann. Sci. Univ. Jassy (2. Partie) 15 (2): 1-13.
MOTAŞ, C. & TANASACHI, J. (1944a): Hydracariens des Carpates Roumaines (Massif de Bucegi et Alpes de Transylvanie). Anal. Acad. Romane, Mem. seçt. stiinţ. (s. 3) 19 (7): 241-256.
MOTAŞ, C. & TANASACHI, J. (1963c): Hydrachnellae freatice din bazinul Vîrghisului. (Hydrachnellae phréatiques du bassin du Vîrghis Région de Mures-Autonome Hongroise). Lucrarile Inst. Speol. "Emile Racovitza", 1-2 (1962-1963): 311-340.
MUCKLE, R. (1942): Beiträge zur Kenntnis der Uferfauna des Bodensees. Beitr. naturkundl. Forsch. Oberrheingeb., Karlsruhe 7: 1-109.
MULLEN, G.R. (1974): Acarine parasites of mosquitos. II. Illustrated larval key to the families and genera of mites reportedly parsasitic on mosquitos. Mosquito News 34 (2): 183-195.
MULLEN, G.R. (1977): Acarine parasites of mosquitos. IV. Taxonomy, life history and behavior of *Thyas barbigera* and *Thyasides sphagnorum* (Hydrachnellae: Thyasidae). J. Med. Ent. 13 (4-5): 475-485.
MULLEN, G.R. (1979): Aquatic mites parasitic on stoneflies in North America. Recent Advances Acarology 2: 481-484.
MULLEN, G.R. & VERCAMMEN-GRANDJEAN, P.H. (1980): The larval stage of Stygothrombiidae THOR, 1935 and Wandesiinae SCHWOERBEL, 1961 and erection of a new superfamily, Stygothrombioïdea. Int. J.Acarol. 6 (1): 25-28.
MÜNCHBERG, P. (1935a): Eine *Arrhenurus robustus* KOENIKE nahestehende neue Milbenart aus Nord-italien. Zool. Anz. 110 (3-4): 66-71.
MÜNCHBERG, P. (1935b): Über die bisher bei einigen Nematocerenfamilien (Culicidae, Chironomidae, Tipulidae) beobachten ektoparasitären Hydracarinenlarven. Z. Morph. Ökol. Tiere 29 (5): 720-749.
MÜNCHBERG, P. (1936a): Zur Morphologie der *Arrhenurus*- und *Georgella*-Larven und – Nymphen, mit besonderer Berücksichtigung der Libellenparasiten. Arch. Naturg. (N.F.) 5 (1): 93-115.
MÜNCHBERG, P. (1936b): Zur Kenntnis der Larvenparasitismus der Thyasinae (Hydracarina), zugleich ein Beitrag über Schmarotzer der Limnobiinae (Diptera). Intern. Rev. Hydrobiol. Hydrographie 33: 313-326.
MÜNCHBERG, P. (1956): Über die an Limoniidae und Tipulidae parasitierenden Milbenlarven. Z. Parasitenkde 17: 330-338.
NEUMAN, C.J. (1880b): Om Sveriges Hydrachnider Svensk. Vetensk.-Akad. Handlingar, Stockholm (N.F.), 1879 17 (3): 1-123.
NEWELL, I.M. (1959): Acari. pp. 1080-1116. In: EDMONDSON, W.T. (ed.), Ward & Whipple, Freshwater Biology, 2nd edition.
NIELSEN, B.O. (1998): The water mite *Thyas barbigera* VIETS (Hydrachnellae: Thyasidae) parasitizing mosquitoes. European Mosquito Bulletin 2: 10-12.
NOCENTINI, A.M. (1960a): Hydrachnellae del Lago di Mergozzo. Mem. Ist. Ital. Idrobiol. 12: 245-287.
NORDENSKIÖLD, E. (1905): Hydrachniden aus dem Sudan. In: JÄGERSKIÖLD, A. (ed.), Results Swedish zool. Exped. Egypt and White Nile 1901, 2: (20A): 1-12.
ÖZDIKMEN, H. (2006): Necessary changes of names in Hydrachnellae (Arachnida: Acari). Munis Entomology & Zoology 1: 51-54.
ÖZKAN, M. (1981): [A systematic study on *Georgella*-species (Acari: Hydrachnellae)]. Atatürk Üniversitesi Fen Fakültesi Fen Dergisi Cilt 1 (2): 1-8. (in Turkish)
ÖZKAN, M. (1982): Wassermilben (Acari, Actinedida) aus der Türkei. Entomologia Basiliensia 7: 29-60
ÖZKAN, M. (1988): *Hydryphantes* (s.str.) *crassipalpis* KOENIKE, 1914 (Hydryphantidae, Hydrachnellae, Acari) üzerine bir araştirma. Doğa, Tr. J. of Zoology 12 (1): 86-109.
ÖZKAN, M. (1988a): *Tadjikothyas* cinsi (Hydrachnellae, Acari) ve türkiyeden yeni bir alttürü hakkinda. Doğa, Tr. J. of Zoology 12: 53-61
ÖZKAN, M. (1989): [Systematic studies on the water mites (Acari, Hydrachnellae) of eastern Turkey]. IV. Doga Türk Zooloji Dergisi 13 (2): 88-108. (in Turkish)
ÖZKAN, M. & BADER, C. (1988): Zwei neue Arten nebst Ergänzungen zur türkischen Wassermilben-Fauna (Acari, Actinedida, Hydrachnellae). Arch. Hydrobiol. 114: 133-145.
ÖZKAN, M. & BOYACI, Y.Ö. (1990): Dogu Anadolu su Akarlari (Acari, Hydrachnellae) üzerine sistematik Arastirmalar V.X. Ulussi Biyoloji Kongresi 18-20 Temmuz 1990, Erzurum: 63-75.

ÖZKAN, M. & SOYSAL, Z. (1989): [A new species of genus *Nilotonia* from Turkey]. Doğa, Tr. J. of Zoology 13: 15-27. (in Turkish)

ÖZKAN, M., ERMAN, O. & BOYACI, Y.O. (1995): [Some New Water Mite Species (Acari, Hydrachnellae) for the Turkish Fauna from the Sultan Sazligi (Kayseri) - II.]. Tr. J. Zoology 19: 77-118. (in Turkish)

ONRAT, S.T., AŞÇI, F. & ÖZKAN, M. (2006): A cytogenetics study of *Hydrodroma despiciens* (Müller, 1776) (Acari: Hydrachnellae: Hydrodromidae). Genetics and Molecular Research 5(2): 342-349.

PANESAR, A.R. (2004): Evolution in water mites (Hydrachnellae, Actinedida, Acari). A revision of the Anisitsiellidae KOENIKE, 1910. Bonner Zoologische Monographien 52: 1-144.

PEŠIĆ, V. (2002a): Water mites (Acari, Actinedida) of the stagnant waters from the Skadar Lake drainage basin (Crna Gora, Yugoslavia). Acta entomologica serbica 5: 131-152.

PESIC, V. (2002e): *Hydrodroma reinhardi* sp.n., a new species of water mites (Acari, Actinedida, Hydrodromidae) from the Mediterranean area. Aquatic Insects 23 (4): 317-323.

PEŠIĆ, V. (2005): New records of water mites (Acari: Hydrachnidia) from springs and running waters in Macedonia. Arch. Biol. Sci., Belgrade 57: 23-24.

PESIC, V., GERECKE, R. & SMIT, H. (2009): A redefinition of *Iranothyas* BADER, 1984 with the description of a new species from Oman. Zootaxa 2290: 59-64.

PESIC, V. & SMIT, H. (2007a): Water mite species of the genus *Hydrodroma* KOCH (Acari: Hydrachnidia, Hydrodromidae) from Australasia. Part I. Zootaxa 1389: 31-44.

PEŠIĆ, V. & SMIT, H. (2007b): Water mite species of the genus *Hydrodroma* KOCH (Acari: Hydrachnidia, Hydrodromidae) from Australia. Part II. Zootaxa 1509: 41-50.

PIERSIG, R. (1892a): Beitrag zur Hydrachnidenkunde. (Vorläufige Mittheilung). Zool. Anz. 15 (389): 151-155.

PIERSIG, R. (1896a): Einige neue Hydrachniden-Formen (Vorläufige Mittheilung). Zool. Anz. 19 (515): 438-441.

PIERSIG, R. (1896c): Beiträge zur Kenntnis der in Sachsen einheimischen Hydrachniden-Formen. S.B. naturf. Ges. Leipzig 22/23: 33-103.

PIERSIG, R. (1898b): Hydrachnidenformen aus den deutschen Mittelgebirgen. Zool. Anz. 21 (565): 451-452.

PIERSIG, R. (1896-1899): Deutschlands Hydrachniden. Zoologica, 1896-1899, Stuttgart, 19 (22): 7+1-601.

PIERSIG, R. (1901b): Acarina, Hydrachnidae. In: PIERSIG, R. & LOHMANN, A.: Acarina, Hydrachnidae und Halacaridae. Tierreich, Berlin (Friedländer & Sohn). 13. Lief.: 18+1-336.

PIERSIG, R. (1902): Arthropoda. Arachnida. Zool. Centrbl. 53: 849.

PÍSAŘOVIC, K. (1896): Zur Kenntnis der Hydrachniden Böhmens. (Ein Vorläufiges Verzeichnis). S.B. böhm. Ges. Wiss, Prag, math.-naturw. Cl. 17: 1-11.

PRASAD, V. & COOK, D.R. (1972): The taxonomy of water mite larvae. Mem. Amer. Entom. Inst. 18: I-II+326 pp.

PROTZ, A. (1896b): Beiträge zur Kenntnis der Wassermilben. Zool. Anz. 19 (513): 407-411.

PROTZ, A. (1923): Eine neue *Thyas*-Species. Zool. Anz. 55 (11-13): 237-239.

RACK, G. (1976): Milben (Acarina) von europäischen Limoniidae (Diptera, Nematocera). Mitt. Hamburg. Zool. Mus. Inst. 73: 63-85.

RACK, G. (1977): Weitere Untersuchungen über Milben (Acarina) an Nematocera (Diptera). Acarologia 19: 20-25.

RENZ, A., GERECKE, R. & MARTIN, P. (2004): Parasitic mites (Acari: Hydrachnidia) on pupae and adults of Simuliidae (Insecta: Diptera). Deutsche Gesellschaft für allgemeine und angewandte Entomologie, Nachrichten 18 (3), 121-122.

SCHECHTEL, E.P. (1912): [Eine neue Hydrachniden-Gattung aus der polnischen Tatra, *Wandesia* n.g.]. Bull. Acad. Sci. Cracowie. Cl. Sci. math.-natur., Sér. B, Sci. natur.: 463-468. (in Polish)

SCHWOERBEL, J. (1957a): Zur Kenntnis der Wassermilbenfauna des südlichen Schwarzwaldes (Hydrachnellae, Acari). 3. Beitrag. Mitt. Bad. Landesver. Naturkde. Naturschutz, (N.F.) 7 (1): 41-52.

SCHWOERBEL, J. (1958a): Neue Wassermilben (Acari, Hydrachnellae) aus dem Schwarzwald. Zool. Anz. 160 (9-10): 193-204.

SCHWOERBEL, J. (1959): Ökologische und tiergeographische Untersuchungen über die Milben (Acari, Hydrachnellae) der Quellen und Bäche des südlichen Schwarzwaldes und seiner Randgebiete. Arch. Hydrobiol., Suppl. 24, 3 (3-4): 385-546.

SCHWOERBEL, J. (1961a): Subterrane Wassermilben (Acari: Hydrachnellae, Porohalacaridae und Stygothrombiidae), ihre Ökologie und Bedeutung für die Abgrenzung eines aquatischen Lebensraums zwischen Oberfläche und Grundwasser. Arch. Hydrobiol., Suppl. 25 (2-3): 242-306.

SCHWOERBEL, J. (1963): Süßwassermilben aus Mazedonien. Acta Mus. Macedon. Sci. Nat. 9 (4): 51-75.

SEMUSHIN, R.D. (1981): [Water mites (Sperchontidae), parasites of black flies (Simuliidae)]. Parasitologiya 15 (1): 27-30. (in Russian)

SMIT, H. (1992): The water mites (Acari, Hydrachnellae) described by A.J. BESSELING from The Netherlands. Ent. Ber., Amsterdam 52 (11): 165-168.

SMIT, H. (1995): New Records of Water Mites from Morocco, with the Description of one New Subspecies (Acari, Hydrachnellae). Aquatic Insects 17 (1): 17-24.

SMIT, H. (2002): Two new water mite species from New Zealand (Acari: Hydrachnidia). Zootaxa 61: 1-10.

SMIT, H. & VAN DER HAMMEN, H (1992): New and rare water mites from The Netherlands (Acari: Hydrachnellae). Ent. Ber., Amsterdam 52: 144-146.

SMIT, H. & VAN DER HAMMEN, H. (2000): [Atlas van de Nederlandse watermijten (Acari: Hydrachnidia)]. Nederlandse Faunistische Mededelingen 13: 1-272. (in Dutch)

SMITH, B.P. (1998): Loss of larval parasitism in parasitengonine mites. Exp. Appl. Acarol. 22: 187-199.

SMITH, I. M. (1979): A review of water mites of the family Anisitsiellidae (Prostigmata: Lebertioidea) from North America. Can. Ent. 111: 529-550.

SMITH, I. M. (1982): Larvae of water mites of the genera of the superfamily Lebertioidea (Prostigmata: Parasitengona) in North America with comments on phylogeny and higher classification of the superfamily. Can. Ent. 114: 901-990.

SMITH, I. M. (1991): Descriptions of new species representing new or unreported genera of Lebertiodea (Acari: Hydrachnida) from North America. Can. Ent. 123: 811-825.

SMITH, I. M. & COOK, D. R. (1999): North American species of *Parathyasella* VIETS stat. nov. and *Amerothyasella* gen. nov. (Acari: Hydrachnida: Hydryphantidae). Int. J. Acarol. 25: 43-50.

SMITH, I. M. & COOK, D. R. (2009): Three new Euthyadine water mites representing previously unreported or inadequately known genera in North America (Acari: Hydrachnidiae: Hydryphantidae). Int. J. Acarol. 35: 385-395.

SMITH, I. M. & OLIVER, D. R. (1976c): The parasitic associations of larval mites with imaginal aquatic insects, especially Chironomidae. Can. Ent. 108: 1427-1442.

SMITH, I. M. & OLIVER, D. R. (1986): Review of parasitic associations of larval water mites (Acari:Parasitengona:-Hydrachnida) with insect hosts. Can. Ent. 118: 407-472.

SMUKALLA, R. & MEYER, E. (1988): Insect emergence from a shallow southern West German lake, with special reference to the parasitic host-associated water mite larvae. Hydrobiologia 169: 149-166.

SOAR, C. D. (1917a): Two new species of Hydracarina or water-mites. J. Quekett micr. Cl. (s. 2) 13 (80): 277-282.

SOKOLOW, I. (1925b): [Zur Kenntnis der Hydracarinenfauna des Oka-Flusses nebst Ergänzungen zur Hydracarinenfauna der Umgebung der Stadt Murom (Gouv. Wladimir)]. Arb. biol. Oka-Stat. Murom 3 (2-3): 89-105. (in Russian)

SOKOLOW, I. (1926a): Eine neue Hydracarinen-Gattung aus Russisch-Karelien. *Kondia* nov. gen. Zool. Anz. 66: 32-40.

SOKOLOW, I. (1926b): Zwei neue Hydracarinenarten aus der Gattung *Atractides* C. L. KOCH. Zool. Anz. 68 (1-2): 72-76.

SOKOLOW, I. (1927c): Beitrag zur Kenntnis der Hydracarinenfauna vom Kaukasus. Trav. Stat. biolog. Caucase Nord, Gorsky Inst. Agronom., Vladicaucase 5 (2, 1): 43-72.

SOKOLOW, I. (1930a): Die Hydracarinen von Russisch-Karelien. Zool. Jb. Syst. 59: 139-232.

SOKOLOW, I. (1930b): Beiträge zur Kenntnis der Hydracarinen Sibiriens. Arch. Hydrobiol. 22: 306-350.

SOKOLOW, I. (1934c): Eine neue Hydracarinengattung *Rutripalpus* n. gen. Zool. Anz. 106 (11-12): 318-325.

SOKOLOW, I. (1936b): Über die Hydracarinen der Quellen und Quellbäche des Leningrader Gebietes. Arch. Hydrobiol. 30: 463-496.

SOKOLOW, I. (1940a): Hydracarina (1.re partie: Hydrachnellae). In: SERNOW & STACKELBERG (eds): Faune de l'URSS. Arachnides, 5 (2): 24+1-51, Inst. zool. Acad. Sci. URSS. (n. s.), N. 20, Moscow, Leningrad.

SPARING, I. (1959): Die Larven der Hydrachnellae, ihre parasitische Entwicklung und ihre Systematik. Parasitologische Schriftenreihe, Jena 10: 1-168.

STEINER, G. (1911): Biologische Studien an den Seen der Faulhornkette im Berner Oberland. Intern. Rev. Hydrobiol. Hydrogr., Biol. Suppl. (s. 2) 4: 1-72.

SZALAY, L. (1936): Zwei neue Hydracarinen aus der Gattung *Protzia* PIERSIG. Zool. Anz. 115 (1-2): 51-54.

SZALAY, L. (1937): Eine neue Hydracarine aus der Gattung *Lebertia* NEUMAN. Zool. Anz. 119 (1-2): 40-43.

SZALAY, L. (1944a): Weitere Süßwassermilben (Hydrachnellae, Acari) aus unterirdischen Gewässern in Ungarn. Fragm. faun. Hungar. 7 (2-3): 33-39.

SZALAY, L. (1947): Einige Atractides-Formen (Hydrachnellae) aus unterirdischen Gewässern des Karpathenbeckens. Ann. Hist. natur. Mus. nation. Hungar. 40 (7): 289-303.

SZALAY, L. (1953a): Neue Hydrachnellae-Formen aus dem Karpatenbecken. Ann. Hist.-natur. Mus. Hungar. 1952 (Budapest 1953) (n. s.) 3: 213-224.

SZALAY, L. (1956a): Wassermilben (Hydrachnellae) aus der Umgebung des Balatons. Acta Zool. Acad. Sci. Hung. 2 (1-3): 269-300.

THON, K. (1899a): Ein neues Hydrachnidengenus aus Böhmen, nebst einigen Bemerkungen über böhmische *Hydryphantes*-Formen. (Vorläufige Mittheilung). Zool. Anz. 22 (581): 100-102.

THOR, S. (1897c): Andet bidrag tilkunskaben om Norges Hydrachnider - 'Norske Hydrachnider II'. Arch. Math. Naturv. 20 (3): 1-40.

THOR, S. (1899a): En ny Hydrachnide-slegt og andre nye arter fundne i Norge sommeren 1899. Foreløbig meddelelse. Kristiania, O. Norli, 1-5 (Fig. 166-173).

THOR, S. (1899c): Tredie bidrag til lundskaben om Norges Hydrachnider. 'Norske Hydrachnider III'. Arch. Math. Naturv. 21 (5): 1-64.

THOR, S. (1901a): Hydrachnologische Notizen IV-VIII. Nyt Mag. Naturv. 38 (4): 369-389.

THOR, S. (1905c): *Lebertia*-Studien I. Zool. Anz. 28 (26): 815-823.

THOR, S. (1906a): *Lebertia*-Studien VI-VIII. Zool. Anz. 29 (25-26): 761-790.

THOR, S. (1906d): *Lebertia*-Studien XI-XIV. Zool. Anz. 30 (15): 463-484.

THOR, S. (1906e): Über zwei neue in der Schweiz von Herrn C. WALTER (Basel) erbeutete Wassermilben. Zool. Anz. 31 (2-3): 67-71.

THOR, S. (1911c): Eine neue *Neolebertia*-Art und eine neue *Pilolebertia*-Art aus Sachsen nebst Bemerkungen über eine wahrscheinlich neue *Neolebertia*-Art aus Irland. Zool. Anz. 38 (13): 326-331.

THOR, S. (1912b): *Lebertia*-Studien XXVI-XXVIII. Zool. Anz. 39 (17-18): 529-536.

THOR, S. (1913a): Neue Acarinenformen aus Vättern (Schweden), gesammelt von Dr. Sven EKMAN. Zool. Anz. 41 (4): 164-172.

THOR, S. (1914b): *Lebertia*-Studien XXXIII-XXXV. Zool. Anz. 45 (1): 27-33.

THOR, S. (1914c): Prispevky k seznáni voduli fauny kavkazské die sbirky p. Jul. Komárka Rozprav. Ceské Akad., Praze (s.2) 23 (26): 1-10.

THOR, S. (1923): Neue Acarinasammlung vom Wolgadistrikt, welche im Grundwasser der Wolga und einiger ihrer Nebenflüsse vom Direktor der Biologischen Wolgastation Dr. A. BEHNING gesammelt wurde - Arb. biol. Wolgastat., Saratow 7 (1-2): 44-57.

TUZOVSKIJ, P. V. (1975): [The nymph of the watermite *Parathyas thoracata* (PIERSIG, 1896) (Hydryphantoidea, Acariformes)]. Biol. Vnutrenn. Vod, Inform. Bjull. 27: 31-35. (in Russian)

TUZOVSKIJ, P. V. (1977b): [Postembryonic development of water mite *Sperchon clupeifer* (Acariformes, Sperchonidae)]. Zool. Žurnal 56 (11): 1616-1628. (in Russian)

TUZOVSKIJ, P. V. (1978c): [Die Nymphe der Wassermilbe *Oxus angustipositus* VIETS, 1908 (Lebertiidae, Acariformes)]. Biol. Vnutrenn. Vod, Inform. Bjull. 37: 48-51. (in Russian)

TUZOVSKIJ, P. V. (1980): [A new species of the genus *Thyas* (Hydrypantidae, Acarina) from the north Caucasus]. Parazitologia 14 (1): 18-25. (in Russian)

TUZOVSKIJ, P. V. (1981): Postembrionalnye stadii vodjanogo klesca *Torrenticola amplexa* (KOENIKE, 1908) (Acariformes, Torrenticolidae). Entomol. Obozrenie 60 (4): 205-212.

TUZOVSKIJ, P. V. (1982): Opisanie dvuch novych vidow vodjanych klescej nadsem. Lebertioidea (Acarina, Trombidiformes) iz Magadanskoj Oblasti. Entomol. Obozrenie 61 (3): 644-656.

TUZOVSKIJ, P. V. (1987): [Morphology and postembryonal development of water mites]. Nauka, Moskwa: 1-176. (in Russian)

TUZOVSKIJ, P. V. (1990): [Key to deutonymphs of water mites]. Nauka, Moskow: 1-238. (in Russian)

TUZOVSKIJ, P. V. (2000a): A new species of *Thyas* from Russia. Zoosyst. Rossica 8: 243-250.

TUZOVSKIJ, P. V. (2000b): [Description of two new water mite species of the genus *Monatractides* (Acariformes, Torrenticolidae) from the Northern Caucasus]. Zoologicheskii Zhurnal 79 (1): 111-119. (in Russian)

TUZOVSKIJ, P. V. (2001b): Description of larva of the water mite *Oxus (Gnaphiscus) ekmani* (Acariformes, Oxidae). Acarina 9: 121-124.

TUZOVSKIJ, P. V. (2001g): A new species of water mites of the genus *Pseudohydryphantes* (Hydryphantoidea, Acariformes). Acarina 9: 247-252.

TUZOVSKIJ, P. V. (2002): [The larva of the water mite *Frontipoda musculus* (MÜLL.) (Acariformes, Oxidae)]. Biol. Vnutr. Vod. 1: 23-27. (in Russian)

TUZOVSKIJ, P. V. (2003): [A new genus of water mites of the family Lebertiidae (Acariformes, Hydrachnidia)]. Zoologicheskii Zhurnal 82(7): 872-879. (in Russian)

TUZOVSKIJ, P. V. (2005): The morphology of larva of the water mite *Panisus major* TUZOVSKIJ (Acariformes, Hydryphantidae). Acarina 14: 93-97.

TUZOVSKIJ, P. V. (2006a): Larval morphology of the water mite *Lebertia densa* (Acariformes, Lebertiidae) and redescription of adults. Acarina 13: 199-202.

TUZOVSKIJ, P. V. (2006b): The water mite *Frontipoda carpenteri* HALBERT, 1911 from Karelia (Acariformes: Oxidae). Zoosystematica Rossica 15 (1): 19-26.

TUZOVSKIJ, P. V. (2007): Water mites of the genus *Thyas* KOCH, 1835 (Acariformes: Thyadidae) in Russia. Acarina 15 (2): 173-259.

TUZOVSKIJ, P. V. (2008): A new record of the water mite *Hydryphantes tenuipalpis* THON (Acariformes, Hydryphantidae) for Russia, with description of its developmental stages. Acarina 16: 57-64.

TUZOVSKIJ, P. V. (2009): A new water mite species of the genus *Acucapito* (Acariformes: Acucapitidae) from Vietnam. Zoosystematica Rossica 18 (2): 212-217.

TUZOVSKIJ, P. V., BENFATTI, D. & GERECKE, R. (2001): The water mite family Acherontacaridae COOK, 1967 nov. stat. and diagnosis of the superfamily Hydrovolzioidea THOR, 1905 (Acariformes, Hydrachnidia). Acarologia 41: 451-473.

UDALZOW, A. D. (1907b): [Zur Fauna und Biologie der Hydrachniden des Gouv. Moskau]. Arb. hydrobiol. Stat. Seer Glubokoje 2: 216-274.

ULLRICH, F. (1976): Biologisch-ökologische Studien an rheophilen Wassermilben (Hydrachnellae, Acari), unter besonderer Berücksichtigung von *Sperchon setiger* (THOR 1898). Thesis, Univ. Kiel, 241pp.

ULLRICH, F. (1978): Biologisch-ökologische Studien an den Larven rheophiler Wassermilben (Hydrachnellae, Acari). Arch. Hydrobiol., Suppl. 54: 189-255.

ULLRICH, F. (1978a): Interspezifische Beziehungen bei den Hydrachnellae eines Mittelgebirgsbaches. Verh. Ges. Ökol. 7: 253.

VAN DER HAMMEN, H. & SMIT, H. (1996): The water mites (Acari:Hydrachnidia) of streams in The Netherlands: distribution and ecological aspects on a regional scale. Netherlands Journal of Aquatic Ecology 30 (2-3): 175-185.

VAN RENSBURG, C. A. J. (1971): Potamophreatic mites (Acari, Trombidiformes) from the Jura and the Tessin, Switzerland. Verhandl. Naturf. Ges. Basel 81: 319-367.

VIETS, K. (1913c): Drei neue Wassermilben-Arten aus den Gattungen *Thyas*, *Hydrachna* und *Arrhenurus*. Abh. naturw. Ver. Bremen 21 (2): 407-411.

VIETS, K. (1920b): Hydracarinen aus norddeutschen und schwedischen Quellen. Arch. Hydrobiol. 12 (4): 805-814.

VIETS, K. (1922d): Hydracarinen aus Quellen in den Weserbergen (Vogler und Ith.). Arch. Naturg. 88 (A. 9): 53-76.

VIETS, K. (1923a): Hydracarinen aus Quellen. (Systematische und biologisch-faunistische Untersuchungen). Arch. Hydrobiol., Suppl. 3 (1): 156-204, 3 (2): 205-384.
VIETS, K. (1925g): Beiträge zur Kenntnis der Hydracarinen aus Quellen Mitteleuropas. Zool. Jbch. Syst. 50 (4-6): 451-596.
VIETS, K. (1925h): Nachträge zur Hydracarinen-Fauna von Kamerun. (Sammlungen aus Kamerun, Span. Guinea und Fernando Poo). Arch. Hydrobiol. 16: 197-242.
VIETS, K. (1926g): Hydracarinen aus Bulgarien. Zool. Anz. 67(1-2): 7-27.
VIETS, K. (1926l): Versuch eines Systems der Hydracarinen. Zool. Anz. 69 (7-8): 188-199.
VIETS, K. (1928a): Zur Mikrofauna einer Quelle auf der Insel Herdla bei Bergen (Norwegen). Bergens Mus. Årbok , 1927; Naturvid. Rekke 5: 1-16.
VIETS, K. (1928c): Außerdeutsche europäische Hydracarinen-Literatur der letzten Jahre. Arch. Hydrobiol. 19: 175-186.
VIETS, K. (1928f): Wassermilben, Hydracarina. 57 pp. In: BROHME, EHRMANN & ULMER (eds): Die Tierwelt Mitteleuropas. Leipzig (Quelle & Meyer), 3, Abt.8 (4).
VIETS, K. (1929h): Fünfte Mitteilung über Hydracarinen von den Sunda-Inseln. Zool. Anz. 86 (3-4): 49-56.
VIETS, K. (1930d): Zur Kenntnis der Hydracarinen-Fauna von Spanien. Arch. Hydrobiol. 21 (2): 175-240, 21 (3): 359-446.
VIETS, K. (1934a): Fünfte Mitteilung über Wassermilben aus unterirdischen Gewässern. (Hydrachnellae und Halacaridae). Zool. Anz. 105 (5-6): 133-141.
VIETS, K. (1934c): Siebente Mitteilung über Wassermilben aus unterirdischen Gewässern. Zool. Anz. 106 (5-6): 118-124.
VIETS, K. (1935c): Über Wassermilben. Zool. Anz. 110 (11-12): 273-279.
VIETS, K. (1936b): Hydracarinen aus Jugoslawien. (Systematische, ökologische, faunistische und tiergeographische Untersuchungen über die Hydrachnellae und Halacaridae des Süßwassers). Arch. Hydrobiol. 29: 351-409.
VIETS, K. (1936d): Wassermilben oder Hydracarina (Hydrachnellae und Halacaridae). In: DAHL, F.: Tierwelt Deutschlands, Jena (G. FISCHER) 31: 10+1-288, 32: 289-574.
VIETS, K. (1938k): Über die verschiedenen Biotope der Wassermilben, besonders über solche mit anormalen Lebensbedingungen und über einige neue Wassermilben aus thermalgewässern. Verh. Intern. Ver. Limnol. 8 (3): 209-224.
VIETS, K. (1939a): Wassermilben (Hydrachnellae und Porohalacaridae, Acari) aus den französischen Pyrenäen. Zool. Anz. 125 (1-2): 1-15.
VIETS, K. (1949b): Nomenklatorische und taxonomische Bemerkungen zur Kenntnis der Wassermilben (Hydrachnellae, Acari). I-X. Abh. naturw. Ver. Bremen 32 (2): 292-327.
VIETS, K. (1950d): Water mites (Hydrachnellae) from Greece. J. Quekett microsc. Cl. (s. 4) 3 (3): 150-161.
VIETS, K. (1951d): Hydrachnellae (Acari) aus den französischen Alpen und aus Afrika. Zool. Anz. 147 (11-12): 285-303.
VIETS, K. (1952c): Weitere Hydrachnellae (Acari) aus den französischen Alpen und aus Afrika. Zool. Anz. 149 (11-12): 245-253.
VIETS, K. (1954e): Wassermilben aus dem Amazonasgebiet (Hydrachnellae, Acari). (Systematische und ökologische Untersuchungen). Bearbeitung der Sammlungen Dr. R. BRAUN, Arau, und Dr. H. SIOLI, Belém. Schweiz. Zschr. Hydrologie 16: (1): 78-151, 16 (2): 161-247.
VIETS, K. (1956b): Die Milben des Süßwassers und des Meeres. Hydrachnellae et Halacaridae (Acari). II. und III.Teil: Katalog und Nomenklator. Jena. G. FISCHER, 870 pp.
VIETS, K.O. (1955): Neue deutsche Wassermilben. Zool. Anz. 154 (3-4): 72-78.
VIETS, K.O. (1956): Die europäischen *Frontipoda*-Arten (Hydrachnellae, Acari). Zool. Anz. 156 (9-10): 243-248.
VIETS, K.O. (1957): Wassermilben aus der Salmoniden-Region von Harzbächen. Abh. naturwiss.Ver. Bremen, 35 (1): 135-161.
VIETS, K.O. (1967): Hydracarina. In: ILLIES, J. [ed.]: Limnofauna Europaea: 124-148.
VIETS, K.O. (1976b): Über *Lebertia (Pilolebertia) longiseta* (WALTER) BADER 1955, eine seltene Milbenart aus dem Bodensee (Acari, Hydrachnellae). Nachr. Naturw. Mus. Aschaffenburg 84: 17-24.
VIETS, K.O. (1978): Hydracarina. In: ILLIES, J. (ed.): Limnofauna Europaea: 154-181.
VIETS, K.O. (1987): Die Milben des Süßwassers (Hydrachnellae und Halacaridae [part.], Acari). II.: Katalog. Sonderbände des Naturwiss. Vereins Hamburg 8: 1-1012.
WAINSTEIN, B.A. (1980): [The water mites larvae]. Inst. Biol. Vnutrenn. Vod; Nauka 1980: 1-238. (in Russian)
WAINSTEIN, B.A. & TUZOVSKIJ, P.V. (1974a): [The body chaetom of water mites, its ontogenesis and evolution]. Inst. Biol. Vnutrenn. Vod; Trudy 25 (28): 230-269. (in Russian)
WALTER, C. (1906b): Neue Hydrachnidenarten aus der Schweiz. Zool. Anz. 30(17-18): 570-575.
WALTER, C. (1906a): Hydrachniden aus der Tiefenfauna des Vierwaldstätter Sees. Zool. Anz. 30 (10): 322-326.
WALTER, C. (1907b): Die Hydracarinen der Schweiz. Rev. Suisse Zool. 15: 401-573.
WALTER, C. (1912b): Beiträge zur Hydracarinen-Fauna der Umgebung von Lunz (Niederösterreich) III. Zool. Anz. 40 (4-5): 145-150.
WALTER, C. (1915a): Notizen über die Entwicklung torrentikoler Hydracarinen. Zool. Anz. 45 (10): 442-456.
WALTER, C. (1918): Hydracarinen, pp. 436-438. In: STEINMANN & SURBECK, Die Wierking organischer Verunreinigungen auf die Fauna schweizerischer fließender Gewässer. Bern.
WALTER, C. (1919a): Hydracarinen aus den peruanischen Anden und aus Brasilien. Rev. Suisse Zool. 27 (2): 19-59.
WALTER, C. (1922a): Hydracarinen aus der Alpen. Rev. Suisse Zool. 29 (7): 227-411.

WALTER, C. (1922c): Die Hydracarinen der Alpengewässer. Denkschr. schweizer. naturf. Ges. 58 (2): 11+ 60-251.

WALTER, C. (1924c): Quelques espèces nouvelles d'Hydracariens du Maroc et du Soudan. Bull. Soc. Sci. natur. Maroc, Rabat et Paris 4 (3-4): 61-71.

WALTER, C. (1925a): Hydracariens de l'Algérie et de la Tunisie (collections de M.H. GAUTHIER). Première liste). Bull. Soc. Hist. natur. Afrique Nord, Alger 16: 189-238.

WALTER, C. (1928c): Zur Kenntnis der Mikrofauna von British Indien. II. Hydracarina. Rec. Indian Mus., Calcutta 30 (1) : 57-108.

WALTER, C. (1935): Hydracarina. In: Voyage de Ch. ALLUAUD et P.A. CHAPPUIS en Afrique occidentale française (Déc. 1930–Mars 1931). Arch. Hydrobiol. 28 (1): 69-136.

WALTER, C. (1944): Die Hydracarinen der Ybbs. – 1. Teil. Intern. Rev. Hydrobiol. Hydrogr. 43 (4-6): 281-367.

WALTER, C. (1947): Neue Acari (Hydrachnellae, Porohalacaridae, Thrombidiidae) aus subterranen Gewässern der Schweiz und Rumäniens. Verh. naturf. Ges. Basel 58: 146-238.

WIGGINS, G.B., MACKAY, R.J. & SMITH, I.M. (1980): Evolutionary and ecological strategies of animals in annual temporary pools. Arch. Hydrobiol., Suppl. 58 (1-2): 97-206.

WILES, P.R. (1982): A note on the watermite *Hydrodroma despiciens* feeding on chironomid egg masses. Freshwater Biology 12: 83-87.

WILES, P.R. (1985): The systematics of the British Hydrodromidae VIETS, 1936. Arch. Hydrobiol., Suppl. 70 (3): 365-403.

WILES, P.R. (1987): Observations on the parasitic biology of the watermite *Hydrodroma despiciens pilosa* BESSELING (Acari: Hydrodromidae). Arch. Hydrobiol., Suppl. 76 (4): 369-392.

WILES, P.R. (1996): A new family, genus and species of watermite (Acari: Hydrachnidia, Lebertioidea) from Brunei. Quekett Journal of Microscopy 37: 692-695.

WILES, P.R. (1997b): Asian and Oriental Torrenticolidae PIERSIG, 1902 (Acari: Hydrachnidia: Lebertioidea): a revision of the family and description of new species of *Torrenticola* PIERSIG and *Pseudotorrenticola* WALTER, from Southeast Asia. J. Natur. Hist. 31: 191-236.

WOLCOTT, R.H. (1905b): A review of the genera of the water-mites. Stud. zool. Labor., Nebraska No. 66; Trans. Americ. microsc. Soc. 26: 161-243.

ZAWAL, A. & DYATLOVA, E.S. (2008): Parasitizing on damselflies (Odonata: Coenagrionidae) by water mite (Acari: Hydrachnidia) larvae from Odessa Province (Southwestern Ukraine). Natura Montenegrina, Podgorica 7: 453-462.

Taxonomic index Acari II

(underlined: accepted taxa dealt with in this book)
(bot.) = plant taxa
slight differences in spelling are possible between text and index
(e. g., „chironomid“ = „Chironomidae“)

aberrata, Lebertia 122-124, 126, 129, 130-131
aberrata, Lebertia cuneifera 126
aberrata, Torrenticola 181
aberratus, Atractides 181
aberratus, Monatractides 180-182
abnormis, Hydryphantes 50
Acadiosperchon 155
Acarus 53
Acerbitas 24
Acucapitidae 90-91
acuta minor, Lebertia 109
adoratus, Monatractides 178-179
affinis, Lebertia 106
africana, Lebertia 106
Africoxus 145
akguli, Sperchonopsis 172
alba, Nymphaea 11
algeriensis, Hydryphantes 50-51
algeriensis, Lebertia 142
algeriensis, Monatractides 183
algeriensis, Sperchon 163, 165
Allowandesia 74
Almuerzothyas 37
Alnus (bot.) 113
alpina, Protzia 63
Amasis 89
Amphipoda 58
amplexa, Torrenticola 188-189, 191, 194, 198
amplexa minutivalvata, Torrenticola 200
amplexus, Atractides 194
andrei, Torrenticola 187
angusta, Partnunia 56-58
angusta, Thyas 56, 58
angustipositus, Oxus 147-148
angustirostris, Atractides 195
Anisitsiellidae 79-80, 90, 177
Anisitsiellinae 80
anisitsipalpis-group (*Bandakia*) 81
Ankelothyadinae 12
annularis, Protzia 66
anomala, Torrenticola 77, 190, 192, 194-195
anomalus, Atractides 185, 194
Apataniidae (Trichoptera) 64
Apeltosperchontinae 154
Apheviderulicidae 207
Apheviderulix 210
Apolebertia 133, 136
apposita, Lebertia 109
apsteini, Georgella 43
aquaticus ruber, Acarus 53
areolata, Lebertia 98
armata, Trichothyas 38
armentarius, Hydryphantes 45-46, 48
arndti, Lebertia 109
Arrenuroidea XIII, 1, 76, 209
Arrenurus 165, 206, 210-211
aspera, Lebertia 119
asteroidea, Georgella 43
Atractideidae 177
Atractides 181, 185-187, 194-196, 198-199, 201-202, 210
Aturus 206, 211
auriculatus, Hydryphantes hellichi 53
baderi, Hydryphantes 48, 51
Balcanohydracarus 211
balcarica, Lebertia 109
Bandakia 80-83
Bandakiinae 80
Bandakiopsidae 152
Bandakiopsis 152
barbigera, Parathyas 25-28
barbigera, Thyas 26, 28
barsica, Lebertia 134
barsica, Protzia invalvaris 60
barsica, Torrenticola 190-191, 193, 195, 199
barsica tisialis, Torrenticola 187
barsicus, Atractides 195
bayeri, Calonyx 65
bayeri, Hydryphantes 45
bayeri nonundulatus, Hydryphantes 52
beetles (Coleoptera) 14
berlesei, Georgella 42, 44
berlesei, Lebertia 136
Bharatonia 80
Bharatoniinae 80
bicincta, Torrenticola 182-183
bicinctus, Atractides 182
bicinctus, Monatractides 182
bieberi, Bandakia 82
bisbisetosa, Lebertia 122-124, 126
borneri, Nilotonia 84-86
Brachycentridae (Trichoptera) 59
Braconidae (Hymenoptera) 23
bracteata, Lebertia 125-127, 136, 139
Bradybates 15
Brentalebertia 92-93, 95-97
brevipalpis, Protzia 61-62
brevipora, Lebertia 117-119, 127-128
brevirostris, Atractides 196
brevirostris, Sperchon 158, 161, 167

brevirostris, Torrenticola 189-190, 192, 196
brigantina, Lebertia 94, 99, 103, 111
brunnea, *Lebertia* 98
bruzelii, Parathyas 25, 27-28
bruzelii, *Thyas* 27
bucculenta, *Thyas dirempta* 29
cabardinicus, *Calonyx* 67
Calonyx 59, 65-68
canadiensis, *Rutripalpus* 152
cancellata, Thyopsis 36-37, 205
cancellata maderensis, *Thyopsis* 37
cancellatomima, *Thyopsis* 36
Capniidae (Plecoptera) 75
carpaticus, *Sperchon* 156
carpenteri, Oxus 146, 148-150
castalia, Lebertia 78, 103-105
cataphractus, *Panisus* 21-22
caucasica cubanica, *Lebertia* 144
caucasica, *Protzia* 68
ceratopogonids (Diptera) 88
Chaoboridae (Diptera) 11
Chaoborus flavicans (Diptera) 11
Charoelia 154, 169
Chimerathyadinae 12
Chimerathyas 12
Chironomidae (Diptera) 10-11, 18-19, 24-25, 32, 58-59, 64, 72, 82, 88, 106-107, 109, 114-116, 129-130, 135, 137, 139, 146, 162-163, 166, 168-170, 172, 175, 184
Chironominae (Diptera Chironomidae) 64, 106, 115, 137, 146
Chloroperlidae (Plecoptera) 75
circumclusa, *Lebertia sefvei* 137
clupeifer, *Sperchon* 155, 157-158, 162, 164
clypeatus, *Hydryphantes* 45
clypeolata, *Panisus* 24
clypeolata, *Thyas* 23
Coenagrionidae (Odonata) 42
cognata, *Lebertia* 100-101, 105, 114
cognatus, *Atractides dudichi* 195
Coleoptera 14
Collembola 18
colligera, Parathyas 26, 28
colligera, *Parathyas rivalis* 28
cometes, *Hygrobates* 175-176
cometes, Teutonia 77, 174-177
compacta, *Lebertia* 98
compactilis, Sperchon 158-159, 162-163
compactilis, *Sperchon papillosus* 162
concreta, Bandakia 81-82
contracta, Lebertia 118-119, 128, 133
corsica, *Bandakia* 82
corsica, *Lebertia* 116
Corynoneura (Diptera Chironomidae) 64
costata, *Lebertia* 119
Cowichania 12
Cowichaniinae 12
crassipalpis, Hydryphantes 47, 50
crassipalpis lacustris, *Hydryphantes* 50
crenophila, Lebertia 119, 124-125, 128
Cricotopus sylvestris (Diptera Chironomidae) 11
Crustacea 52, 58
Ctenothyadidae 1
cubanica, *Lebertia caucasica* 144
Culicidae (Diptera) 16, 18, 21, 25, 27, 35, 44, 53,
cuneifera, Lebertia 122-124, 126, 129, 131
cuneifera aberrata, *Lebertia* 126
curvifrons, Panisopsis 6, 19-20
curvifrons, *Thyas* 19
curvifrons, *Walterothyas* 19
curvipalpis, *Lebertia* 140
curvipalpis, *Protzia* 67
Cyclohydryphantes 69
Cyclothyadinae 12
cylindrata, *Lebertia* 119
cylindrica, *Lebertia rufipes* 116
Dacothyas 4, 37
dalmatica fragilis, *Lebertia* 98
damselflies (Odonata) 56
danielei, *Lebertia* 97
danuviensis, *Hydrodroma* 12
danuviensis, *Hydrodroma torrenticola* 12
Dartia 83-88
Dartiella 83-85, 87-88
Dartonia 83
dauphinensis, *Georgella* 44
dauphinensis, *Hydryphantes* 44
densa, *Lebertia* 130
dentata,*Thyas* 34-35
dentatus, Thyasides 7, 35
denticulatus, Sperchon 156-157, 163-164, 167
denticulatus hibernicus, *Sperchon* 163
depressostriata, Lebertia 120-121, 130-131
despiciens, Hydrodroma 8-11
despiciens pilosa, *Hydrodroma* 11
Diamesinae (Diptera Chironomidae) 59
diaspora, *Torrenticola* 196
Dicranomyia (Diptera Limoniidae) 27
Diplocalonyx 59
Diplodontinae 3, 13
Diplodontus 2-3, 12-14
Diptera 10-11, 16, 18-19, 21, 24-25, 27, 32, 35, 37-38, 44, 50, 53, 58-59, 64, 72, 75, 82, 88, 106-107, 109, 114-116, 129-130, 135, 137, 139, 146, 153, 162-163, 165-166, 168-170, 172, 175, 184
dirempta, Parathyas 25, 28-29
dirempta bucculenta, *Parathyas* 29
dirempta rostrata, *Parathyas* 29
diremptellus, *Parathyas* 29
dispar, Hydryphantes 47, 50
distincta, Protzia 62-64
distincta, *Protzia eximia* 63
distinctus, *Thyas* 28-29
Dixidae (Diptera) 88
dresdensis, *Lebertia* 98

dubia, Lebertia 119-120, 130, 139
dubiaeformis, Lebertia 133-134
dudichi, Atractides 196
dudichi, Torrenticola 190, 192-193, 196
dudichi cognatus, Atractides 195
dudichi oraviensis, Torrenticola 201
duplicata, Trichothyas 38
duricoria, Lebertia 119
ekmani, Gnaphiscus 151
ekmani, Oxus 151-152
elephantula, Euthyas 16
Elliptera (Diptera Limoniidae) 25
elliptica, Lebertia 98
elliptica, Torrenticola 190-191, 193, 196-197, 201
elliptica microphallus, Torrenticola 200
ellipticus similis, Atractides 201
elongata, Lebertia 106
elsteri, Lebertia 92, 96-97
Empididae (Diptera) 58, 64, 75
Eolebertia 92-93, 96-97
Ephydridae (Diptera) 50
Erioptera (Diptera Limoniidae) 25, 27
erlangensis, Monatractides latissimus 182
Estelloxus 91
Eupatrella 13
Eupatrellinae 12-13
Euthyadidae 1
Euthyadinae 1, 3, 15, 44, 56, 72
Euthyas 6, 15-16, 208
Euthyasinae 15
Euwandesia 72
excellens, Lebertia 120-121, 130-131
eximia, Protzia 2, 59, 62-64
Eylais 208, 210-211
Eylaoidea XIII, 207
eximia distincta, Protzia 63
eximia macrognatha, Protzia 63
eximia rugosa, Protzia 68
extendens, Lebertia 131
extendens, Parathyas 25
faber, Protzia 66
fagei, Atractides 186
fagei, Torrenticola 185-186
Fagus (bot.) 20
felix, Protzia 64
feuerborni, Lundbladia 37
fibulata, Tadjikothyas 4
fimbriata, Georgella 43
fimbriata, Lebertia 98, 103-106, 116-117
Flabellifrontipoda 145
flavicans, Chaoborus (Diptera Chaoboridae) 11
flexuosus, Hydryphantes 45, 49, 54-55
flexuosus thoni, Hydryphantes 55
flumenia, Lebertia 98
fontana, Lebertia 123-124, 129, 131-133
fonticola, Vietsthyas 72
fontinalis, Hydryphantes 46, 48, 51, 54
fragilis, Lebertia dalmatica 98
frici, Hydryphantes 45
frigida, Parathyasella 32-34
frigida, Thyasella 33
Frontipoda 145
gallica, Nilotonia 88
Georgella 3-4, 42-44
giardinai, Lebertia 123-124, 127, 131-133, 137
glabra, Lebertia 98, 100-101, 107-108, 115
gladiator, Lebertia 98, 103
glandulosus, Sperchon 158, 160, 168, 170
glandulosus thienemanni, Sperchon 168
globifera, Lebertia schechteli 98
Glossosomatidae (Trichoptera) 64
Gnaphiscus 145, 151
gomerae, Torrenticola 197
Gonomyia (Diptera Limoniidae) 25
gracilipes, Lebertia 132
gracilis, Psammotorrenticola 79-80
grandicollis, Hydrochus (Coleoptera) 14
Gripopterygidae (Trichoptera) 75
groenlandica, Lebertia 103-104, 107, 110
guadarramensis, Torrenticola 195
guignoti, Hemisphaera (Coleoptera) 14
guttata, Lebertia 98-99, 107, 109
halberti, Calonyx 65
halberti, Lebertia 118-119, 132
halberti, Protzia 61-62, 65, 67
halberti, Torrenticola 187
haliki, Lebertia 134
halophilus, Oxus 145-146
handschini, Hydryphantes 48
harenarum, Lebertia 103-104, 108
harrisi, Dartia 86
harrisi, Nilotonia 84-87
Heleniella (Diptera Chironomidae) 64
hellichi, Hydryphantes 53
hellichi auriculatus, Hydryphantes 53
helocrenica, Lebertia 125, 133
helvetica, Georgella 4, 42-44
helvetica, Hydrodroma 42-43
helvetica, Lebertia 119-120, 133, 136
helvetica, Lebertia oudemansi 133
helvetica, Wandesia 74
Hemisphaera (Coleoptera) 14
Heterothyas 37
Hexalebertia 117, 126-132, 134-138
hibernica, Torrenticola 182
hibernicus, Monatractides 182
hibernicus, Sperchon 156-157, 163, 166
hilaris, Hydryphantes 44, 48
hirtipalpis, Lebertia 98
hispanica, Torrenticola 195
Hispidosperchon 156, 162-167
hispidus, Sperchon 155-157, 160, 162, 164
histriensis, Torrenticola 196
holsatica, Lebertia 78, 119, 121, 134-135, 138
hupalupu, Lebertia 96

husmanni, Lebertia 109
Hydrachna 210-211
Hydrachnidia 202-211
Hydrachnoidea XIII, 207
Hydrellia (Diptera Ephydridae) 50
Hydrobaderia 44, 48
Hydrochus (Coleoptera) 14
Hydrodroma 1-2, 8-12, 42, 54, 165
Hydrodromidae 1, 8
Hydrophilidae (Coleoptera) 14
Hydropsychidae (Trichoptera) 64
Hydrovolzia 206
Hydrovolzioidea XIII, 207
Hydryphantes 3-4, 42, 44-56, 208, 210-211
Hydryphantidae 1, 12, 16, 40-41, 72, 154, 212
Hydryphantinae 1, 3, 12
Hydryphantoidea 1, 12, 15, 72, 209
Hygrobates 142, 175, 211
Hygrobatidae 185
Hygrobatoidea XIII, 1, 76, 209
hygropetrica, Lebertia 92, 95
Hymenoptera 23
hyporheica, Torrenticola 189-190, 192, 198-199
hystrix, Sperchon 156
Ignacarus 6-7, 72
ignatowi, Lebertia 112
ilicaensis, Hydrobaderia 50
imperfecta, Georgella 42
inaequalis, Hygrobates 142
inaequalis, Lebertia 140-143
incognita, Lebertia 119
inepta, Parathyas 26, 30-31
inepta, Thyas 30
inflexa, Lebertia 119
insignis, Lebertia 94, 140-144
insignis, Sperchon 158-159, 164-165
insignis, Sperchon setiger 164
integer, Oxus 146
intermedia, Georgella 44
intermedius, Hydryphantes 44, 50
interposita, Lebertia sefvei 137
intronata, Lebertia 106, 116
invalvaris, Protzia 2, 59-60, 62, 65-66
invalvaris barsica, Protzia 60
invalvaris multispina, Protzia 60
inversa, Lebertia 105
Iranothyas 6-7
ischnophallus, Torrenticola 188-189, 191, 198
jadrensis, Lebertia 116
Javathyas 37
jeanneli, Atractides 199
jeanneli, Torrenticola 189-190, 192, 198-199
Kashmirothyas 37
koenikei, Georgella 43-44
koenikei, Oxus 149149
komareki, Lebertia 98
kuzini, Lebertia 144
Laccobius (Coleoptera Hydrophilidae) 14
lacustris, Hydryphantes crassipalpis 50
lacustris, Lebertia 98
larrieui, Protzia 60
laskai, Torrenticola 77, 191, 193-194, 199-200
lata, Partnunia 66
lata, Protzia 2, 59-61, 63, 66
lata, Torrenticola 183
laticoxalis, Lebertia 98
latissima latissima, Torrenticola 182
latissimus, Monatractides 181-182
latissimus erlangensis, Monatractides 182
lativalvata, Torrenticola 199
lativentris, Lebertia 120-121, 130-131, 135
latus, Calonyx 66
Lebertia 78, 91-144, 210
Lebertia s. str. 92, 94, 97-117, 140
Lebertiidae 76, 78, 90-91, 145, 152, 212
Lebertioidea 1, 76, 80, 154, 209
Lepidoptera 15, 24
Lepidostomatidae (Trichoptera) 11, 59
Leptoceridae (Trichoptera) 11
limicola, Rutripalpus 77, 152-153
Limnephilidae (Trichoptera) 64
Limnesia 87, 165
Limnesiidae 79-80
Limnochares 205, 208, 210-211
Limnolegeria 77, 173-175
Limonia (Diptera Limoniidae) 27, 38
Limoniidae (Diptera) 25, 27, 32, 37-38, 58
limula, Torrenticola 189-190, 192, 197
lineata, Lebertia 107
lineatus, Oxus 146, 150
Ljania 206
lobata, Torrenticola 183
longicornis, Hydryphantes 54
longiepimerata, Lebertia 116
longipalpis, Lebertia 130, 135
longipalpis-group (*Bandakia*) 81
longipes, Lebertia 119
longipora, Dartia 87
longipora, Nilotonia 84-85, 87-88
longirostris, Sperchon 158-159, 168, 170
longirostris, Torrenticola 194
longiseta, Lebertia 140-143
longiseta, Limnolegeria 77, 173-175
longiseta, Pseudomarica 149
longisetus, Oxus 147, 149, 151
longissimus, Sperchon 158, 160, 169
loricata, Nilotonia 83
loricata, Teutonia 83
lucifuga, Wandesia 75
lunata, Teutonia 177
lundbladi, Panisus 23
Lundbladia 37-39
lusitanica, Lebertia 98-99, 108-109
lutea, Nuphar (bot.) 11
lychnobia, Wandesia 73
macilenta, Lebertia 120-121, 136

macrognatha, *Protzia* 63-64
macrognatha, *Protzia eximia* 63
maculosa, *Lebertia* 98-100, 108-109, 115
maderensis, *Thyopsis* 37
maderensis, *Thyopsis cancellata* 37
madericola, *Lebertia* 97, 140
maderigena, *Lebertia* 97, 113, 115, 140
madritensis, *Monatractides* 178, 180, 182-183
magadanensis, *Parathyas* 28
maglioi, *Lebertia* 102, 104, 107-110, 113
maglioi, *Torrenticola* 187
major, *Panisus* 22
Mamersinae 12
Mamersonia 83
mandibularis, *Thyas* 33
mandibularis, *Thyasella* 5, 33-34
mandibularis torrenticola, *Thyasella* 34
Manotonia 83
marismortui, *Iranothyas* 6
mediterranea, *Lebertia* 127
Megapalpis 178-179, 185-187
meridianus, *Sperchon* 165
meridionalis, *Torrenticola* 197
Mesowandesia 75
Metriocnemini (Diptera Chironomidae) 115
michaeli, *Panisus* 5, 22-24
Microcrustacea 25
micrommata, *Tartarothyas* 70, 72
microphallus, *Torrenticola* 191, 193, 199
microphallus, *Torrenticola elliptica* 200
Micropsectra (Diptera Chironomidae) 72
Mideopsis 210
minimus, *Amasis* 89
minor, *Lebertia acuta* 109
minuta, *Lebertia* 98
minuticornis, *Lebertia* 140
minutipalpis, *Lebertia* 92, 95-96
minutivalvata, *Torrenticola* 188, 191, 200
minutivalvata, *Torrenticola amplexa* 200
Mixolebertia 93, 95, 117-139
Mixosperchon 162
mollis, *Lebertia* 98
Molophilus (Diptera Limoniidae) 25, 27, 32
monardi, *Lebertia* 98
Monatractides 178-184
monstruosus, *Sperchon* 163
montanus, *Sperchon* 156
moraguesi, *Laccobius* (Coleoptera) 14
mucronata, *Torrenticola* 195
multicincta, *Lebertia* 98
multipilosa, *Lebertia* 110
multipora, *Protzia* 68
multispina, *Protzia invalvaris* 60
muscicola, *Trichothyas* 38
musculosa, *Protzia* 67
musculus, *Frontipoda* 149
musculus, *Oxus* 146, 149
mutilus, *Sperchon* 78, 158, 161, 169
Mycetophilidae (Diptera) 16
nana, *Torrenticola* 195, 201
natans, *Lebertia* 94, 99, 110
Nemouridae (Plecoptera) 75
Neoatractides 179
Neoatractidinae 177
Neocalonyx 56
neocomensis, *Lebertia* 142
Neolebertia 109-112, 116, 143
Neothyas 37
nilotica, *Sigthoria* 79, 89-90
niloticus, *Amasis* 89
Nilotonia 78, 80, 83-89
Niphargus (Crustacea Amphipoda) 58
nitida, *Lebertia* 134, 137
nodigerus, *Oxus* 147, 150
nonundulatus, *Hydryphantes* 52
nonundulatus, *Hydryphantes bayeri* 52
novatus, *Hydryphantes ruber* 53
novipalpis, *Lebertia* 119
Nuphar (bot.) 11
Nymphaea(bot.) 11
obesa, *Lebertia* 105
oblonga, *Lebertia* 94, 99, 103, 111
oblonga, *Zschokkea* 2, 40-41
oblongus, *Oxus* 145
Octohydryphantes 54
octopora, *Protzia* 59
octoporus, *Hydryphantes* 45, 55-56
Odonata 42, 56
olonensis, *Lebertia* 112
oraviensis, *Torrenticola* 190, 193, 195, 201
oraviensis, *Torrenticola dudichi* 201
ornatus, *Sperchon* 156
Orthocladiinae (Diptera Chironomidae) 64, 106-107, 109, 115, 130, 135, 137, 146
oudemansi, *Lebertia* 119-120, 134, 136
oudemansi helvetica, *Lebertia* 133
ovalis, *Hydrachna* 150
ovalis, *Oxus* 147, 150-151
oxa, *Lebertia* 137
Oxidae 90-91, 145
Oxopsis 9
Oxus 77, 145-152
Pachygaster 91
pachygnatha, *Protzia* 62, 66
pachystoma, *Parathyas* 26, 30-31
pachystoma pannonica, *Thyas* 30
pachystoma paucispina, *Thyas* 30
pallida, *Lebertia* 112
Palpisperchon 155
palustris, *Parathyas* 7, 26, 29-31
palustris, *Thyas* 31
Panisellus 5, 17-18, 39
Panisoides 18, 20
Panisopsis 6-7, 18-22, 206
Panisus 5, 7-8, 17, 21-24
pannonica, *Thyas pachystoma* 30

Papilloporus 44, 48, 54
papillosus, Sperchon 158-159, 162-163, 165
papillosus compactilis, Sperchon 162
Paracalonyx 56
Paradelphomyia (Diptera Limoniidae) 32
Paralebertia 90-91
Parathyas 7-8, 24-32, 208
Parathyasella 7, 32-33
parmulatus, Hydryphantes 47, 51
Parochlus (Diptera Chironomidae) 32
Partnunia 2-3, 56-59, 66
Partnuniella 74
parva, Dartia 88
parva, Nilotonia 84-85, 88-89
Parvadartia 87
parvipalpis, Atractides 182
parvipalpis, Monatractides 182
parvipalpis, Torrenticola 182
parvulus, Pseudohydryphantes 2, 69
paucipora, Protzia squamosa 68-69
paucispina, Thyas pachystoma 30
Pediciidae (Diptera) 37, 58
peltatus, Hydryphantes 51
pennata, Thyas 37
Perlodidae (Plecoptera) 75
peroviensis, Hydryphantes 45
petrophila rutae, Trichothyas 38-39
petrophila, Lundbladia 38
petrophila, Trichothyas 38-39
petrophilus, Thyas 38
Philopotamidae (Trichoptera) 59
phreatica, Sperchonopsis 172
pictus, Oxus 149
Piersigia 208, 211
Pilaria (Diptera Limoniidae) 27, 32
Pilolebertia 94, 110, 128, 136, 140-144
pilosa, Hydrodroma 8-11
pilosa, Lebertia 140-141, 143-144
Pinquicola 178-180
piriformis, Torrenticola 194
placationis, Hydryphantes 46, 52
planus, Hydryphantes 48-49, 52-53
plauta, Lebertia 140
Plecoptera 58, 75
Plesiothyas 37
plumifera, Lundbladia 38-39
polonica, Lebertia 100-101, 111
Polycentropodidae (Trichoptera) 64
Polyhydryphantes 45, 48, 54-56, 208
Pontarachnidae 205
porosa, Lebertia 140-141, 143-144
primaria, Teutonia 176
procera, Sperchonopsis 161, 171-172
procerivalvata, Torrenticola 187
Prodiamesinae (Diptera Chironomidae) 107
prolongata, Hydryphantes ruber 53
propinqua, Wandesia 74
Protzia 2-3, 59-69, 211
Protziinae 56
Psammotorrenticola 79-80
Pseudatractides 89
Pseudohydryphantes 2-3, 69-70
Pseudohydryphantinae 3, 69
Pseudolebertia 91, 107-109, 114-116, 139
Pseudomarica 149
Pseudothyas 1
Pseudotorrenticola 179
Pseudotorrenticolella 179
pseudotuberosa, Lebertia 109
Pseudowandesia 73-74, 211
Psychodidae (Diptera) 64
Psychomyiidae (Trichoptera) 64
Ptychopteridae (Diptera) 25, 153
pulchella, Lebertia 98
pullinidentis, Protzia 66
pusilla, Lebertia 101-102, 112-113
pustulosa, Thyas 25, 33
pycnica, Protzia 66
pyrenaica, Lebertia 119
pyrenaica, Nilotonia 84-85, 88-89
pyrenaica, Protzia walteri 65
pyrenaicus, Hydryphantes 48
pyriformis, Pseudatractides 89
quadripora, Lebertia 142
quinquescutata, Pseudotorrenticola 184
racovitzai, Wandesia 73
raetica, Partnunia 58
raetica, Protzia 63
raetica, Tartarothyas 71
ramiger, Atractides 198
ramigera, Torrenticola 198
ramosus, Hydryphantes 48
rarus, Parathyas 31
reducta, Lebertia 133
reducta, Sperchonopsis verrucosa 172
reinhardi, Hydrodroma 10, 12
remyi, Torrenticola 178, 185-187
resupinus, Sperchon 158, 160, 169
reticulata, Lebertia 99-100, 105, 112-114
reticulata, Lebertia rufipes 112
Rheolimnesia 76, 89
Rhyacophilidae (Trichoptera) 64
Rhynchohydracaridae 1
rhynchota, Pseudotorrenticola 178-179, 184
rivalis colligera, Thyas 28
rivalis, Parathyas 31
rivalis, Pseudohydryphantes 70
rivulorum, Lebertia 101-102, 113
robusta, Lebertia 131
robusta, Nilotonia 87
robusta, Torrenticola 182
robustus, Atractides 182
robustus, Monatractides 182
romanica, Tartarothyas 2, 71
rostrata, Thyas dirempta 29
rothae, Thyopsis 37

rotunda, Protzia 59, 61, 65, 67
rotundus, Calonyx 65, 67
rubella, Lebertia 115
ruber mozleyi, Hydryphantes 53
ruber novatus, Hydryphantes 53
ruber prolongata, Hydryphantes 53
ruber tenuipalpis, Hydryphantes 54
ruber, Acarus aquaticus 53
ruber, Hydryphantes 4, 44, 48-49, 52-54
rufipes cylindrica, Lebertia 116
rufipes reticulata, Lebertia 112
rufipes, Lebertia 97, 100-101, 105, 113-114
rugosa, Protzia 62, 68
rugosa, Protzia eximia 68
rugosus, Sperchon 170
Rusetria 179, 185
Rusetriella 179
rutae, Trichothyas petrophila 38-39
Rutripalpidae 76, 152
Rutripalpus 77, 152-153
salarius, Ignacarus 6
salebrosa, Lebertia 98-100, 114-115
sandalensis, Torrenticola 187
sandozi, Sperchon 164
sarasini, Panisus 22
savulescui, Dacothyas 4
saxicola, Lebertia stackelbergi 144
saxonica, Lebertia 140, 143
scapularis, Diplodontus 2, 13-14
schechteli, Lebertia 98-100, 108-109, 112, 115
schechteli globifera, Lebertia 98
schischkoffi, Lebertia 109
schloethi, Charoelia 169
schmidtii, Lebertia 132
sculptopetiolata, Protzia 65
scutata, Thyas 39-40
scutata, Vietsia 5, 39-40
scutellata, Lebertia 109
Scutolebertia 90-91
sefvei, Lebertia 119, 124-125, 129, 137
sefvei circumclusa, Lebertia 137
sefvei interposita, Lebertia 137
sefveoides, Lebertia 132
semiperforatus, Diplodontus 14
semireticulata, Lebertia 120-121, 134, 137-138
separata, Lebertia 125-126, 138
separata, Lebertia stigmatifera 138
Septlebertia 92
serratipalpis-group (Torrenticola) 197
setiger, Sperchon 18, 78, 158-159, 165
setiger insignis, Sperchon 164
setipes, Panisoides 20
setipes, Panisopsis 18-20
setipes, Thyas 20
setosus, Gnaphiscus 151
setosus, Oxus 147, 149, 151
shadini, Lebertia 111
Shivatonia 80
Siboneyacarus 89
sigthori, Lebertia 119
Sigthoria 79-80, 89-90
silesiaca, Lebertia 112
similis, Torrenticola 190-191, 193, 196
Simuliidae (Diptera) 64, 165-166
Sindacoides 18, 20
sinuata, Lebertia 119-120, 139
slovenica, Lebertia 114
somnambula, Wandesia 73-74
sparsicapillata, Lebertia 103-106, 114, 116
speciosa, Bandakia 81-83
Sperchon 78, 154-171, 206, 211
Sperchon s. str. 156, 167-171
Sperchonidae 154
Sperchonopsis 154-155, 161, 171-172
Sperchontidae 79, 152, 154
Sperchontinae 154
Sperchontoidea 154
sphagnorum, Thyasides 35
spinipes, Hydryphantes 53
spinirostris, Torrenticola 187
Sporadoporus 59
squamosa, Protzia 59-60, 68-69
squamosa paucipora, Protzia 68-69
squamosus, Calonyx 68
squamosus, Sperchon 154-155, 158, 160, 168, 170
stackelbergi saxicola, Lebertia 144
stadleri, Atractides 183
stadleri, Monatractides 180-181, 183-184
stadleri stenostoma, Torrenticola 184
stadleri stenostomus, Monatractides 184
steinmanni, Partnunia 2, 57-58
stenostoma, Torrenticola stadleri 184
stenostomus, Atractides 184
stenostomus, Monatractides 181, 184
stenostomus, Monatractides stadleri 184
stigmatifera, Lebertia 125-127, 136, 138-139
stigmatifera separata, Lebertia 138
strigata, Hydrachna 150
strigatus, Oxus 145, 147, 150
Stygomamersopsis 79-80
stygophila, Wandesia 73-74
Stygothrombioidea XIII, 207
Stygothrombium 208, 210
Stygotoniidae 76
subalpina, Protzia 66
subalpina, Teutonia 174-177
sublitoralis, Lebertia 111
Subteutonia 175
subtilis, Lebertia 101-102, 110, 117
suecica,Tartarothyas 71
sylvestris, Cricotopus 11
Synaptia 179
tabernaculata, Partnunia 57-59
Tadjikodartia 83
Tadjikothyas 4-5

Taeniopterygidae (Plecoptera) 75
Tanypodinae (Diptera Chironomidae) 64, 82
Tanytarsini (Diptera Chironomidae) 106-107, 114-116, 139, 169
Tartarothyadinae 3, 70
Tartarothyas 2-3, 70-72
Tartarothyasinae 70
tauinsignita, Lebertia 98
tauinsignita, Pachygaster 91
tazzecensis, Euthyas 16
tegulata, Nilotonia 83-85
tenuabilis, Hydryphantes 48
tenuicollis excellens, Lebertia 131
tenuipalpis, Hydryphantes 47, 49, 54
tenuipalpis, Lebertia 98
tenuirostris, Torrenticola 186-187
tenuisetis, Oxus 148, 151
tenuistriata, Lebertia 106
Teratothyadidae 1
Testudacarinae 177
Testudacarus 177
Teutonia 77, 83, 173-177
Teutoniidae 77, 152, 173
Thermacaridae 1
Thermacarus 1
thienemanni, Panisellus 5, 17-18
thienemanni, Sperchon 158-159, 161, 168, 170
thienemanni, Sperchon glandulosus 168
Thienemanniella (Diptera Chironomidae) 64
thoni, Hydryphantes 55
thoni, Hydryphantes flexuosus 55
thoracata, Parathyas 7, 23-25, 32
thoracatus, Thyas 32
thori, Atractides 187
thori, Panisopsis 19-21
thori, Torrenticola 185-187
thori, Wandesia 72-75
Thyadidae 1, 15
Thyadinae 15
Thyas 15, 20, 24, 33
Thyasella 5-6, 33-34
Thyasides 7-8, 34-35
Thyasinae 15
Thyopsis 5, 36-37, 205
ticinensis, Sindacoides 20
Tipulidae (Diptera) 11, 16, 25, 37
tisialis, Torrenticola barsica 187
Todothyas 24
Torrenticola 77, 178-202
Torrenticola s.str. 185, 187-202
torrenticola, Hydrodroma 2, 8-12
torrenticola danuviensis, Hydrodroma 12
torrenticola, Thyasella mandibularis 34
Torrenticolidae 77, 90-91, 177
torrenticolus, Panisus 22-24
Trichoptera 11, 25, 58-59, 64, 75, 163
Trichothyas 4, 37-39, 208
truncata, Euthyas 5, 16
truncatus, Bradybates 15, 16
tuberculata, Limnesia 87
tuberosa, Lebertia 115
turcica, Nilotonia 86
turgidus, Sperchon 158, 160, 171
umbrata,Hydrodroma 9
ungeri, Atractides 202
ungeri, Torrenticola 178, 188, 191, 202
unguiculatus, Atractides 183
unguiculatus, Monatractides 183
uniscutata, Torrenticola 179
uniscutatus, Monatractides 179
Utaxatax 79-81
vaginosus, Sperchon 156-157, 160, 163, 166
valvata, Parathyas 25
variolata, Lebertia 110
vastipalpis, Hydryphantes 54
vegacabrerae, Lebertia 106, 116
ventriplax, Utaxatax 79, 81
verrucosa procera, Sperchonopsis 172
verrucosa reducta, Sperchonopsis 172
verrucosa, Sperchonopsis 155, 171-172
verrucosus, Sperchon 172
vesnae, Sperchon 168-169
viedmai, Stygomamersopsis 79
Vietsia 5-6, 39-40
vietsi-group (*Bandakia*) 81
Vietsthyas 72
vigilans, Panisopsis 18-19, 21
vigilans, Thyas 21
violaceus, Sperchon 155-157, 163, 166-167
wackeri, Lebertia 142
walteri, Lebertia 98
walteri, Protzia 60, 65
walteri pyrenaica, Protzia 65
Walterothyas 18
Wandesia 1, 3, 72, 206, 211
Wandesiidae 72
Wandesiinae 1, 3, 72
Wandesioidea 72
westfalica, Lebertia 116
xylurgella, Torrenticola 197
yalvaci, Hydryphantes 48, 51
zachariasi, Lebertia 112
Zelandothyadidae 1
zermattensis, Lebertia 109
Zschokkea 2-3, 40-41
zschokkei, Lebertia 109

www.ingramcontent.com/pod-product-compliance
Lightning Source LLC
Chambersburg PA
CBHW082006130926
51338CB00021B/49
* 9 7 8 3 8 2 7 4 1 8 9 4 4 *